Glacier–Permafrost Interactions

The Cryosphere Science Series

Permafrost, sea ice, snow, and ice masses ranging from continental ice sheets to mountain glaciers, are key components of the global environment, intersecting both physical and human systems.The study of the cryosphere is central to issues such as global climate change, regional water resources, and sea level change, and is at the forefront of research across a wide spectrum of disciplines, including glaciology, climatology, geology, environmental science, geography and planning.

The Wiley-Blackwell Cryosphere Science Series comprises volumes that are at the cutting edge of new research or provide a focused interdisciplinary review of key aspects of the science.

Series Editor

Peter G. Knight, senior lecturer in geography, Keele University

Related Titles

Till: A Glacial Process Sedimentology
David J A Evans, 2018
9781118652596

Recent Climate Change Impacts on Mountain Glaciers
Mauri Pelto, November 2016
978-1-119-06811-2

Remote Sensing of the Cryosphere
Marco Tedesco, December 2014
978-1-118-36885-5

Glacier–Permafrost Interactions

Dr Richard I. Waller
Keele University
United Kingdom

 The Cryosphere Science Series

Series Editor: Dr Peter Knight, Keele University

WILEY

Registered Office(s)
John Wiley & Sons, Inc., 111 River Street, Hoboken, NJ 07030, USA
John Wiley & Sons Ltd, The Atrium, Southern Gate, Chichester, West Sussex, PO19 8SQ, UK

For details of our global editorial offices, customer services, and more information about Wiley products visit us at www.wiley.com.

Wiley also publishes its books in a variety of electronic formats and by print-on-demand. Some content that appears in standard print versions of this book may not be available in other formats.

Library of Congress Cataloging-in-Publication Data

Names: Waller, Richard I., author.
Title: Glacier–permafrost interactions / Richard I. Waller.
Description: Hoboken, NJ : Wiley, 2024. | Includes bibliographical
   references and index.
Identifiers: LCCN 2023053438 (print) | LCCN 2023053439 (ebook) | ISBN
   9781118620984 (hardback) | ISBN 9781118620977 (adobe pdf) | ISBN
   9781118620960 (epub)
Subjects: LCSH: Glaciers. | Permafrost. | Global warming.
Classification: LCC GB2403.2 .W35 2024 (print) | LCC GB2403.2 (ebook) |
   DDC 551.31/2–dc23/eng/20231220
LC record available at https://lccn.loc.gov/2023053438
LC ebook record available at https://lccn.loc.gov/2023053439

Cover Design: Wiley
Cover Image: Courtesy of Richard Waller

Set in 9.5/12.5pt STIXTwoText by Straive, Pondicherry, India
SKY10074263_050424

Dedication

This book is dedicated to the memory of my father, John Waller, a fellow academic and eternally supportive father who would have enjoyed discussing the contents of this book.

Contents

Author Biography

Richard I. Waller is a Senior Lecturer in Physical Geography at Keele University in the United Kingdom. Following the completion of a PhD degree in 1997 at the University of Southampton, he spent five years working as a Postdoctoral Research Fellow in Glaciology at Greenwich University prior to moving to Keele in 2001. His research interests focused initially on the formation, deformation and mechanisms of debris transfer associated with debris-rich basal ice layers, subsequently broadening to encompass subglacial processes, the development of glacial landsystems and the processes and products of glacier–permafrost interactions that form the focus of this book. Being largely field based, this research has provided opportunities to work extensively on modern and ancient glacial and permafrost environments in Alaska, Canada, Iceland, Greenland and Scandinavia as well as the United Kingdom. This work has led to the publication of papers within several peer-reviewed journals as well as book chapters, although this is his first sole-authored research text. He has been actively involved with the Quaternary Research Association, the International Glaciological Society and the International Permafrost Association and has presented at a number of events including the International Conference on Permafrost and the INQUA Congress. He is a passionate believer in the importance of outreach and widening participation and has worked with the British Science Association, the Geographical Association and the Royal Geographical Society to create a range of educational resources designed to foster a wider appreciation and interest in the Earth's cold environments and the impacts of climate change. In his spare time, he enjoys any time spent in the 'great outdoors' whether by foot, by bike or by kayak.

Preface

In a year that has witnessed record-breaking temperatures around many parts of the planet, there has been an increasing public consciousness of the accelerating impacts of climate change in the form of heat waves, wildfires and floods. Associated media interest has extended to the global cryosphere and as I write, there is recent news of the lowest recorded extent of Antarctic sea ice since satellite observations commenced in 1979. Against this backdrop, it is surprising that so little research attention has focused on the extent, nature and significance of the interactions between glaciers and permafrost as two of the most significant components of the cryosphere. This is in part a legacy of a longer-term disciplinary divide that has seen those studying glaciers and those studying permafrost working largely if not entirely in isolation. The primary aim of this book is consequently to showcase the distinctive processes and products associated with glacier–permafrost interactions that have not previously received the attention they deserve.

From the perspective of research priorities, one of the main reasons why glacier–permafrost interactions have received limited attention is that glaciers and permafrost have traditionally been considered to be largely mutually exclusive with significant thicknesses of glacier ice, active basal processes and the production of meltwater considered inimical to the preservation of subglacial permafrost. Recent field studies however have indicated that subglacial permafrost can extend for several kilometres beneath ice margins. Meanwhile, modelling studies have indicated that subglacial permafrost can persist for millennia beneath advancing ice sheets due to its thermal inertia whilst also suggesting that it can re-form and aggrade during periods of recession and downwasting.

With the thermal regime of a glacier or ice sheet having long been considered a key determinant of their dynamic behaviour, another reason for the limited research interest has been the assumption that glaciers overlying permafrost are by definition 'frozen to their beds' such that basal processes are inactive and ice flow is limited to internal creep. With research agendas focusing very much on the states of fast ice-flow driven by basal processes, cold-based glaciers associated with very low velocities have been considered of limited intrinsic interest. Again, more recent research has however indicated that these ice masses can be much more dynamic than previously thought with the potential for basal processes to persist at temperatures well below the bulk freezing point.

More recently, the glaciological literature has increasingly focused on hydrology as a key driver of various facets of glacier behaviour. Whilst this work has largely targeted temperate glaciers in environments lacking permafrost, the few studies undertaken in permafrost environments have revealed complex connections between glacier thermal regimes, hydrological pathways and velocities capable of driving significant seasonal variations in the generation, storage and release of meltwater. Looking at polar catchments more broadly, following the pioneering work of Olav Liestøl, ongoing research has demonstrated the landscape-scale hydrological interconnectivity of glaciers and permafrost and the consequent need to treat them as coupled components of a wider system. This has become apparent in the explanation of one of the more paradoxical impacts of polar climate change whereby climate warming has led to glacier recession and cooling, a reduction in the extent of warm-based ice critical for groundwater recharge and the shutdown of spring systems.

When viewed from the perspective of geomorphologist and geologists, cold-based glaciers associated with permafrost have traditionally been associated with a lack of geomorphic activity, providing another reason for the limited prior research interest. However, it is precisely this ability to preserve preglacial features – including biologically viable vegetation – that is arguably one of the most enigmatic and consequential aspects of their behaviour that is worthy of further investigation. Glacier–permafrost interactions can nonetheless be associated with the creation of a variety of distinctive landforms, sediments and landsystems, although these are often markedly different from the well-known 'textbook' examples associated with temperate glaciers. With the landscape record containing an inherent bias towards warm-based conditions and the potential for misinterpretations of landforms and sediments, elucidating their landscape impacts and expression can be considered a key priority. Looking to the future, it is also likely that glacier–permafrost interactions will play an influential role in determining landscape responses to climate change in high mountain and high latitude regions, with the potential for major rockfalls and related hazards in the former and the exposure and subsequent melt-out of massive ground ice in the latter.

The book has been designed to provide a systematic overview of these themes, developments and the fundamental underlying concepts:

- **Chapter** 1 provides a short introduction to the research context of the book and the significance of glacier–permafrost interactions.
- **Chapter** 2 provides an introduction to permafrost considering its thermal and physical properties, its formation and extent and its rheological properties.
- **Chapter** 3 considers the distinctive characteristics of glaciers found within permafrost environments and the influence of glacier–permafrost interactions on subglacial and ice-marginal processes.
- **Chapter** 4 provides an integrated overview of the hydrological and dynamic behaviour of glaciers that interact with permafrost before considering their implications for glacial sediment transfer.
- **Chapter** 5 addresses the landscape expression of glacier–permafrost interactions, considering the concept of glacial protectionism, the broad range of associated landforms, sediments, structures and landsystems and finally, the paraglacial landscape adjustments in which they play a prominent role.
- **Chapter** 6 concludes by exploring a range of potential avenues for future research.

Ultimately, it is hoped that this book will encourage those studying cryospheric environments to view glaciers and permafrost not as separate entities, but as part of a thermally, mechanically, hydrologically and geologically integrated system. At the same time, it is hoped that this recognition will encourage more collaborative research activity between glaciologists and permafrost scientists. This can only be to the benefit of both disciplines and to the collective and pressing need to fully comprehend the accelerating impacts of climate change in those parts of the world that continue to be dominated by ice.

Richard I. Waller

Acknowledgements

With this book focusing on the interactions between glaciers and permafrost, I would like to express my sincere gratitude to all those who have stimulated my interest in glaciers, in permafrost and more recently, in the interactions between the two that have culminated in this contribution to the research literature. Special thanks in particular to Jane Hart, who provided me with the opportunity to develop my nascent interest in glaciers, to Julian Murton who introduced me to permafrost and the 'wonderful world of underground ice' and to Peter G. Knight without whose encouragement this book would never have happened.

Since the late 1980s, I have been privileged to travel to a range of spectacular Arctic and sub-Arctic locations to undertake field work with some remarkable people and to benefit from their company, knowledge, expertise and enthusiasm. In addition to those already mentioned, these include Dave Evans, Matthew Bennett, Colin Whiteman, Mike Hambrey, Zoe Robinson, Brian Moorman, Emrys Phillips amongst others who I've undoubtedly forgotten to recognise. They have all contributed to some of the most enjoyable and memorable experiences of my life that continue to shape my views of the natural world. I'm especially grateful to the individuals and organisations who enabled these opportunities and provided the financial support necessary.

I would also like to acknowledge the pioneers whose work on cold-based glaciers and glacier–permafrost interactions has provided the foundation and inspiration for this book. In addition to those mentioned above, the research findings, insights and conceptual frameworks provided by Wilfried Haeberli, Bernd Etzelmüller, Sean Fitzsimons, Cliff Atkins and Art Dyke have proven invaluable.

Thanks to all those at Wiley who have been supportive and patient over the years this book has taken to complete. I'd like to thank Mandy Collison and Indirakumari Siva in particular, without whose guidance this book would never have come to fruition. Special thanks also to Sean Fitzsimons, Jason Gulley and Mike Hambrey who kindly allowed the reproduction of their images within the book.

Finally, I would like to thank my wife for her patience as I've disappeared for hours or days at a time whilst writing this book. Rest assured - it'll be some time before I attempt another writing project of this magnitude! I would also like to thank my wonderful mother to whom I owe so much. I hope you get the chance to finally see this book in print.

1

Introduction

1.1 Research Context and Academic Background

Glaciers and permafrost constitute two of the most important elements of the global cryosphere, a dynamic component of the Earth system dominated by cold climatic conditions and water in its solid form. Modern-day glaciers and ice sheets alone are estimated to cover a total area of over 16 million km^2 with the majority occurring in the form of the Antarctic and Greenland Ice Sheets (covering c. 14 million and 1.7 million km^2, respectively; NSIDC, 2019), with the remaining ice being distributed between 198,000 smaller glaciers and ice caps (Pfeffer *et al.*, 2014). Meanwhile, permafrost regions are estimated to occupy an additional area of almost 23 million km^2 in the Northern Hemisphere alone, which equates to almost 24% of the Earth's exposed land area (Zhang *et al.*, 2003). In combination therefore, the global cryosphere can be estimated to currently cover a total area of almost 40 million km^2, occurring primarily within the polar regions and high-altitude continental interiors.

In spite of their obvious geographical associations within similar climatic settings, surprisingly little research activity has focused on an examination of the nature and potential significance of the interactions between glaciers and permafrost. Part of the reason for this limited interest in glacier–permafrost interactions within the glaciological research community in particular relates to two widely held assumptions. First, that glaciers and permafrost are largely mutually exclusive with substantial thicknesses of glacier ice insulating any underlying permafrost from the prevailing climatic conditions and the heat generated by basal processes rapidly degrading any permafrost present. Second, where glaciers are observed to rest on permafrost, the resulting cold-based thermal regime is thought to preclude the operation of processes such as basal sliding and subglacial-sediment deformation that are in turn associated with dynamic ice flow and significant landscape modification. In addition to explaining the lack of targeted research activity, both assumptions emphasise the central importance of basal thermal regimes to any consideration of the connections between glaciers and permafrost (Section 3.2).

Cold-based glaciers occurring in areas of permafrost are therefore commonly conceptualised as being frozen to rigid and undeformable beds and to be both slow moving

Glacier–Permafrost Interactions, First Edition. Richard I. Waller.
© 2024 John Wiley & Sons Ltd. Published 2024 by John Wiley & Sons Ltd.

and geomorphologically impotent. They have consequently been disregarded as being of limited intrinsic research interest. In combination with their tendency to occur within remote regions that are typically difficult and costly to access, this has resulted in glacier scientists historically tending to focus primarily on temperate or warm-based glaciers, inducing a research bias that has resulted in our understanding of the behaviour of cold-based glaciers remaining limited in comparison. The same situation also applies to ancient glacial environments where glacial geomorphologists and geologists have tended to study areas glaciated by warm-based ice or characterised by hard bedrock, for example the Fennoscandian and Canadian Shields. In contrast, only a handful of studies have been published in international research journals that consider the potential interactions between cold-based ice sheets and the permafrozen sediments that underlie huge swathes of western Siberia and Arctic Canada and where permafrost is believed to have persisted throughout entire glacial cycles (e.g. Astakhov *et al.*, 1996, section 5.4.3).

Some have argued that the limited interest in glacier–permafrost interactions is a reflection of a more fundamental and deep-seated dichotomy that has developed and persisted within the cryospheric sciences. Harris & Murton (2005) note that whilst the term 'geocryology' had originally been introduced to encompass both glaciers and permafrost, the introduction of the term 'periglacial' by Lozinski in the early 20th century resulted in a split between those interested in the study of glaciers and those interested in the study of frozen ground. These two communities have subsequently worked largely if not entirely in isolation, such that studies into the processes and products that occur at the interface between glaciers and permafrost have rarely received the focused and systematic research attention they deserve. They have instead been largely limited to isolated studies that have for example considered the potential role played by permafrost in promoting the development of large push moraines (e.g. Rutten, 1960; section 5.3.2). This enduring schism between the two disciplines is further illustrated by the usage of the term glaciology. Whilst this literally means the study of ice, it is widely regarded as referring more exclusively to the study of glaciers whilst the study of glaciers and ice sheets has been seen to be beyond the remit of permafrost scientists.

As a rare example of someone who has worked extensively across this research divide, Haeberli (2005) has argued that this schism has been to the detriment of both disciplines, resulting in terminological confusion and a hampering of research progress in both fields that have ultimately limited the credibility of cryospheric research as a whole. The lack of collaborative research appears particularly perverse when one considers shared research interests in the behaviour of ice-sediment mixtures for example (e.g. Waller *et al.*, 2009a). Glacier scientists recognise the potential influence of the debris-rich ice that commonly occurs at the base of glaciers and ice sheets on their dynamic behaviour, sediment transport and geomorphic impact. At the same time, permafrost researchers have made significant progress in describing the nature, origin and engineering properties of different types of ground ice commonly found within permafrost regions. However, in spite of these overlapping areas of mutual interest relating to the study of essentially identical materials using similar techniques, the amount of collaborative research has remained limited until relatively recently (Section 2.5.1).

The resolution of a series of major research challenges relating for example to hazard mitigation in cold-climate regions, the secure burial of radioactive waste in cold regions and the potential impacts of future climate change on the global cryosphere have led to the recognition of a pressing need for a more integrated approach to the study of the cryosphere and renewed calls for the two research communities to collaborate more actively. This provides the opportunity to open up a 'new scientific land' (Haeberli, 2005, p36) in which workers in both communities can recognise and actively incorporate rather than ignore and exclude the findings of the other discipline. Such collaborative approaches are essential for the resolution of a range of interdisciplinary research questions that span the two subject areas. These include the accurate interpretation of buried ice within permafrost environments hypothesised to constitute buried glacier ice that has been preserved within the permafrost since deglaciation (Section 5.3.4). Similarly, recent work exploring the link between the dynamic behaviour of ice streams, basal freezing and till rheology has benefitted from the application of pre-existing models of frost heave originally devised by permafrost engineers (Section 4.3.3). These two brief examples provide an insight into the potential benefits that could be enabled by a more integrated approach. Most importantly, the fostering of a more interdisciplinary approach can prevent the promulgation of misconceptions and theoretical shortcomings that could have devastating implications in the context of natural hazards in high mountain areas (Section 5.6.1).

Whilst there are signs of a growing awareness of the importance of glacier–permafrost interactions, coupled with attempts to bridge the existing gap between glacier science and permafrost research, the fact remains that 'the geological and geomorphological processes at the interface between glaciers and permafrost have received less attention than they warrant, and the influence of the one on the other has been largely neglected' (Haeberli, cited by Harris & Murton, 2005, p2). It is hoped that this book will go some way to appreciating, emphasising and promoting the importance of glacier–permafrost interactions through a consideration of the historical roots of these interactions and its attempts to review recent developments, the state of current knowledge, the key gaps in our understanding and profitable avenues for future research.

1.2 Overview of the Significance of Glacier–Permafrost Interactions

As mentioned in the previous section, one of the principal reasons behind a lack of research interest in glacier–permafrost interactions has stemmed from the assumption that glacier–permafrost interactions are of limited extent and that glaciers resting on permafrost are slow moving and geomorphologically ineffectual. Research over the past 30 years in particular has however demonstrated that these assumptions are by no means universally applicable. The rebuttal of these assumptions has in turn stimulated renewed interest in the nature and significance of glacier–permafrost interactions that are briefly reviewed within this section prior to their more detailed examination in the subsequent chapters.

Numerical ice-sheet modelling and field observations relating primarily to Pleistocene Ice Sheets have indicated that glacier–permafrost interactions are likely to have been more extensive and of longer duration than was previously thought (Section 2.4.1). This is particularly the case during the growth phases of continental ice sheets when they are most likely to have advanced over areas of pre-existing and potentially thick permafrost with a considerable thermal inertia. Conversely and more topically, ongoing glacier recession resulting from climate change in high-latitude regions such as Svalbard is resulting in an increase in the extent of subglacial permafrost as the glaciers thin and decelerate and the insulating effect of the ice decreases, suggesting that glacier–permafrost interactions can also be important during deglacial phases (Section 2.4.2).

The basal thermal regime of a glacier or ice sheet is widely regarded as one of the principal controls of its dynamic behaviour (Section 3.2). With glaciers resting on permafrost by definition being cold-based in character, this establishes clear process-related connections between permafrost, basal thermal regimes, subglacial hydrology and dynamic behaviour. Cold-based glaciers resting on permafrost are commonly referred to as being 'frozen to their beds', which suggests that they are only able to move via internal creep and are very slow moving as a consequence (Section 3.4). This is in marked contrast to warm-based glaciers where the active production of meltwater enables the basal processes such as basal sliding and subglacial-sediment deformation that are central to states of fast ice flow and flow instabilities. Recent work in both modern and ancient glacial environments has demonstrated however that these basal processes are not restricted to warm-based glaciers as was previously thought and that these processes can remain active at temperatures well below the pressure-melting point due to the presence of 'premelted water' (Section 3.5). This means that glaciers interacting with permafrost may flow more rapidly than was previously assumed to be the case, although the magnitude and significance of any increase in velocity remains as yet unclear. In addition, the identification of deep-seated permafrost deformation beneath former cold-based ice sheets raises the intriguing possibility of glaciers mechanically coupling with permafrost to create an integrated dynamic system (Sections 3.5.5, 5.4.3).

Further consideration of the basal thermal regime of glaciers and their position within hydrological catchments highlights the hydrological implications of glacier–permafrost interactions both for the glacier and the wider catchment (Section 4.2). Glaciers occurring in permafrost areas exhibit distinctive hydrological regimes and experience far less basal melting than glaciers in more temperate environments. Cold-based glaciers are commonly associated with deeply incised supraglacial drainage systems that are reactivated during the melt season to feed lateral meltwater systems. Larger polythermal glaciers with ice thicknesses sufficient to produce areas of warm-based ice are capable of generating perennial subglacial drainage which can lead to the development of proglacial icings during the winter months. These zones of basal melting beneath polythermal glaciers are often associated with taliks or unfrozen zones in the permafrost. Their connection with wider catchment groundwater systems means that they constitute one of the few sources of groundwater recharge in areas of extensive permafrost. This can in turn drive perennial spring systems and lead to the formation of distinctive permafrost landforms such as pingos. Ongoing recession and thinning of glaciers in Svalbard and the shrinking of areas of warm-based ice has as a result been observed to lead to a reduction in groundwater recharge and the drying up of springs, once again highlighting the importance of viewing glaciers and permafrost as part of a connected system.

The conceptual framework used to understand glacier–permafrost interactions also has important implications for the geomorphological processes and products associated with glaciers occurring in permafrost environments. With active glacial erosion traditionally being associated with warm-based glaciers and processes such as basal sliding, cold-based glaciers resting on permafrost have typically been characterised as settings featuring limited geomorphological activity and landscape change. As such they have been associated largely with the preservation of preglacial landforms rather than with the active generation of distinctive glacial landforms, with some arguing their principal geomorphic impact is limited to the isostatic effects associated with crustal loading (Section 5.2). Some studies have however recognised that glacier–permafrost interactions can lead to the creation of a range of distinctive landforms (Section 5.3), for example by enabling the transmission of glacier-induced stresses into a permafrozen foreland and facilitating the development of large push moraine complexes (Section 5.3.2). Others have suggested that the lateral meltwater flows characteristic of cold-based glaciers can create distinctive flights of meltwater channels that can provide a rather different landscape record of ice-margin recession to the moraine sequences more commonly employed within palaeoglaciological reconstructions (Section 5.3.1).

The recognition that basal processes can remain active at temperatures below the pressure melting point has been central to a more recent reappraisal of their continued ability to erode, entrain, transport and deposit sediment (Section 3.5). In this regard, the association of polythermal glaciers terminating in permafrost regions with thicker debris-bearing basal ice layers and higher glacial and fluvioglacial sediment loads suggests that the enhanced availability of sediment in permafrost areas may in fact more than compensate for their lower ice velocities (Section 4.4). Greater interest in and scrutiny of the forelands of cold-based glaciers in permafrost regions has demonstrated that even those glaciers occurring in the coldest regions of Earth, such as the Dry Valleys in Antarctica, are associated with some degree of geomorphological activity and the development of a range of subtle yet distinctive landforms (Section 5.5.4). Work in past glacial environments where glaciers have advanced over and coupled with permafrost have also revealed distinctive sedimentological and structural signatures relating to the deformation and mobilisation of permafrost (Section 5.4). Therefore, rather than being characterised by an absence of geomorphological and geological activity, it is becoming increasingly clear that glaciers interacting with permafrost are associated with a range of distinctive landform-sediment assemblages the nature and diversity of which are only starting to become clear (Section 5.5). This remains a nascent area of enquiry where much work is required to establish the diagnostic links between process and product that provide the fundamental foundations for geomorphological inverse models and palaeoglaciological reconstructions.

Finally, the consideration of glaciers and permafrost as complex, coupled systems is central to an understanding of the geomorphological consequences of climate change in modern-day glaciated permafrost regions and the risks they pose to the resident populations. The Kolka-Karmadon rock/ice slide that occurred in the Caucasus Mountains in 2002 claiming an estimated 120 lives provides a dramatic illustration of the large-scale and catastrophic slope failures that can result from a combination of glacier recession, slope debuttressing and permafrost degradation. High mountain areas featuring glaciers and permafrost are therefore experiencing a transitional phase of rapid change associated with profound landscape disequilibria and the development of complex process chains (Section 5.6.2). In addition, a significant increase in the frequency and intensity of summer storm events has resulted in the exposure and degradation of buried ground ice within large areas of the western Canadian

Arctic and Siberia and the formation and dramatic expansion of 'megaslumps' (Section 5.6.3). With much of this ice being considered to comprise relict Pleistocene glacier ice preserved by permafrost within ice-cored moraine complexes, recent climate change has resulted in a renewed phase of deglaciation and thermokarst activity. This has significant regional implications for regional water quality and transport infrastructure and global implications in terms of the stored carbon these processes are helping to release.

1.3 Explanation of the Book Structure

Chapter 2 considers the distinctive characteristics and properties of permafrost within the specific context of their occurrence of modern and ancient glacial environments. The chapter focuses initially upon thermal and physical properties of permafrost paying particular attention to the presence and significance of water as both a liquid and a solid (Section 2.2). It then explores the conditions and processes associated with its formation, preservation and degradation in both subaerial and subglacial environments (Section 2.3) before considering the spatial extent of glacier–permafrost interactions and the temporal changes related to modern-day and ancient glacier fluctuations (Section 2.4). Finally, detailed consideration is given to the highly variable mechanical and rheological properties of permafrost and the key causes the complex behaviours that have been observed (Section 2.5).

Chapter 3 considers the influence of permafrost on the operation of a diverse range of glacial processes. The chapter starts by reviewing the thermal regimes associated with glaciers occurring in permafrost environments (Section 3.2) before considering the basal boundary conditions associated with both rigid and soft-bed glaciers (Section 3.3). It then examines the traditional glaciological assumptions regarding basal conditions and processes associated with cold glaciers with 'frozen beds' (Section 3.4) before considering the evidence that has led to a recent reappraisal of the activity of subglacial processes at temperatures below the pressure melting point (Section 3.6). The final section examines a range of ice-marginal and proglacial processes that have been specifically related to glacier–permafrost interactions (Section 3.7).

Chapter 4 examines the hydrology, dynamic behaviour and sediment fluxes associated with modern and ancient non-temperate glaciers that are central to understanding the varied roles and potential glaciological and geomorphological impacts of glacier–permafrost interactions. Section 4.2 focuses on the hydrological characteristics of non-temperature glaciers considering the distinctive flow pathways and their seasonal variability before considering their impacts on suspended-sediment and solute fluxes. The impacts of non-temperature glaciers on groundwater fluxes within the broader permafrost catchments are also explored. In view of the emerging connections between glacier hydrology and ice flow, this provides an important precursor to the consideration of the dynamic behaviour of non-temperature glaciers within Section 4.3. This explores the potential implications of glacier–permafrost interactions on glacier dynamics in relation to the velocities of modern-day, non-temperate glaciers, flow instabilities and ice streaming in ice sheets and the longer-term behaviours of Pleistocene Ice Sheets. Finally, Section 4.4 integrates the considerations of glacier hydrology and dynamics in considering their impacts on sediment pathways and fluxes within glaciated permafrost catchments.

Chapter 5 provides a detailed review of the distinctive products and landscape expressions of glacier–permafrost interactions. Section 5.2 starts by considering the broader debates concerning the geomorphic impacts of glaciers and the glacial protectionism and preservation of preglacial features thought to be characteristic of cold-based glaciers. Section 5.3 reviews a range of specific landforms that have been explicitly related to glacier–permafrost interactions before Section 5.4 considers the possible sedimentological features and structural signatures. Section 5.5 integrates these elements within a consideration of the distinctive glacial landsystems found within a range of contrasting glacial environments including those found on Mars. Section 5.6 concludes the chapter by examining the processes of paraglacial landscape adjustment in which glacier–permafrost interactions play a prominent role, including large rock slope failures and the degradation of ground ice.

Chapter 6 concludes by reflecting on the current state of knowledge and on the outstanding research gaps and areas of ongoing controversy whilst suggesting priority research themes and directions for the future.

References

Astakhov, V.I., Kaplyanskaya, F.A. & Tarnogradsky, V.D., 1996. Pleistocene permafrost of West Siberia as a deformable glacier bed. *Permafrost and Periglacial Processes*, 7, 165–191.

Haeberli, W., 2005. Investigating glacier-permafrost relationships in high-mountain areas: historical background, selected examples and research needs. In: Harris, C. & Murton, J.B. (Eds.), *Cryospheric Systems*, Special Publications, 242. Geological Society of London, London, 29–37.

Harris, C. & Murton, J.B., 2005. Interactions between glaciers and permafrost: an introduction. In: Harris, C. & Murton, J.B. (Eds.), *Cryospheric Systems, Special Publications*, 242. Geological Society of London, London, 1–9.

NSIDC, 2019. State of the Cryosphere: Ice Sheets. https://nsidc.org/cryosphere/sotc/ice_sheets.html. Date accessed: 5 November 2019.

Pfeffer, W.T., Arendt, A.A., Bliss, A., *et al.*, 2014. The Randolph Glacier Inventory: a globally complete inventory of glaciers. *Journal of Glaciology*, 60, 537–552.

Rutten, M.G., 1960. Ice-pushed ridges, permafrost and drainage. *American Journal of Science*, 258, 293–297.

Waller, R.I., Murton, J.B. & Knight, P.G., 2009a. Basal glacier ice and massive ground ice: different scientists, same science? In: Knight, J. & Harrison, S. (Eds.), *Periglacial and Paraglacial Processes and Environments*. Geological Society, London, Special Publication, 320, 57–69.

Zhang, T., Barry, R.G., Knowles, K., Ling, F. & Armstrong, R.L., 2003. Distribution of seasonally and perennially frozen ground in the Northern Hemisphere. In: *Proceedings of the 8th International Conference on Permafrost* (Vol. 2). AA Balkema Publishers, Zürich, Switzerland, 1289–1294.

2

Permafrost in Modern and Ancient Glacial Environments

2.1 Introduction

Permafrost is a term that encompasses a diverse range of Earth surface materials that can display a remarkable range of characteristics and behaviours. The presence of water in materials with ambient temperatures close to the thermal transition between its solid and liquid phases results in a high degree of sensitivity to any thermal perturbations in particular. Seemingly minor changes in climatic variables (temperature or precipitation) or surface boundary conditions can sometimes elicit a striking non-linear response in the form of dramatic landscape changes. In addition, variations in pressure (depth), porewater salinity and grain size can add an additional level of complexity to unconsolidated sediments that commonly comprise dynamic mixtures of ice, water and sediment. This can result in highly variable rheological behaviours that can change rapidly over space and time.

This chapter provides an introduction to permafrost as encountered in both modern and ancient environments, which underpins any appreciation and understanding of its interactions with glaciers. The chapter focuses initially on the distinctive thermal and physical properties of permafrost (Section 2.2), before considering its formation within both subaerial and subglacial settings (Section 2.3), associated spatial and temporal variations in its extent (Section 2.4), and finally its complex rheological properties and behaviours (Section 2.5).

2.2 Thermal and Physical Properties

2.2.1 Defining Characteristics

Permafrost is defined as 'ground (soil or rock and included ice or organic material) that remains at or below 0 °C for at least two consecutive years' (International Permafrost Association, 2020). Its modern-day definition is therefore dependent solely on temperature and time in contrast to earlier definitions that used the term as an abbreviation for 'permanently frozen ground' (Muller, 1943) and which therefore inferred the presence of water in a frozen state as a defining characteristic.

Glacier–Permafrost Interactions, First Edition. Richard I. Waller.
© 2024 John Wiley & Sons Ltd. Published 2024 by John Wiley & Sons Ltd.

When viewed as a purely thermal and temporal condition, permafrost can in theory encompass any type or combination of Earth surface materials. Dry bedrock can therefore be considered to be just as much a permafrost material as an ice-rich sediment or peat deposit. In considering this book's focus on glacier–permafrost interactions, it is important to note that glaciers and ice sheets can themselves be considered a subset of permafrost as they constitute Earth surface materials that have by definition been at or below 0 °C for a significant period of time. Whilst most cryospheric scientists would see permafrost and glaciers as separate entities, the ability to consider both as part of the same definition illustrates the shared attributes, processes and features that will become increasingly apparent both in this chapter and the book as a whole.

This commonality is particularly evident when one considers the physical composition of glaciers and permafrost which typically comprise varied mixtures of ice, sediment, water and air. Where glaciers occur in permafrost environments, the two can become virtually indistinguishable on account of their very similar compositions (Figure 2.1). This has resulted in some confusion within the literature as to whether such mixtures should be considered to be part of the glacier (i.e. basal ice) or part of the glacier bed (i.e. frozen subglacial sediment). In specific relation to the formation of ice-pushed ridges, Kupsch (1962) for example concluded that 'the permafrost layer in essence constituted the lower part of the glacier, the bottom of which was not the top of the bedrock but the lower limit of the ground ice sheet' (p591). This association is made most explicitly by Hughes (1973), who coined the term 'glacial permafrost' to refer to the 'regolith-charged basal ice layer of a glacier or ice sheet' (p213), thereby illustrating what at times are indivisible connections between glaciers and permafrost. Whilst this might appear to be a question of semantics, it does have broader implications. If some basal ice comprises pre-existing permafrost entrained by an overriding glacier (Waller *et al.*, 2000), then this highlights the importance of glacier–permafrost interactions for the processes of debris entrainment and basal ice formation (Section 3.5.4). It also has important implications for our conceptualisation of the location and nature of the ice-bed interface that is associated with a variety of influential dynamic and geomorphological processes (Section 3.5.5).

Although ice and liquid water are not defining features, they are commonly found within permafrost areas and are central to many of the more distinctive processes and landforms for which these environments are renowned. Therefore, in considering the properties of permafrost, it is vitally important at the outset to differentiate between the thermal condition of the substrate and the state of any constituent water. The terms 'cryotic' (0 °C) and 'non-cryotic' (>0 °C) are generally used within permafrost literature to refer explicitly to the thermal condition in relation to the freezing/melting point of pure water at atmospheric pressure. These terms differ from the more widely used terms 'frozen' and 'unfrozen' that refer to the state of any constituent water (e.g. Williams & Smith, 1989).

Unfortunately, these seemingly subtle yet fundamental distinctions between temperature and state are frequently conflated or ignored within published literature, leading to the simplistic assumption that the freezing and thawing of water in bedrock or sediments invariably occur at a single bulk freezing temperature (typically 0 °C or the pressure melting point when considering subglacial environments). If this were the case, then water-saturated substrates would in essence constitute a binary system displaying two distinct states separated by a clearly defined thermal threshold, i.e. unfrozen, weak and permeable

(a)

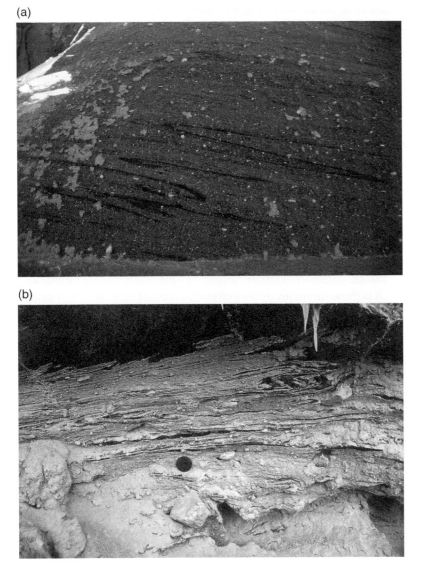

(b)

Figure 2.1 (a) Massive ground ice exposed at Mason Bay in the western Canadian Arctic and (b) basal ice exposed at the western margin of the Greenland Ice Sheet. Note the similar compositions in the form of folded debris laminations.

vs. frozen, rigid and impermeable. In reality, as discussed in more detail in the next section, the connection between the temperature of permafrost and the state of any water present is both highly complex and dynamic in nature. The thermal boundaries between cryotic/ non-cryotic and frozen/unfrozen states can in fact differ by several degrees Celsius due to the freezing-point depression caused by a range of controls such as the presence of dissolved salts. Additionally, the freezing of water within permafrost substrates commonly occurs progressively over a broad range of temperatures rather than at a single temperature, resulting in a spectrum of states and behaviours rather than a convenient binary system with a clear boundary between frozen and unfrozen states.

The need to distinguish between temperature and state is illustrated for example by the concept of 'seasonally active permafrost', which is used to refer to a near-surface layer that seasonally thaws and which can be subject to mass movement but which remains cryotic and is therefore part of the permafrost layer. Similarly, a warm-based glacier could overlie entirely unfrozen ground with a temperature that exceeds the pressure melting point but is defined as permafrost as it has remained at or below 0 °C, in which case the presence of permafrost is of little consequence. The importance of this distinction has fundamental implications for the understanding of allied processes that will be explored further in Chapter 3 and consequently in relation to the hydrological and dynamic characteristics of glaciers in Chapter 4 and the generation of distinctive landforms, sediments and landscapes in Chapter 5.

2.2.2 Water in Permafrost

Permafrost frequently contains water as a liquid, a solid (ice) or in both states simultaneously. As such, permafrost can be 'unfrozen', 'partially frozen' or 'frozen' depending on the state of any water present. It is these variations in the presence, state and relative concentration of ice and liquid water that result in permafrozen substrates displaying such varied and complex behaviours, particularly in relation to their mechanical and rheological properties (Section 2.5). In contrast, if no water is present, the permafrost can be referred to as 'dry' and its characteristics are likely to be very similar to the non-cryotic substrates found in more temperate environments (French, 2017).

Water within porous media such as sediment and bedrock can occur in two forms: (i) 'bulk water' or 'free water' that occurs within the pore spaces and (ii) 'interfacial water' that occurs at the interfaces with ice or mineral particles for example. Pure bulk water at atmospheric pressure will freeze at 0 °C, which is therefore commonly considered to be the specific temperature at which changes in state occur. However, there are a number of factors that can result in a depression of the freezing point such that water can become 'supercooled' – in other words, remain in a liquid state at temperatures below the bulk freezing point. Within the context of glacial environments, glacier scientists have long appreciated that the freezing (and melting) point can be depressed beneath thick ice as a result of increasing confining pressures. Whilst the rate of freezing point depression with increasing pressure is relatively low (0.0074 °C per atmosphere), it can nevertheless lead to significant freezing point depressions under thick ice masses. Consequently, glaciological literature almost always refers to the pressure melting point rather than 0 °C as the temperature at which phase changes occur. For example, at the Byrd Station in Antarctica where the ice thickness is 2164 m (equating to 200 atmospheres), the pressure melting point at the base of the ice sheet is estimated to be −1.6 °C (Gow *et al.*, 1979).

The freezing temperature of bulk water can also be depressed through the addition of impurities. The addition of 2% by weight of sodium chloride for example will depress the freezing point by approximately 1 °C, and it is this influence that results in the widespread gritting of roads in countries such as the United Kingdom in order to prevent the formation of ice at cryotic temperatures (Davis, 2001). Within permafrost environments, the rejection of solutes during the freezing process can lead to the development of layers or lenses of highly mineralised porewater that form taliks or unfrozen zones within otherwise frozen substrates. Marine sediments in coastal permafrost environments commonly feature high levels of dissolved salts that can result in them containing significant amounts of unfrozen bulk water at cryotic temperatures. This can in turn compromise the structural integrity of buildings constructed in these settings. Site investigations associated with the construction of a coal mining facility in central Spitsbergen in an area characterised by marine clays with porewater salt contents of $60\,\mathrm{g\,l^{-1}}$ for example indicated that 40% of the porewater remained unfrozen in spite of a mean annual surface temperature of $-6\,°C$ (Gregersen *et al.*, 1983). A geomorphological consequence of the increased unfrozen water content of marine sediments is the predominance of push moraines on Svalbard below the Holocene marine limit, with these sediments facilitating the development of the décollement horizons required for the formation of these landforms (Etzelmüller *et al.*, 1996; see section 5.3.2 for more detail).

Whilst pressure and impurities can explain freezing-point depression, it is the influence of molecular forces associated with interfaces within ice–water–sediment mixtures that is central to explaining the ability of liquid water and ice to coexist at temperatures below $0\,°C$. It is this ability for liquid water and ice to coexist in varying fractions over a range of cryotic temperatures that is one of the most fundamental and distinctive attributes of many permafrost substrates and a key reason for their complex rheological behaviours (Section 2.5).

When the temperature drops below the bulk freezing point, the 'free' porewater is the first to freeze. As freezing continues, the remaining liquid water is confined to progressively smaller spaces between the mineral particles and the growing ice crystals such that it becomes increasingly subject to capillary effects (Figure 2.2). Capillarity is an effect associated with confined interfaces. It refers to the migration or suction of liquids within narrow spaces that results from the combined influence of adhesive forces between the water and the nearby solid surfaces and the surface tension of the liquid itself. The molecular attraction between

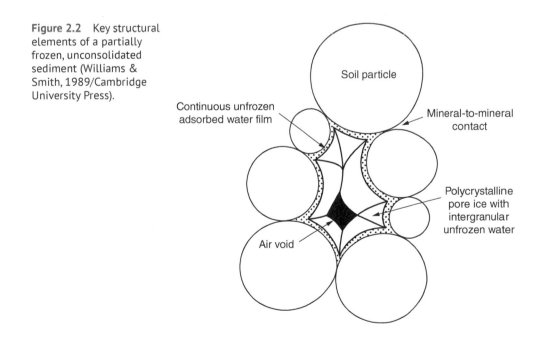

Figure 2.2 Key structural elements of a partially frozen, unconsolidated sediment (Williams & Smith, 1989/Cambridge University Press).

the water and the solid surfaces can also result in the water molecules adhering to the surface of constituent particles to produce thin, interfacial films through a process known as adsorption. Both capillarity and adsorption cause the free energy of the remaining liquid water to fall, resulting in a further depression of the freezing point such that progressively lower temperatures are required for the remaining unfrozen fraction to freeze. The associated development of free energy gradients within a substrate can result in the migration of water to areas of lower free energy that are typically the coldest regions. If these regions are associated with ice within the substrate, the combination of capillary effects and free-energy gradients can result in cryosuction and the migration of water to these ice lenses enabling their progressive growth within what are referred to as 'frost susceptible' substrates. This can however generate a negative feedback with the growth of ice within the capillaries reducing the hydraulic conductivity and limiting the flow of both water and heat (Watanabe & Flury, 2008). It is this bulk transfer of liquid of water within a 'frozen fringe' rather than the volumetric expansion of water upon freezing that is key to understanding the process of frost heave (Rempel, 2011) that is well known for its role in bedrock weathering in subaerial permafrost environments (e.g. Walder & Hallet, 1985; Murton *et al.*, 2006). As a further illustration of the complex process associations between glacial and permafrost environments, it is worth noting that the application of theories of frost heave to subglacial settings has provided new and novel insights into the processes of sediment entrainment beneath ice streams and in turn their dynamic behaviour (see Sections 3.5.4 and 4.3.3, respectively).

Ongoing research into frost heave within partially frozen porous media has further developed our understanding of the processes and circumstances in which unfrozen, or what is commonly referred to as 'pre-melted water', can persist at temperatures up to a few tens of degrees below the bulk freezing temperature (Dash *et al.*, 2006). The necessary depression of the freezing temperature occurs for two reasons (e.g. Rempel, 2007, 2010). First, curvature-induced pre-melting at a solid-melt interface (e.g. ice-water) that has a centre of curvature within the solid results in the development of supercooled water within the pore spaces. Second, interfacial pre-melting associated with the aforementioned molecular forces results in adhesion between the water and any constituent particles and the development of thin interfacial liquid films that encompass the particles. These films are typically only a few nanometres thick, although their thickness is temperature dependent with them becoming thinner as the temperature falls and vice versa.

These pre-melting effects result in porous substrates within permafrost regions displaying significant variations in the relative amounts of water and ice depending on both their temperature and their particle-size characteristics (Williams & Smith, 1989). The influence of interfacial effects within coarse-grained sediments such as sands and gravels is relatively minor. Most of the porewater is therefore bulk water that will tend to freeze rapidly once the temperature drops below the freezing point (Figure 2.3). Conversely, interfacial effects are much more pronounced in fine-grained materials in general and clays in particular, due to their high surface areas and the greater adhesive forces between the particles and the water molecules. As such, fine-grained sediments are capable of retaining significant quantities of unfrozen water content at temperatures well below the bulk freezing point.

A key point to note here is that fine-grained sediments can retain liquid water and therefore remain partially frozen over considerable cryotic temperature ranges. The bulk freezing point simply marks the temperature at which the freezing process can commence with the process continuing as temperatures fall further, resulting in a progressive change in the

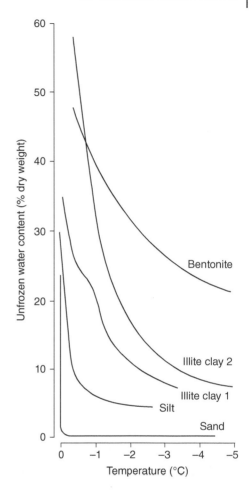

Figure 2.3 Influence of particle size on the unfrozen water content of unconsolidated sediments (Williams & Smith, 1989/Cambridge University Press).

relative concentration of ice and water and a reduction in the amount of residual liquid water. This conception of the freezing process is in marked contrast to the commonly held misconception that melting and freezing occur at a specific temperature and that materials display two states (frozen and unfrozen) above and below this thermal boundary. The implications of this complexity in terms of permafrost rheology and glacier flow are considered further in Sections 2.5 and 3.5, respectively.

More detailed discussion of the temperature-dependency of ice and liquid water in soil can be found in Williams and Smith (1989), with more recent summaries of pre-melting microphysics in porous media being provided by Rempel *et al.* (2004), Dash *et al.* (2006) and Rempel (2007, 2010, 2011).

2.2.3 Ground Ice

Ground ice is a general term employed to refer to the different types of ice located within frozen or partly frozen ground. As the name implies, this important element of the cryosphere occurs below the surface and is therefore typically obscured from view by a layer of sediment and/or vegetation, except where seen in natural exposures. It is its hidden nature that has resulted in this distinctive and increasingly influential element of the cryosphere not receiving the same attention as surface ice types. In a classic early paper on 'underground ice', Mackay (1972) noted that,

> *Remote in the Arctic regions, below ground and thus concealed from view, much of this ice has persisted for millennia, unseen and unsuspected by generations who have lived and travelled only a few feet above it* (p2)

As with liquid water, its presence is not a defining characteristic of permafrost. However, in a number of Arctic and high mountain landscapes, it comprises an important and unique aspect of these terrains whose formation and degradation are associated with a range of distinctive processes, ice-rich facies and sedimentary structures and landforms. The quantity of ice, its characteristics and its distribution are highly variable and can vary from small amounts of ice confined to a substrate's pore spaces to large layers of 'massive ice' up to tens of metres in thickness (Figure 2.4). In terms of the ice content of the ground, a key distinction relates to the concept of excess ice. Excess ice is said to be present if thawing results in the production of a volume of water that exceeds the available pore space (French, 2017). If excess ice is present, the permafrost is often referred to as being ice rich or icy. If not, the ground is referred to as ice poor.

Ground ice occurs in highly variable amounts and can occur in a variety of different forms. The amount and type of ground ice present are determined by the properties of the substrate along with the availability of different water sources, the mechanisms of water

Figure 2.4 Exposure of massive ice in a coastal exposure at Crumbling Point, Summer Island, western Canadian Arctic. Figure for scale.

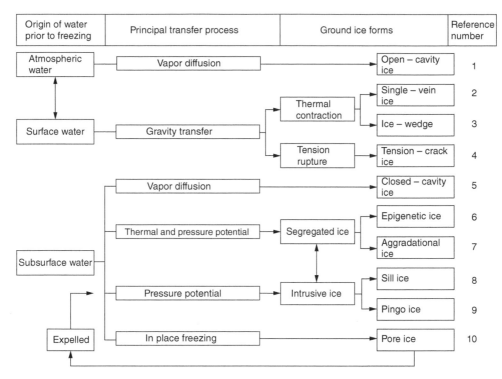

Figure 2.5 Ground ice classification scheme of Mackay (1972)/Taylor & Francis.

transfer to the freezing front and the freezing process. Ground ice can form within pre-existing materials that have already been deposited in which case it is referred to as epigenetic or intrasedimental. Alternatively, syngenetic ice forms whilst deposition of the host sediment is ongoing. Ground ice can also form as a result of the burial of ice that originally formed at the surface. This diversity has been described and explained within a number of classification schemes, the most widely used of which was developed in the aforementioned paper by J. Ross Mackay in 1972 (Mackay, 1972; Figure 2.5). The ten forms of ground ice included in this classification scheme can be synthesised into four main types that include pore ice, segregated ice, intrusive ice and vein ice. The following section provides a brief summary of these key types with more detailed descriptions of the distinctive cryofacies and cryostructures characteristic of permafrost regions being provided by Murton & French (1994) and French & Shur (2010).

1) Pore ice (also known as interstitial ice) is limited to the pore spaces within sediments and therefore occurs as a cement that bonds the constituent grains (Figure 2.6a). Substrates with pore ice are typically ice poor and the ice type therefore typically has no distinctive landform expression.

2) Segregated ice commonly refers to more ice-rich ground that contains excess ice. This excess ice is commonly visible to the naked eye as segregated ice lenses aligned parallel to the freezing front (Figure 2.6b). These can vary in thickness from a few millimetres to tens of metres. Segregated ice forms within fine-grained and frost susceptible substrates

(a) (b)

(c) (d)

Figure 2.6 Images of the principal types of ground ice: (a) Pore ice in the rounded toe of a sand wedge, Crumbling Point, Summer Island, NT, Canada. (b) Segregated ice lens (c. 10 cm thick) located within peat at Nuiqsut, Alaska. Note the tubular bubbles and the discoloured central portion. (c) Intrusive ice dyke located at Peninsula Point near Tuktoyaktuk, western Canadian Arctic (lens cap for scale). Note the glassy texture of the ice and the lack of bubbles. (d) Ice wedges emplaced within massive ice exposed at Crumbling Point, Summer Island, NT, Canada (figure for scale).

capable of generating capillary forces and cryosuction (see previous section). These forces can result in the migration of liquid water to a freezing front or frozen fringe in which the ice lenses progressively develop. The growth and decay of segregated ice is central to a broad range of permafrost processes including frost heave, thaw consolidation, cryoturbation, freeze-thaw weathering and solifluction.

3) Intrusive ice is associated with the injection of liquid water into a layer of permafrost. Consequently, it is ordinarily associated with the development of very high porewater pressures within unfrozen taliks that exceed the overburden pressure and therefore result in the hydrofracturing of the overlying permafrost. This results in the formation

of ice dykes and sills that are conceptually similar to igneous intrusions and which share common features such as chilled margins with very small ice crystals (Figure 2.6c).

4) Vein ice forms when surface water drains into cracks that penetrate into a permafrost layer. The cracks form through tensile fracture associated with rapid winter cooling, and it is the repetition of this process with surface melting in the summer that results in repeated vein-forming events and the development of ice wedges (Figure 2.6d). The distinctive surface manifestation of this ice type in the form of ice-wedge polygons suggests that it is the most extensive ice type comprising between 20–35% of the total volume of ground ice occurring within the upper 5–10 m of permafrost (French, 2017).

Massive ice is a specific type of ground ice that has been widely documented within the western Canadian Arctic (Figure 2.4; e.g. French & Harry, 1990; Mackay & Dallimore, 1992; Murton *et al.*, 2005), western Siberia (e.g. Vtyurin & Glazovskiy, 1986; Astakhov & Isayeva, 1988) and Alaska (e.g. Kanevskiy *et al.*, 2013). Massive ice is defined by an ice content in excess of 250% by weight (French, 2017) and typically occurs as large tabular bodies that can exceed 10 m in thickness. Whilst its distribution, stratigraphic setting and physicochemical properties have been increasingly well constrained, its origin has been the subject of debate. Two principal hypotheses have been advocated, both of which involve glaciers and permafrost. This type of ground ice can therefore be viewed as an explicit product of glacier–permafrost interactions, and as such further detail on its nature and debated origin is provided in Section 5.3.4. In addition, the exposure, ablation and landscape destabilisation associated with the exposure of these massive ice bodies represents one of the most dramatic impacts of Arctic climate change. The recent and ongoing research demonstrating its importance in this context is explored further in Section 5.6.2.

2.3 The Formation and Preservation of Permafrost

2.3.1 Subaerial Permafrost

With permafrost involving the establishment of a thermal regime that is required to persist for a given period of time, its formation is dependent on the exchanges of heat and moisture in combination with the thermal properties of the ground itself. With the majority of permafrost being subaerial in origin, the ground thermal regime is driven primarily by the energy exchanges that occur between the atmosphere and the ground surface. These are described by the surface energy balance, which comprises the net exchange of radiation (Q^*), the transfer of sensible heat (Q_H) and latent heat (Q_{LH}) and the conduction of heat into the ground (Q_G) (Williams & Smith 1989):

$$Q^* = Q_H + Q_{LH} + Q_G$$

Any change in the surface energy balance will result in a change in the ground surface temperature. During daytime for example, Q^* is positive (i.e. more radiation is absorbed than is reflected). This results in an increase in surface temperature that can in turn cause an increase in sensible and latent heat fluxes into the atmosphere and the conduction of heat into the ground such that the sum of any changes is zero (i.e. energy is conserved).

Longer-term seasonal changes in the surface energy balance drive changes in the ground thermal regime that can penetrate to a depth of up to 10–20 m. The point at which seasonal changes are no longer discernible is the point at which the mean annual ground temperature (MAGT) is recorded. For permafrost to occur therefore, the MAGT must be at or below 0 °C for at least two years.

With surface energy balances and temperatures being primarily determined by net radiation (Q*), there is a general tendency for both atmospheric and surface temperatures to decrease with increasing latitude as the receipt of solar radiation decreases. As a general rule of thumb, permafrost can generally be found in climatic regimes where the mean annual air temperature (MAAT) is below −1 °C. In reality however, local variations in ground surface conditions lead to complex and spatially variable relationships between the MAAT and the MAGT, which can vary by several degrees Celsius. As a consequence, the distribution and thickness of permafrost are typically more complex and variable than one would expect if climate was the sole driver. In a study of the connections between microclimatic variations and permafrost distribution in the Mackenzie Delta (Canadian NWT), Smith (1975) for example identified vegetation and snow cover as the most influential controls that in combination led to differences in temperature at a depth of 1 m of up to 6 °C over distances of only 12 m. An increase in vegetation cover was generally associated with a decrease in ground surface temperature such that the removal of only 10 cm of organic material led to an increase in the mean daily temperature of 3 °C at a depth of 10 cm. This is important in that it highlights the thermal sensitivity of permafrost environments to what might seem like comparatively minor surface disturbances. Spatial variations in snow depths led to recorded variations in ground surface temperatures of up to 20 °C in March that were still identifiable as 10 °C variations in late summer. An increase in the thickness of the snow cover increased the insulation from the cold winter air temperatures such that the areas associated with the thickest snow accumulations were associated with taliks (no underlying permafrost).

Large water bodies that are commonly found in permafrost and glacial environments can also have a profound impact on local ground thermal regimes and the distribution of permafrost. Lakes and rivers with depths exceeding the thickness of winter ice are characterised by bottom temperatures above 0 °C that promote the formation of subjacent taliks. Measurements of lake bottom temperatures on Richards Island (Mackenzie Delta) by Burn (2002) recorded values of between 1.5 and 4.8 °C beneath the central pools, with subsequent geothermal modelling predicting that the combined thermal influence of these lakes is sufficient to create a talik beneath 10–15% of the island's total area. Conversely, one of the most dramatic field experiments undertaken, involving the artificial drainage of a large lake at Illisarvik in 1978, demonstrated the rapidity of permafrost aggradation and ice wedge formation once the lake was deliberately drained (Mackay, 1986). Interestingly, the initially rapid development of the ice wedges was found to have ceased over much of the lake bed by 1990 as a consequence of vegetation development and increased snow entrapment that increased the winter ground temperatures and limited the potential for thermal contraction cracking (Mackay & Burn, 2002).

Focusing on the near-surface materials below this surface boundary layer, the equilibrium depth of permafrost (Z_p) can be determined according to the long-term mean annual ground surface temperature (T_s) and the geothermal gradient, which is in turn controlled by the heat flux (Q_G) and the thermal conductivity of the substrate (K) (Williams & Smith, 1989):

$$Z_p = T_s K / Q_G$$

This equation indicates that under steady-state conditions, the thermal conductivity of the substrate is equally as important as surface temperatures in determining permafrost thickness, with thick permafrost being promoted by low surface temperatures, low heat flows and high thermal conductivities. In practice, permafrost thickness is complicated by additional factors, most notably the geological history (e.g. Allen *et al.*, 1988) and the presence of ice and water close to their phase transition temperature within the substrate. The presence and state of porewater can also significantly modify the thermal conductivity of a material, an influence further complicated by the volumetric fraction of ice and liquid water also being temperature-dependent (Section 2.2.2). In addition, the high heat capacity of water and the release of latent heat upon freezing can substantially slow the rate of permafrost aggradation. Finally, temperature gradients in frozen sediments drive the migration of unfrozen water which results in additional heat fluxes. Recent summaries of the thermal regime of permafrost that address these mechanisms in greater detail are provided by Burn (2004, 2007) and French (2017).

2.3.2 Subglacial Permafrost

Whilst the majority of permafrost is formed in subaerial settings in response to surface-atmosphere energy exchanges and the prevailing climatic regime, permafrost can also grow and persist within subaqueous and subglacial environments so long as the interface between the permafrost and the adjacent medium remains cryotic for at least two years. In glacial environments, this basic requirement suggests that active subglacial permafrost will only occur beneath cold-based ice, where the temperature of the basal ice remains below the pressure melting point for at least two years. Strictly speaking, subglacial permafrost can also occur beneath warm-based ice in a situation where the temperature of the basal ice is below 0 °C but above the pressure melting point. This situation would however be no different to warm-based ice resting on a non-permafrost substrate. Focusing on the basal conditions and processes associated with cold-based ice is therefore of greater relevance to a consideration of the distinctive nature and impacts of glacier–permafrost interactions.

In considering the basal heat budgets of glaciers, Lorrain & Fitzsimons (2011) define three basic scenarios in relation to basal thermal regimes that differ according to the basal temperature gradient and the rate at which heat energy can be removed from the glacier bed.

1) More heat is supplied to the basal ice than can be conducted upwards through the glacier. In this case, the surplus heat energy will increase its temperature and/or cause basal melting if it is already at the pressure melting point.
2) The heat supplied is equal to that which can be conducted.
3) The heat supplied is lower than the amount that can be conducted. In this case, which is of most relevance to cold-based glaciers, the glacier will tend to freeze to its bed.

Hooke's (1977) qualitative review of the basal temperatures of polar ice sheets provides a more detailed insight into the glaciological conditions required for cold-based conditions to occur and consequently for subglacial permafrost to form and persist. Starting with the simplest scenario of a stagnant ice mass that is below the pressure melting point throughout, the steady-state temperature distribution is determined by the conditions at the upper and lower boundaries (Figure 2.7). At the lower boundary (the glacier bed), there is a geothermal heat flux from the Earth, the magnitude of which determines the temperature gradient within the ice mass. A typical geothermal heat flux for the Canadian Shield for example will result in temperature gradient of $0.017\,°C\,m^{-1}$. The surface temperature then determines the distribution of temperatures within the ice mass that vary according to the prescribed geothermal gradient. In an ice sheet 1000 m thick for example with a temperature gradient of $0.017\,°C\,m^{-1}$, the temperature difference between the surface and the bed will be 17 °C. A mean surface temperature of −27 °C will therefore correspond with a basal temperature of −10 °C. For cold-based thermal conditions to occur, the surface temperature must therefore be low enough to result in a basal temperature that is below the pressure melting point. As pure ice has a substantially higher thermal conductivity than snow and is therefore associated with low thermal gradients (Williams & Smith, 1989), cold-based conditions and subglacial permafrost can in theory be maintained beneath substantial ice thicknesses. This is supported by empirical observations of the temperatures beneath thick polar ice sheets. Herron & Langway (1979) for example recorded a temperature of −13 °C at the base of the Greenland Ice Sheet at Camp Century where the ice thickness was 1375 m.

Hooke (1977) goes on to consider the influence of the vertical velocities generated by surface accumulation and ablation. Within the accumulation area, the velocity is downward. It is also at its maximum at the surface and zero at the bed with the most rapid rate of decrease occurring near the surface. In order to maintain a steady-state temperature distribution, cold surface ice moving downwards must therefore warm up. The heat required is removed from the (geothermal) heat flux moving upwards resulting in a reduction in the temperature gradient that therefore decreases upwards. The resulting profile is likened to an elastic band that has been stretched downwards with colder ice being advected downwards from the surface. Conversely, the *upward* vertical velocity in the ablation area requires ice to be *cooled* as it rises. This augments the geothermal heat flux as the temperature gradient

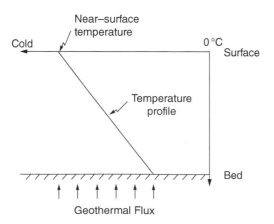

Figure 2.7 Theoretical temperature distribution in a stagnant ice mass (Hooke, 1977/Cambridge University Press).

increases towards the surface. With both the vertical velocity and temperature gradient increasing upwards, this results in an exponential decrease in temperature towards the surface that again is likened to an elastic band being stretched upwards.

The significance of these vertical velocities for the predicted basal temperatures is clearly illustrated through the calculation of a series of temperature profiles for both the ablation zone and the accumulation zone. These involve the same boundary conditions (surface temperatures and geothermal heat fluxes) and therefore highlight the influence of changes in ice thickness and vertical velocity (in turn related to accumulation and ablation rates). In the case of the ablation zone profiles (Figure 2.8a), the basal temperature is observed to increase markedly as the vertical velocity (v) increases. Where $v = 0.5 \, \mathrm{m \, yr^{-1}}$, the temperature at the bed reaches the pressure melting point such that only 46% of the geothermal heat flux can be conducted upwards through the ice resulting in basal melting. A reduction in ice thickness also has a significant influence with a reduction in ice thickness from 500 to 350 m resulting in a 7.4 °C reduction in basal temperature where $v = 0.25 \, \mathrm{m \, yr^{-1}}$. In the case of the accumulation zone profiles (Figure 2.8b) calculated for a thicker ice sheet,

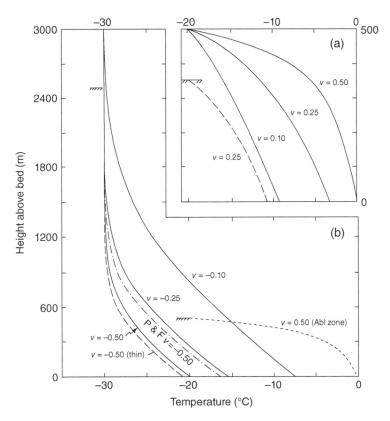

Figure 2.8 Theoretical temperature profiles for the ablation zone (a) and accumulation zones (b) for different vertical velocities (v = vertical velocity at the glacier surface in $\mathrm{m \, yr^{-1}}$ with positive being upward and vice-versa). Horizontal conduction and advection and internal heat generation are ignored. In each case, the basal temperature gradient is 0.017 °C m^{-1} and the surface temperature is −20 °C (Hooke, 1977/Cambridge University Press).

an increase in the vertical velocity (i.e. accumulation rate) results in a significant decrease in the basal temperature. With the surface ice being effectively isothermal, the influence of a reduction in ice thickness from 3000 to 2500 m is comparatively minor, resulting in a decrease in temperature of only 0.9 °C. Hooke (1977) cautions therefore that an increase in ice thickness in the accumulation area of an ice mass will not necessarily result in a significant increase in basal temperature and may actually cause a reduction in temperature as colder ice is advected downwards from the surface.

The downwards advection of cold firn created in the central accumulation areas of ice masses in cold climatic regimes therefore provides an important mechanism through which subglacial permafrost can be maintained. Even at relatively low accumulation rates, this can result in the penetration of very low surface temperatures to a considerable depth beneath the centre of the ice cap that can promote the preservation of subglacial permafrost even in situations where the glacier or ice sheet is much thicker than the permafrost (Robin, 1955).

In examining the influence of thermal regime on ice flow at Storglaciaren, a polythermal glacier resting on a bed of unconsolidated sediment (Section 4.3.1), Moore *et al.* (2011) consider the heat balance required to actively freeze sediments onto the base of the glacier, a process that if sustained over a sufficient period of time will lead to the subglacial aggradation of ice-rich permafrost. As identified by Hooke (1977), the thermal character of the overlying glacier is of primary importance. If the glacier is cold throughout its thickness, then the heat flux upwards from the bed (Q_c) is described by the following equation (where k is the thermal conductivity of the ice, T is the temperature and z is the vertical distance):

$$Q_c = -k \, \mathrm{d}T / \mathrm{d}Z$$

If the heat conducted upwards away from the bed exceeds that supplied by heat sources such as the geothermal heat flux, then additional heat must be derived from alternative sources. These can include frictional heat associated with active basal processes, dissipated energy associated with water flow or latent heat release associated with basal freeze-on. In recording relatively high basal thermal gradients suggestive of basal heat fluxes four times higher than those expected for geothermal heat fluxes and low basal velocities, the authors suggest that the latent heat release from the freezing of porewater within the subglacial sediments provides the only energy source capable of providing the increasing amounts of heat required by the vertical temperature gradient that is observed to steepen towards the margin as the ice thins. Using a one-dimensional ice flow model, the authors calculate that a rate of subglacial freeze-on of a few centimetres to a few decimetres per year is capable of providing the heat flux required.

When one considers the longer timescales associated with the formation and decay of ice sheets (see Section 2.4.2), rather than initially forming underneath a glacier, subglacial permafrost may alternatively occur as a result of a glacier's development within a landscape characterised by pre-existing permafrost. In examining the likely evolution of basal thermal conditions beneath Late Wiconsinan ice caps in the Canadian High Arctic, Dyke (1993) considers their thermal evolution and the related impacts on the presence and thickness of subglacial permafrost over time. An ice cap initially forming over thick permafrost would inevitably be entirely cold based. Progressive thickening of the ice mass will subsequently result in a concomitant thinning of the subglacial permafrost, although the approximation of the thermal profile to the subglacial geothermal gradient means that it is likely to persist as long as the ice thickness is less than the thickness of the pre-existing permafrost. Empirical evidence for this general rule of thumb is provided by Koerner (1989) who recorded temperature profiles beneath the accumulation areas of the Devon and Agassiz Ice Caps that indicated cold-based conditions. However, as noted earlier, the downwards advection of cold ice in the accumulation area can limit the warming of the basal ice and promote the continued persistence of subglacial permafrost even in situations where the ice is thicker than the permafrost.

Continued thickening of an ice mass and the associated increase in basal shear stress have the potential to result in strain heating that can provide an important additional source of heat energy, particularly in the basal ice where the temperatures and related strain rates are highest. A metre per year of ice movement at the base in this respect is estimated to generate $2.8\,\mathrm{mJ\,m^{-2}\,s^{-1}}$ of heat energy (Robin, 1955; cited in Dyke 1993, p229). If this warming results in active basal sliding, then the frictional heat generated can lead to further warming and potentially basal melting. A sliding rate of only $1\,\mathrm{m\,a^{-1}}$ for example is capable of generating sufficient frictional heat to raise the temperature of a 0.05 m basal ice layer by $1\,^{\circ}\mathrm{C}$. The combination of these additional sources of heat with the geothermal heat flux is typically thought to result in an increased temperature gradient within the base of the ice mass (Hooke, 1977).

However, if subglacial permafrost is present, Dyke (1993) suggests that this heating within the basal ice can create a temperature inversion at the glacier bed (Figure 2.9). Heat energy will consequently migrate both upwards and downwards from the inversion. Whilst this will act to limit its magnitude, it will persist as long as the strain heating is sufficient

Figure 2.9 Sequence of subglacial permafrost dissipation below an ice cap generating enough strain heating to create a temperature version (*I*) at the ice-bed interface. *G* is the geothermal gradient (Dyke (1993, p230). Figure 4).

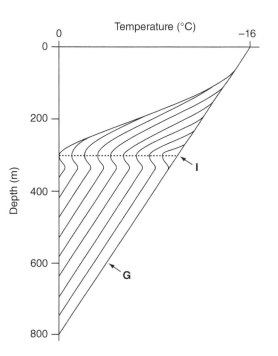

creating an 'inversion trap' that prevents geothermal heat from flowing upwards into the glacier. This heat energy is instead utilised to drive a progressive warming and thinning of the subglacial permafrost as well as a steepening of the temperature gradient within the overlying ice. In this way, the additional heat energy generated by strain heating in the basal layers of more dynamic ice masses is capable of rapidly degrading any permafrost initially present. In calculating the time required to thaw pre-existing permafrost, Dyke (1993) estimates that if a geothermal heat flux of $60\,mW\,m^{-2}$ is focused entirely on the melting of a permafrost layer 600 m thick with a surface temperature of $-11\,°C$ and a volumetric ice content of 6%, it would nevertheless take over 10,000 years for the layer to thaw. However, if the basal ice reached the pressure melting point, then the heat energy associated with more rapid rates of sliding and the production of significant volumes of meltwater is likely to result in its more rapid degradation such that warm-based conditions and subglacial permafrost are typically considered to be mutually exclusive.

With glaciation and an increase in ice thickness and basal shear stress being associated with the degradation and potential destruction of subglacial permafrost, it is worth noting that deglaciation can subsequently result in the re-formation and aggradation of subglacial permafrost. The stagnation and immobilisation of formerly active ice can lead to the dissipation of heat and the migration of the $0\,°C$ isotherm below the ice-bed interface such that the glacier essentially becomes part of the local permafrost (Boulton, 1970b). Further detail on the temporal changes in the extent of subglacial permafrost and their association with the impacts of ongoing climate change in Arctic glacial environments is provided in Section 2.4.2.

2.4 The Distribution and Spatial Extent of Permafrost

2.4.1 The Modern Day and Last Glacial Maximum Distribution of Permafrost

The permafrost region in the Northern Hemisphere has been estimated to occupy up to c. 23.6 million km^2 (c. 24%) of the exposed land area (Zhang *et al.*, 2008). The majority of this region occurs between 60 and 68°N with the most significant areas of permafrost occurring in Northern Canada, Siberia and Alaska. In addition, significant areas of permafrost can also be found at high altitudes in lower latitude regions including the Tibetan Plateau, the mountains of southwest Asia, the Rocky Mountains of North America and the Alps and Caucasus mountains of Europe. Within this broader region, permafrost is classified in terms of its spatial extent, varying – in a transect from colder northern areas to warmer southern areas – from *continuous* (where it underlies 90–100% of the land surface), through *discontinuous* (50–90%), to *sporadic* (10–50%) and *isolated* (0–10%). As a result, the actual area of exposed land in the Northern Hemisphere that is underlain by permafrost is significantly smaller than that of the overall permafrost region, being estimated to be between c. 12 and 17 million km^2 (c. 13–18%) (Zhang *et al.*, 2000). With the majority of permafrost being located in polar and alpine environments where glaciers and ice sheets are also prevalent, glaciers and permafrost are often juxtaposed. For example, the ice-free areas surrounding the ice sheets in Antarctica and much of Greenland are characterised by extensive permafrost (e.g. van Tatenhove and Olesen, 1994; Bockheim, 1995; Ruskeeniemi et al., 2018). Although just 2% of the Antarctic continent is ice-free, this still represents an area of 280,000 km² which is roughly equivalent to the area of the United Kingdom.

During the glacial phases of the Quaternary Period, permafrost is thought to have been much more extensive than it is today, extending for example out onto the currently submerged shallow continental shelves of Eurasia (Vandenberghe & Pissart, 1993; Ballantyne & Harris, 1994). In 2010, the International Permafrost Association established an action group to map the extent of permafrost in the Late Pleistocene, focusing largely on lowland areas (<1000 m elevation). Employing archival data on the distribution of features indicative of permafrost such as ice-wedge pseudomorphs, sand wedges and large cryoturbations, Vandenberghe et al. (2014) constructed a map illustrating the maximum distribution of permafrost in the Northern Hemisphere between 25 and 17 ka BP, a period they term the 'Last Permafrost Maximum' (LPM). This mapping suggests that the southern boundary of permafrost extended to 47–44°N in Europe and further south to 48–36°N in Central Asia and North America (Figure 2.10). In a follow-up study, Lindgren et al. (2016) employed GIS techniques to enhance the compilation and integration of available empirical evidence and thereby quantify the estimated extent of permafrost around the time of the Last Glacial Maximum (LGM). They concluded that permafrost extended over an area of 34.5 million km² with estimated maxima and minima of 35.3 and 32.7 million km², respectively. Taking the highest figure, this represents an extension of almost 11 million km² to encompass a total area roughly 33% larger than its present-day distribution. In view of this book's focus on glacier–permafrost interactions, it is important to note that the LGM and LPM are unlikely to have been synchronous events. The maximum extent of Late Pleistocene permafrost and ice sheets almost certainly occurred at different times in different places. As such, Lindgren et al. (2016) prefer the use of the term 'LGM permafrost' extent with the LGM period itself spanning a broad period of time (26.5–19 ka BP following Clark et al., 2009).

With the thickest permafrost sequences requiring tens of thousands of years to form, the present-day distribution of permafrost remains in part a thermal legacy of this expanded distribution during the LGM. The permafrost in some regions today is still much thicker than would be anticipated to form in relation to the current climatic regime. This *relict permafrost* also persists in some areas where the current MAAT is not sufficiently cold to account for active permafrost development. Szewczyk & Nawrocki (2011) for example have identified a layer of relict permafrost at least 93 m thick at a depth of 357 m in north-eastern Poland that constitutes a deep-seated thermal remnant of the cold climatic regimes of the Late Pleistocene. The sub-sea permafrost occurring on the shallow continental shelves of the Northern Hemisphere comprises the most extensive type of relict permafrost, with an estimated 2.4 million km² estimated to have persisted following their exposure due to eustatic sea-level lowering during the LGM (Lingren et al., 2016).

2.4.2 The Spatial Distribution of Contemporary Permafrost

The distribution of contemporary subglacial permafrost beneath glaciers and ice sheets is poorly understood due to the inherent inaccessibility of subglacial environments. Some direct observations of the basal thermal regime of the major ice sheets have been

LPM permafrost map

Extent of LPM permafrost:

Equilibrium permafrost (climate-controlled)

Land area under relict permafrost

Exposed land under relict permafrost

Exposed land at time of LPM (assumed to be permafrost)

Extent of continental ice sheets at time of LGM

Exposed land without permafrost at time of LPM

Approximate boundary between LPM continuous and discontinuous permafrost

Southern boundary of present-day permafrost

Approximate limit of LPM winter sea ice extent

Rivers

Modern coastline

Elevation contour lines (m a.s.l.)
1000
2000
4000

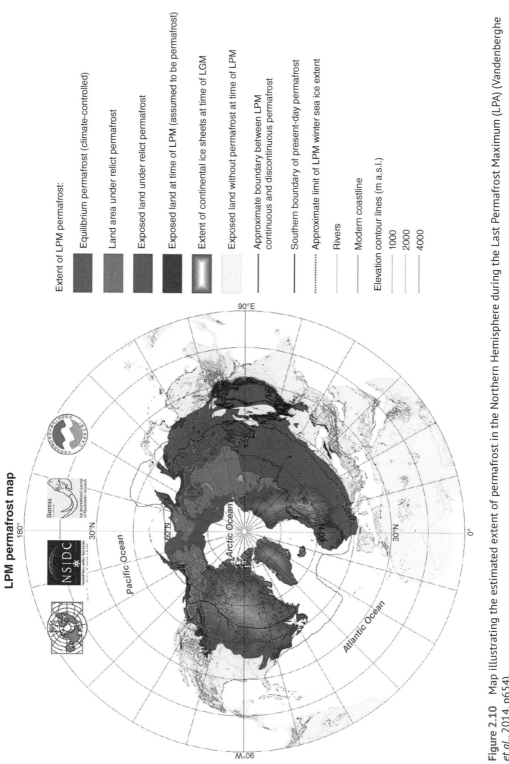

Figure 2.10 Map illustrating the estimated extent of permafrost in the Northern Hemisphere during the Last Permafrost Maximum (LPA) (Vandenberghe *et al.*, 2014, p654).

undertaken as part of broader ice-coring projects. Whilst these only provide occasional 'spot measurements', they have nonetheless demonstrated the presence of cold-based ice and therefore confirmed the potential for subglacial permafrost to occur beneath thick ice masses, with Hansen & Langway (1966) for example recording a basal temperature of $-13\,°C$ beneath the 1375 m thick ice at Camp Century in northwest Greenland. Borehole measurements have also confirmed the presence of cold ice at the base of smaller and thinner ice caps within the High Arctic, with two boreholes close to the summit of the Devon Island Ice Cap in the Canadian Arctic recording basal temperatures of $-18.4\,°C$ beneath the 300 m thick ice (Paterson *et al.*, 1977).

The development of non-invasive geophysical techniques has allowed researchers to image much larger areas of the glacier bed, yielding a more extensive insight into the spatial variations of both basal thermal regimes and the extent of subglacial permafrost. The dielectric differences between liquid water, ice, air and sediment have resulted in electromagnetic techniques being the most frequently employed. In Svalbard, Björnsson *et al.* (1996) highlighted the use of airborne radio-echo sounding as an effective means of mapping the thermal regimes of polythermal glaciers. Radar signals at 320–370 MHz easily penetrated ice at temperatures well below the pressure melting point but were reflected by the interface with the subjacent temperate basal ice that contained liquid water. In contrast, lower-frequency signals of 5–20 MHz were able to penetrate both the warm and cold ice and therefore to image the bed. The application of these techniques at Midre Lovenbreen, a polythermal valley glacier in northwest Spitsbergen, indicated that the glacier was cold based beneath its snout and along its lateral margins but it retained a core of warm basal ice beneath its centreline.

More recently, Ruskeeniemi *et al.* (2018) report on the results of electromagnetic (EM) surveys undertaken as part of the 'Greenland Analogue Project' that provide what is considered to be the first direct evidence for the presence of subglacial permafrost beneath the margin of the west Greenland Ice Sheet near Kangerlussuaq. With the electrical resistivity of bedrock being in part dependent upon the presence of porewater, the immobilisation of charge carriers within permafrost with limited bulk free water can result in a significant resistivity contrast between frozen and unfrozen substrates. Following preliminary EM surveys in the proglacial area that revealed a cryogenic stratigraphy comprising a thin active layer (c. 1 m), a thick permafrost layer (c. 500 m) and a subpermafrost talik, the authors applied the technique to a $3\,km \times 3\,km$ grid beneath the ice margin (Figure 2.11). Under the ice margin, the EM surveys identified a thin (1–10 m thick) and low resistivity layer (300–1 kΩm) immediately below the ice-bed interface that is interpreted as a thin layer of unfrozen till. This was almost always underlain by a thick (c. 500 m) high-resistivity layer (30–40 kΩm) interpreted as a thick layer of subglacial permafrost with resistivity values identical to those characterising the proglacial permafrost. Finally, the subglacial permafrost was in places underlain by another low-resistivity layer (c. 500Ωm) that is interpreted as a subglacial talik containing saline groundwater, previously identified and sampled during borehole investigations (Harper *et al.*, 2017). As the profiles were extended up ice, the subglacial permafrost layer was observed to persist for 2 km after which it was replaced by a layer with values more typical of unfrozen bedrock (c. 5000Ωm). Whilst this provides clear evidence for the persistence of thick permafrost underneath a modern ice sheet, it is important to note that the recorded extent of 2 km under the margin

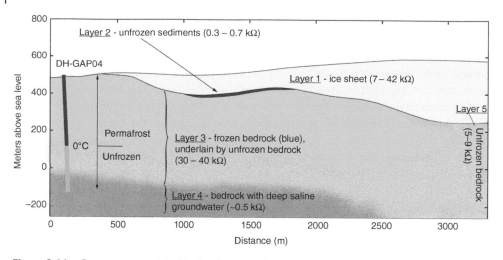

Figure 2.11 Conceptual model of bedrock and its thermal conditions below the ice-margin of the west Greenland ice sheet. Permafrost (frozen bedrock) extends for at least 2.5 km beneath the ice margin (Ruskeeniemi *et al.*, 2018, p182).

provides a minimum estimate. As acknowledged by the authors, the technique is used to differentiate areas of frozen and unfrozen material and therefore focuses on the state of any porewater present rather than the temperature of the substrate that is key to the definition of permafrost. It is therefore highly likely that subglacial permafrost with temperatures persistently below 0 °C but which contains significant amounts of liquid water extends further under the ice-sheet margin.

Ground-based ground penetrating radar (GPR) surveys have also been employed extensively in attempts to detect and map thermal transitions and thereby to elucidate the characteristics of the subglacial materials and the possible presence of subglacial permafrost. Whilst cold ice appears transparent in GPR surveys, the presence of water inclusions within temperate ice leads to backscattering and a more chaotic response allowing their differentiation and systematic mapping (e.g. Gusmeroli *et al.*, 2012a). In a number of cases, these GPR surveys have been combined with borehole observations and measurements that allow the geophysical interpretations to be ground-truthed. Murray *et al.* (2000) for example combined the use of GPR surveys and borehole measurements both to relate a cold-temperate transition in basal thermal regime to the active surge front of Bakaninbreen in Svalbard and also to estimate the thickness (15–22 m) of subglacial permafrost beneath the glacier's cold-based lower reaches (see Section 4.3.1). Repeat GPR surveys on a polythermal glacier on Bylot Island in the Canadian Arctic by Irvine-Fynn *et al.* (2006) have demonstrated the technique's ability to detect temporal changes in a glacier's thermal structure. Repeat surveys revealed seasonal changes in signal-to-noise, attenuation and the appearance and disappearance of internal reflectors suggestive of the temporary generation and storage of englacial meltwater during the ablation season. Gusmeroli *et al.* (2012b) similarly used repeat GPR surveys over a 20-year period (1989–2009) to chart the progressive loss of a cold surface layer at Storglaciären in Sweden in response to ongoing climatic warming. Conversely, systematic GPR surveys of the snout of Midtdalsbreen in south-central Norway have demonstrated the unexpected presence of areas of cold ice in both the ablation and accumulation areas of what was previously assumed to be an entirely temperate outlet glacier (Reinhardy *et al.*, 2019).

Numerical ice-sheet models have also been employed to reconstruct the basal thermal regime of both modern and ancient ice masses over large spatial and long temporal scales. In a seminal contribution, Sugden (1977) used the steady-state model of Budd (1970) to reconstruct the basal thermal regime of the Laurentide Ice Sheet at its maximum extent. The model predicted the presence of four distinct and largely concentric thermal zones extending from the centre of the ice sheet to its periphery (Figure 2.12): (i) an inner zone of warm-based conditions and net basal melting, (ii) a zone of warm-freezing basal ice where the latent heat released by the refreezing of meltwater maintained the ice at the pressure

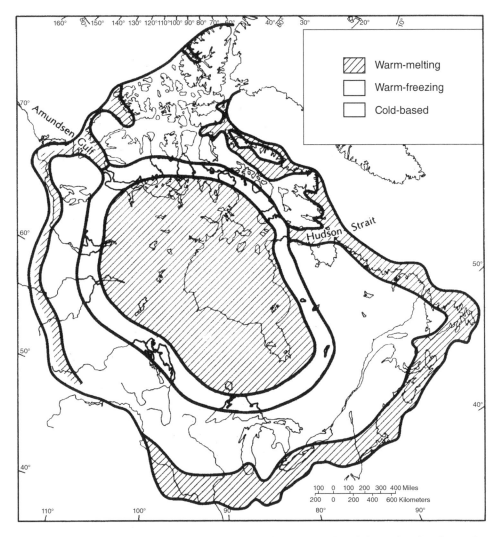

Figure 2.12 Model-based reconstruction of zones of contrasting basal thermal regime beneath the Laurentide Ice Sheet as its maximum (Sugden, 1977, p38).

melting point, (iii) an extensive marginal zone of cold-based ice occurring everywhere except beneath deep troughs (Hudson Strait & Amundsen Gulf) and (iv) a peripheral zone of warm-melting ice.

Subsequent numerical modelling has suggested that cold-based ice may have been even more widespread. Three-dimensional ice-temperature modelling by Marshall & Clark (2002) has suggested that rather than cold-based conditions being limited in extent, the Late Pleistocene North American ice sheet was in fact dominated by this basal thermal regime for the majority of the last glacial cycle with the area of cold-based ice remaining at c. 65% for 70,000 years. It was only after the LGM that the area of warm-based ice increased exponentially in response to the increased insulation promoted by the thick ice and a combination of geothermal and strain heating. This resulted in a concomitant increase in flow velocities that ultimately triggered the ice sheet's rapid collapse. Focusing on modern-day ice sheets, early modelling undertaken by Herterich (1988) that excluded basal processes and ice shelves has also suggested that cold-based conditions are likely to be widespread beneath the modern-day Antarctic Ice Sheet (Figure 2.13). Subsequent flowline modelling incorporating basal processes suggests a more spatially complex picture with Wilch & Hughes (2000) for example suggesting that whilst East Antarctica is dominated by a low thawed fraction (i.e. a predominance of cold-based ice), warm-based conditions can readily occur in subglacial basins (with thicker ice) and in areas of rapid flow convergence (ice streams).

The occurrence of thick permafrost sequences coupled with the observation of geothermal temperature profiles within deep boreholes located in formerly glaciated permafrost regions has been used to argue for the persistence of permafrost beneath Pleistocene ice sheets for one or more glacial cycles. Allen *et al.* (1988) used thermal modelling to suggest

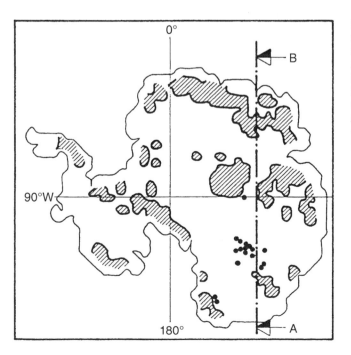

Figure 2.13 Modelled distribution of cold-based ice beneath the modern-day Antarctic Ice Sheet (identified by the hatched areas). Dots illustrate the location of subglacial lakes and the line A–B refers to a modelled profile (Herterich, 1988, p34/ Cambridge University Press).

that the deepest permafrost in the Mackenzie Delta area (western Canadian Arctic) has been aggrading for c. 120,000 years such that it must have persisted through at least two glacial cycles (albeit close to the margin of the Laurentide Ice Sheet). Inversion models have also been developed to convert the vertical temperature profiles recorded in deep boreholes into ground surface thermal histories (e.g. Shen & Beck, 1992). Whilst these techniques are usually employed to inform palaeoclimatic reconstructions, Rolandone *et al.* (2003) used the temperature measurements from six deep (>1600 m) boreholes across Canada to provide empirical evidence for large-scale spatial variations in the basal thermal regime of the last Laurentide Ice Sheet. In contrast to Marshall & Clark (2002), they argued that ground surface temperatures exceeding the melting point of the ice were widespread during and after the LGM with the associated geothermal heat fluxes contributing to the ice sheet's instability during its deglaciation.

Geomorphological inverse models relating distinctive landforms to thermally regulated processes have also been developed by geomorphologists to provide an alternative empirical means of reconstructing the basal thermal regimes of Pleistocene ice sheets that can in turn be used to test the outputs of numerical models (see Sections 5.3 and 5.5 for further details). Various landforms including thrust-block moraines, hummocky moraines and tunnel channels have been used for example to suggest that frozen-bed conditions and ice-marginal permafrost were limited to narrow zones of no more than a few kilometres in width beneath the southern margin of the Laurentide Ice Sheet (e.g. Attig *et al.*, 1989; see section 5.5.2). Sugden (1977) acknowledged the potential existence of an outermost fringe of cold-based ice that could be viewed as an extension of proglacial permafrost beneath the ice margin, although its limited extent meant that it was not depicted within his large-scale numerical model mentioned previously (Figure 2.12). Conversely, in relating the formation of ribbed moraines to thermal transitions from cold to warm-based ice (Section 5.3.3), Kleman & Hättestrand (1999) have suggested that large areas of the interior parts of the Laurentide and Fennoscandian Ice Sheets were associated with frozen-bed conditions and therefore by inference with subglacial permafrost.

Geomorphological inverse models typically rely on a central assumption that basal processes are restricted to areas of warm or wet-based ice (Section 3.4). Areas associated with limited modification and the preservation of preglacial features are consequently identified as a means of inferring the former presence of cold or dry-based ice (see Section 5.5.1 for more detail). Kleman & Glasser (2007) applied this approach to a range of mid-latitude Pleistocene Ice Sheets and the contemporary Antarctic ice sheet, concluding that ice sheets are likely to display a more complex subglacial thermal organisation than has hitherto been recognised within numerical models (Section 4.3.4). Focusing more specifically on the potential location of subglacial permafrost, 'frozen-bed patches' are thought to have occurred at various spatial scales ranging from large areas of the bed (e.g. near continental ice divides) to isolated islands in areas otherwise dominated by warm-based conditions (Section 5.5.2).

A number of authors have however highlighted the challenges and limitations inherent in the use of geomorphological data to infer the basal thermal regime and therefore the potential extent of permafrost beneath former ice sheets. Geomorphological and geological records have an inherent tendency to underestimate the extent of glacier–permafrost interactions, with Kleman & Glasser (2007) noting that,

The wet-bed system is ... responsible for cumulative consummation of the direct evidence for frozen-bed conditions, whereas the frozen-bed system preserves the evidence for wet-bed conditions unaltered. This induces a strong bias towards wet-bed conditions in face-value interpretations of the geomorphological record. Stratigraphic data contained in glacial deposits will be equally biased towards evidence for wet-based conditions (p585).

Moreover, whilst numerous authors have assumed that the boundaries between warm-based (thawed) and cold-based (frozen) zones are clear and distinct, as with any kind of mapping, this is unlikely to be the case in reality. With this in mind, Mooers (1990) for example cautioned that it is,

...unlikely, however, that the boundary between thawed-bed and frozen-bed zones would be abrupt. The transition from thawed to frozen beds is likely to occur as a diffuse zone, possibly with isolated frozen patches separated by areas of basal melting (p857).

In conclusion, whilst there are discrepancies between numerical, geophysical and geomorphological reconstructions of basal thermal regimes, there is a broad agreement with pre-existing research that subglacial permafrost beneath large ice masses is most likely to occur in three principal locations:

1) **Marginal fringes** where sub-freezing basal temperatures are promoted by thin ice cover. With the majority of this work focusing on the mid-latitude margins of Pleistocene ice sheets, it is generally assumed that these fringes reach no more than a few kilometres beneath the ice margin. However, it is reasonable to infer that more extensive cold-based zones existed beneath their high-latitude margins in response to the colder climatic conditions and the presence of thicker and colder pre-glacial permafrost (Astakhov *et al.*, 1996; Murton *et al.*, 2004).
2) **Central areas of ice masses (ice-dispersal centres)** where air temperatures are lowest, cold firn is advected to the base of the ice mass and flow-related strain heating is minimised by the low ice-surface slope (Dyke, 1993; Kleman & Hättestrand, 1999; Marshall & Clark, 2002; Kleman & Glasser, 2007).
3) **Areas of flow divergence** where cold ice is brought into contact with the bed and ice thicknesses and basal shear stresses are reduced (Kleman *et al.*, 1999; Dyke, 1993).

2.4.3 Temporal Changes in the Extent of Subglacial Permafrost

Ongoing research undertaken on both modern and ancient glaciers and ice sheets is starting to reveal the increasingly complex and at times, counterintuitive relationships between climate change, glacier extent, glacier thermal regime and the extent and thickness of subglacial permafrost. The potential for significant changes in the extent of subglacial permafrost to occur over relatively short timescales is particularly well illustrated by ongoing glaciological research in Svalbard that is revealing the connections between climate change and various glaciological characteristics. Numerous investigations on a range of glaciers with contrasting thermal characteristics have demonstrated a strong link between ice thickness and basal thermal regime and therefore the presence or absence of subglacial permafrost. An ice thickness of c. 100 m is seen as an approximate threshold above which the glacier is likely to have a warm bed, and below which the glacier is likely to have a cold bed (Murray *et al.*, 2000). As a consequence, the combination of glacier recession and downwasting relating to recent climate change is resulting in an increase in the extent of cold-based ice and subglacial permafrost aggradation, the glaciological and hydrological implications of which are currently being explored (see Sections 4.2, 4.3 and 6.6).

With glaciers and ice sheets having experienced dramatic changes in their extent and thickness in response to the major climatic oscillations that have characterised the Quaternary Period, it is inevitable that the extent of cold-based ice and subglacial permafrost will have changed significantly during glacial cycles in response to variations in ice thickness, the advection of cold ice from ice-sheet interiors and strain heating associated with active basal processes (e.g. Kleman & Glasser, 2007). In considering this temporal dimension, this section expands upon the previous section in considering the changes likely to accompany periods of ice mass growth and subsequent recession.

Focusing firstly on periods of ice-mass inception and growth, the onset of glaciation occurring through either *in situ* accumulation or ice advance is likely to occur initially as a thin and cold-based ice cover (e.g. Marshall & Clark, 2002), with extensive field evidence for the presence of pre-glacial permafrost leading many authors to conclude that Pleistocene ice sheets would almost certainly have advanced over permafrost during their initial expansion. For example, the initial stages of the Middle Pleistocene glaciation of Britain during Marine Isotope Stage (MIS) 12 experienced an intensely cold and dry climate that persisted for millennia (Jones & Keen, 1993). Consequently, permafrost at least tens of metres thick is likely to have formed prior to the onset of glaciation, especially in the more ice-marginal areas (e.g. northern Norfolk, UK), where the period of time for permafrost growth was longest (Waller *et al.*, 2009b). Modern-day evidence relating the presence of subglacial permafrost to ice-sheet advance over aggrading subaerial permafrost is provided by the EM surveys in western Greenland described in the previous section. In considering the origin of the subglacial permafrost extending beneath the ice-sheet margin, Ruskeeniemi *et al.* (2018) argue that any permafrost that formed in the area prior to the last glaciation would have thawed when it was overridden by a larger LGM ice sheet. Alternatively, it could also have thawed during the warm climatic conditions that occurred during the early deglacial phase during which time the ice sheet reached a minimum position between 6000 and 5000 years BP. They therefore suggest that the permafrost is most likely to have formed within the deglaciated area during a subsequent colder phase (post 4400 years BP) with 3000–3500 years providing sufficient time for the aggradation of the current permafrost layer which is currently >300 m thick. The permafrost was subsequently overridden by the ice sheet during a recent late Holocene advance after which it is predicted to have switched from a growth phase to a steady or possibly degrading state.

As noted in the previous section, face value interpretations of the geomorphological evidence left by large ice sheets and the outputs of steady-state numerical models have led to the presumption that glacier–permafrost interactions are typically limited in extent. However, transient ice-sheet models have indicated that subglacial permafrost is likely to have been far more extensive during phases of ice advance than previously thought.

Numerical modelling of the Green Bay Lobe of the southern Laurentide Ice Sheet for example led Cutler *et al.* (2000) to conclude that its bed was frozen for 60–200 km up-glacier of the margin during the LGM. They also suggest that the thermal inertia of thick permafrost resulted in its persistence for anything between a few hundred and a few thousand years. Earlier work has similarly suggested that subglacial permafrost can persist for millennial timescales, particularly when it is ice rich. Mathews and Mackay (1960) calculated that if a 400 m thickness of saturated, frozen, unconsolidated sediments, with ground temperatures comparable to the current Mackenzie Delta region (c. −10 °C), were overridden by a glacier with a basal temperature of 0 °C, permafrost c. 100–200 m thick should still persist after a period of 10,000 years. Dyke (1993) similarly calculated that 10,000 years would be required to thaw a permafrost layer 600 m thick with a surface temperature of −16 °C and a limited ice content of 6% by volume. This length of time is doubled if the geothermal heat flux is also required to warm the overlying glacier to its pressure melting point.

In considering the evolution of the highly distinctive glaciated landscapes of the Canadian Arctic (Section 5.3.1), Dyke (1993) employs a combination of geomorphological mapping and ice-sheet modelling to reconstruct the changing thermal character of ice caps through the Late Wisconsinan. He suggests that these smaller high-latitude ice masses remained entirely cold based during their initial growth with ice masses that reached no more than 30 km in radius (which includes modern-day remnants such as the Devon Island ice cap; Koerner & Fisher, 1979) remaining in this state. Ice caps that grew beyond this size developed warm-based peripheral zones in response to an increase in strain heating and a reduction in the influence of cold-ice advection. These areas of warm-based ice are associated with evidence of subglacial scouring that can affect up to 75% of the flow line length in areas where flow convergence enhances the influence of strain heating. In contrast to the aforementioned shift to warm-based conditions following the LGM that led to the collapse of the continental-scale Laurentide Ice Sheet (section 2.4.2, Marshall & Clark, 2002), these smaller ice caps switched back to becoming entirely cold-based during deglaciation in response to a concomitant reduction in ice thickness and velocity. Mapping of glacial landforms in southern Norway by Sollid & Sørbel (1994) led the authors to conclude that the central part of the much larger Scandinavian Ice Sheet also remained cold-based throughout the late Weichselian and that deglaciation similarly led to an increase in the extent of cold-based conditions, starting at the highest elevations where ice thicknesses were at their lowest.

Ice-mass decay can therefore result in a return to cold-based conditions and a second phase of permafrost growth during deglaciation. The historical and ongoing recession of glaciers in Svalbard from their Little Ice Age maxima has been observed to result in a reduction in the extent of warm-based ice and a thermal transition from polythermal to entirely cold-based conditions that provide the potential for subglacial permafrost to aggrade and expand (e.g. Lovell *et al.*, 2015). Field research at Midtdalsbreen in southern-central Norway demonstrating the expansion of cold-based ice beneath a supposedly temperate glacier led Reinhardy *et al.* (2019) to suggest that these recession-related thermal changes are not restricted to the Arctic, with temperate glaciers potentially transitioning to polythermal regimes as they recede and downwaste (Section 3.2).

In considering recent changes in the basal thermal regime of Bakaninbreen, Svalbard, Murray *et al.* (2000) utilise a simple model to quantify the thickness of subglacial permafrost that could be expected to develop as a consequence of glacier recession during the late twentieth century. The time (t) taken to form a given thickness of permafrost (H) from unfrozen sediment at 0 °C can be estimated using the following equation,

$$H = \sqrt{2k\Delta Tt} \, / \, \rho Ln$$

where ρ is sediment density (estimated at 2000 kg m^{-3}), n is the porosity (estimated at 0.25) and L is the latent heat of fusion for ice (3.35 × 10^5 J kg^{-1}). This equation estimates that the 12–13 m of subglacial permafrost thought to occur beneath Bakaninbreen (Section 4.3.1) is likely to have formed within 190 years.

Climate warming can therefore result in a series of counterintuitive changes that have included significant impacts on groundwater systems during the course of the twentieth century as the aggradation of subglacial permafrost has limited aquifer recharge (e.g. Haldorsen & Heim, 1999) (Section 4.2.4). If deglaciation is accompanied by permafrost aggradation, then as mentioned previously, the final remains of these glaciers may in fact be preserved for millennia as massive ground ice deposits that Hughes (1973) has referred to as 'glacial permafrost' (Section 5.3.4).

2.5 Permafrost Rheology

2.5.1 The Rheology of Ice, Sediment and Ice–Sediment Mixtures

The mechanics and rheological behaviour of ice–sediment mixtures have been studied for over 50 years by engineers, permafrost scientists and glacier scientists amongst others. A general and commonly held belief is that ice–sediment mixtures are stronger than the end members of pure ice or ice-free granular sediment with their strength being dependent on a combination of the cohesion of the ice matrix and the frictional resistance of the sediment particles (e.g. Ting, 1983). Closer inspection of the literature however reveals a range of seemingly contradictory results with some studies suggesting that ice–sediment mixtures are stronger and more resistant to deformation than pure ice (e.g. Goughnour & Andersland, 1968), whilst others suggest that they are weaker and more easily deformed (Lawson, 1996; Fitzsimons *et al.*, 2001). This reflects the diverse range of interacting variables that can influence the rheological properties and behaviours of ice–sediment mixtures that include the relative concentrations of sediment and ice, the particle size of the sediment, the temperature, the solute concentration and the nature of the applied stress.

Before looking at the specific case of ice–sediment mixtures, it is worth briefly examining the rheological properties of the two constituent elements: ice and unconsolidated sediment. Looking firstly at pure ice, terrestrial ice is characterised by a planar crystal structure which results in an anisotropic behaviour, particularly in response to shear stresses. If the applied stress is aligned with the crystal structure (the basal plane) then the ice will deform relatively easily via dislocation glide. The resistance to shear in other orientations can however be orders of greater in magnitude. Ice in the natural environment typically occurs in a polycrystalline form, the behaviour of which is complicated by the varied alignments of the basal planes of the individual ice crystals. It is further complicated by the influence of processes that operate at the crystal boundaries. Polycrystalline ice consequently displays a distinctive non-linear and time-dependent response to an applied stress. Laboratory experiments that have been highly influential in glaciology involving the application of a constant stress have helped to constrain the complex creep behaviour of polycrystalline ice (e.g. Glen, 1955). These indicated that following an initial elastic response, polycrystalline ice exhibits three phases of creep deformation. During primary creep, the strain rate decreases (strain hardening) as the crystals with differing orientations interfere until a minimum or secondary creep rate is attained. Recrystallisation resulting in the development of crystals with orientations more favourable to deformation results in a subsequent increase to a tertiary creep rate up to an order of magnitude higher than the secondary creep rate in shear (Budd & Jacka, 1989).

Glaciologists have continued to invest considerable research effort in their attempts to devise a flow law for ice as a fundamental control on glacier dynamics (see Cuffey & Paterson, 2010 for a detailed review). The aforementioned laboratory work on ice samples by John Glen is seen as a seminal contribution that led to the development of a power law (Glen, 1955) linking the applied stress (σ) and strain rate (e) in what has subsequently been termed Glen's flow law:

$$e = k\,\sigma^{n}$$

As a power law, this relationship includes an exponent n that in the case of glaciers is most commonly thought to approximate to 3, with the strain rate therefore increasing as the cube of the applied stress. The law also includes a constant k that is referred to as a hardness or fluidity parameter that can be modified for example via the inclusion of dissolved impurities within ice crystals (e.g. Jones & Glen, 1969; Holdsworth & Bull, 1970). Of particular note here is its dependence on temperature such that the strain rate for a given stress is five times higher at $-10\,°C$ than it is at $-25\,°C$ (Cuffey & Paterson, 2010). The ability of ice flow in turn to generate heat through dissipation results in the potential development of self-reinforcing creep instability feedbacks whereby a small perturbation in flow can lead to the development of regions of fast ice flow (Clarke *et al.*, 1977; section 4.3.4). An additional influence of temperature relates to the ability of unfrozen water at grain boundaries to facilitate dislocation creep, causing k to increase linearly with the unfrozen water content (Moore, 2014).

In switching the focus to the rheology of sediment, the ability of unfrozen sediment to resist deformation is determined by the shear resistance of the constituent particles. The yield stress of a granular material (τ) is provided by a combination of cohesion and internal friction as described by the Mohr–Coulomb equation:

$$\tau = C + \sigma \tan\varphi$$

It is important to note that this equation is not a flow law in that it does not define a relationship between the applied stress and the strain rate. Rather, it determines a threshold stress *below which* no permanent deformation occurs, *at which* permanent strain occurs and *above which* accelerating deformation occurs (Moore, 2014). In the context of unfrozen and consolidated sediments, cohesion (C) results from the electrostatic attraction between the constituent particles that is high within sediments containing fine-grained materials (especially clays) and low or non-existent within coarser-grained sands and gravels. Cohesion can also be provided by physical cementation, which explains why sandstones for example are much stronger than their unlithified counterparts. Finally, apparent cohesion can arise as a result of capillary forces within any thin water films adhering to the constituent particles. The frictional element is represented by the component $\sigma \tan\phi$ where σ is the normal stress and ϕ is the friction angle. The friction angle is an inherent property of the sediment which reflects the ability of particles to interlock under stress that is typically high in coarse-grained sediments and low in fine-grained sediments. The inclusion of normal stress in the friction term indicates that frictional strength will tend to increase as the normal stress increases. The role played by water in promoting landslides for example is well known and relates to the ability of high porewater pressures to counterbalance the normal stress such that the effective stress and the associated frictional strength are reduced significantly. In the context of glaciological research, the recognition of the ability of high porewater pressures to substantially reduce the shear strength of subglacial sediments and thereby create a subglacial deforming layer has led to what some consider to be a paradigm shift within glaciology (Boulton, 1986; section 3.3.3).

Returning to the rheology of the ice–sediment mixtures that are commonly found in permafrost and glacial environments, these materials are generally believed to display higher shear strengths than either pure ice or unfrozen sediment. This is widely attributed to the growth of interstitial ice enhancing sediment cohesion and preventing the generation of the high porewater pressures that can reduce the frictional strength (Mathews & Mackay, 1960; Etzelmüller *et al.*, 1996). This conclusion is seemingly supported by a number of widely cited laboratory studies. One of the earliest experimental studies was undertaken by Goughnour & Andersland (1968), who compared the mechanical properties of polycrystalline ice with those of sand–ice mixtures at ice at a range of temperatures between -3.85 and $-12.03\,°C$. They concluded that at low volumetric concentrations, the addition of sand resulted in a linear increase in shear strength according to the relative concentration of sand and ice, surmising that the increasing concentration of sand requires the strain to be accommodated within an ice matrix of diminishing volume. Once the volumetric concentration of sand reached 42%, any further increase resulted in a rapid increase in shear strength as a result of increasing interparticle friction (Figure 2.14). With the cohesion of the ice matrix preventing dilation, this created an effect equivalent to the deformation of unfrozen materials under high effective stresses. Nickling & Bennett (1984) subsequently examined the comparative shear strengths of a wider variety of ice–sediment mixtures at temperatures much closer to the pressure-melting point ($-1\,°C$). In spite of the different testing conditions, both in terms of temperature and the nature of the sediment used (coarse-grained diamicton), their tests produced broadly comparable findings. They found that peak shear strength increases rapidly as the ice content rises from 0 to 25% by volume, reaching a maximum value at 25% by volume (equivalent to the sample porosity) and decreasing thereafter (Figure 2.15). They therefore also concluded that ice–sediment mixtures display higher peak shear strengths, due to the combined influence of the ice matrix on sample cohesion and the increase in internal friction caused by the presence of sediment particles.

Subsequent experimental and analytical studies of ice-cemented granular soils have provided an increasingly detailed insight into the varied controls on their shear strength that are now known to include temperature, ice strength, structural hindrance by the grain skeleton,

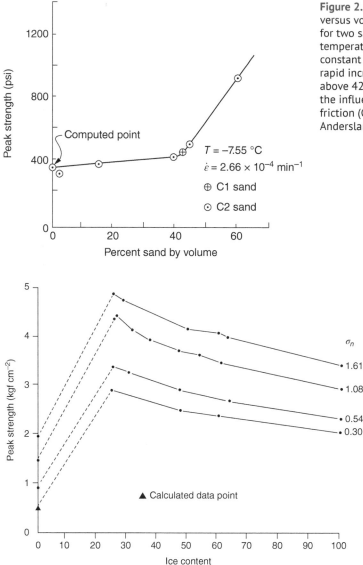

Figure 2.14 Peak strength versus volumetric sand content for two sand samples at a temperature of −7.55 °C and a constant strain rate. Note the rapid increase in peak strength above 42% sand by volume due to the influence of interparticle friction (Goughnour & Andersland, 1968, p930).

Figure 2.15 Variations of peak shear strength of frozen coarse-grained granular debris with increasing volumetric ice contents. Sigma *n* indicates variations in the normal load (Nickling & Bennett, 1984, p355).

crack propagation and unfrozen water content (Eckardt, 1982; Yasufuku *et al.*, 2003; Arenson & Springman, 2005a,b). Arenson *et al.* (2004) and Arenson & Springman (2005a) have demonstrated that the shear strength of ice-rich frozen sediment increases with increased strain rate, with high strain rates leading to brittle deformation and lower strain rates promoting more ductile behaviour. Strain hardening occurs because (i) pressure melting of pore ice at points of high stress concentration can increase the number of grain-to-grain contacts within the sand, thus increasing its frictional strength (Yasufuku *et al.*, 2003); and (ii) refreezing of the intergranular meltwater may lead to 'self-healing' of the sediment and once again an increase in its shear strength (Arenson & Springman, 2005a). However, tri-axial tests carried out by Yang *et al.* (2010a,b) on frozen sand with variable water (and ice) contents and confining pressures, over the temperature range −4 to −6 °C, indicate that strain hardening only occurs during shearing under high confining pressures (≥3 MPa), whereas strain softening occurs during shearing under low confining pressures (0.5–1.0 MPa). Yang *et al.* (2010b) also observed that the strength of frozen sand with a low water content (10% by weight) increased with increasing confining pressure, whereas in frozen sand with a high water content (15–20%), its strength at first increased with increasing confining pressure, and then decreased, probably as a result of pressure melting and grain crushing.

Experimental studies suggest that the highest shear strengths are typically associated with the following conditions: (i) pore ice completely fills the available pore space, maximising both the cohesive effect from ice and the frictional effect from grain-to-grain contacts, and impeding dilatancy of the sediment during deformation; (ii) temperature is as low as possible, minimising the amount of liquid water along ice crystal and sediment grain boundaries and thus maximising the adhesive effects of ice bonding with the sediment (Fish & Zaretsky, 1997); (iii) normal stress is of a moderate value that does not exceed a critical limit whereby the energy needed for dilatation equals the shear strength of individual particles (c. 100 kPa for the coarse granular debris studied by Nickling & Bennett, 1984), beyond which dilatancy is partly replaced by particle fracture and crushing and, as a result, shear strength decreases or remains relatively constant; and (iv) strain rates are high, restricting volume change and accommodation cracking (Goughnour & Andersland, 1968).

As part of a more extensive review of the complex and seemingly contradictory literature on the rheology of ice–sediment mixtures the diverse results of which are illustrated in Figure 2.16, Moore (2014) provides a valuable conceptual framework devised specifically for Earth surface environments that is summarised below:

1) At debris concentrations lower than about 40% by volume (i.e. sample porosity), the rheological behaviour of the ice–sediment mixture is dominated by creep within the ice matrix, which is in turn dependent upon the applied stress and temperature. Isolated debris particles can influence the creep behaviour of the ice by pinning dislocations or influencing ice crystal size for example. The presence of liquid water at the interfaces between particles and the surrounding ice matrix can limit the coupling of the sediment and the ice and localise deformation within these liquid interfaces.

2) At higher debris concentrations, interparticle friction becomes increasingly influential. Little or no deformation occurs when the applied stress is lower than the frictional strength of the particle network. If stresses exceed this threshold, some creep may occur so long as confinement is limited. As the volumetric contribution of the ice matrix is reduced, the frictional component of strength is maintained but the cohesive component is progressively reduced.

3) At any sediment concentration, the rheological properties of the ice–sediment mixture are strongly affected by the abundance and distribution of unfrozen water, which limits the potential for interaction between the sediment particles and the ice matrix. An increase in solute concentration and the quantity of fine-grained sediment can increase the unfrozen water volume at temperatures below the bulk freezing point.

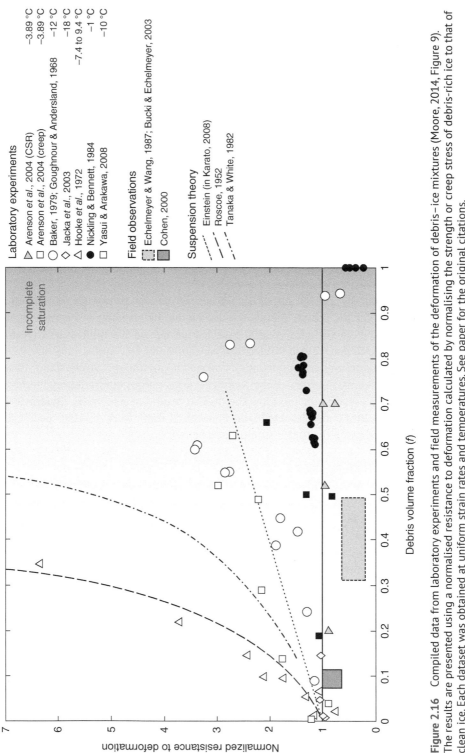

Figure 2.16 Compiled data from laboratory experiments and field measurements of the deformation of debris–ice mixtures (Moore, 2014, Figure 9). The results are presented using a normalised resistance to deformation calculated by normalising the strength or creep stress of debris-rich ice to that of clean ice. Each dataset was obtained at uniform strain rates and temperatures. See paper for the original citations.

Focusing on the final point listed within Moore's (2014) conceptual framework, the nature and the strength of the bonds between the ice and the sediment particles is of particular significance to the mechanical strength and rheology of ice–sediment mixtures. The presence of pre-melted water at the interfaces between the constituent particles and the ice matrix can significantly influence the nature of these bonds, the extent to which these elements can interact and therefore the rheological properties of the substrate in question. Fine-grained materials in particular, can retain significant quantities of liquid water at cryotic temperatures that can in turn have a significant impact on their mechanical and rheological properties. The concept of pre-melted water also helps to explain the strongly temperature-dependent behaviour of ice–sediment mixtures with changes in temperature controlling the thickness of interfacial films and the presence of supercooled water in the pore spaces (see Section 2.2.2).

The importance of particle size in relation to subglacial deformation was recognised as early as 1960 by Mathews & Mackay when, citing field observations from Herschel Island and the Danish Coast, they suggested that permafrost might deform preferentially along clay-rich horizons. In an early experimental study, Andersland & Alnouri (1970) compared the creep characteristics of frozen sand and clay-sized sediment under constant shear stresses and strain rates at −12 °C. The authors demonstrated that at this low temperature, frozen sand displayed a much higher resistance to loading than frozen clay, a behavioural difference they attributed to the clay retaining 12% unfrozen water by volume, whereas the sand was completely frozen. In being performed at low temperatures (−12 °C), these experimental results contrasting the rheological behaviour of frozen sand and clay may not apply at temperatures closer to the pressure melting point. However, the underlying explanation for this difference – that fine-grained sediments contain significant amounts of liquid water at sub-freezing temperatures, whereas coarse-grained materials freeze very rapidly below the pressure melting point - should in theory be applicable at a wide range of temperatures. In this respect, Ogata *et al.* (1983) isolated the influence of unfrozen water more directly by undertaking experiments on samples of sand at −20 °C with varying salinities. They found that increasing the salinity and therefore the unfrozen water content led to a substantial decrease in strength that mirrored the related reduction in pore ice content, an influence considered central to the formation of large push moraines in permafrost terrains in Svalbard (Etzelmüller *et al.*, 1996; section 5.3.2).

This critical importance of water and its state was recognised by Tsytovich (1975) on the basis of numerous engineering studies of the geotechnical properties of permafrost. He consequently developed a conceptual model in which permafrost rheology is represented as a gradation between ground that is 'hard-frozen' and ground that is 'plastic-frozen', with the relative occurrence of ice and liquid water being the key variable determining the rheological behaviour between these two end member states. Hard-frozen ground is firmly cemented by pore ice and has a temperature sufficiently low that most of the water within it is frozen. In sands, the maximum temperature for this condition is about −0.5 °C (Williams & Smith, 1989, p257), whereas in clays it may be several degrees Celsius lower because the large specific surface of clays allows significant amounts of water to remain unfrozen (Section 2.2.2). For example, Tsytovich reported a temperature limit for 'hard frozen' behaviour of −5 °C or even −7 °C for very fine-grained clays, especially those of montmorillonite composition. The deformation of hard-frozen ground is consequently characterised by brittle failure and such ground is considered practically incompressible under loads less than c. 0.5–1 MPa. Plastic-frozen ground, by contrast, is more weakly cemented on account of its relatively high liquid water content. As a result, such ground characteristically displays ductile behaviour and creep deformation, and the material has a relatively high compressibility. It is these materials therefore that have the greatest potential to deform as a consequence of glacier–permafrost interactions in both proglacial and subglacial settings.

2.5.2 The Rheology of Subglacial Permafrost

Current research focusing specifically on the rheological behaviour of subglacial permafrost is sparse in nature on account of the general inaccessibility of subglacial environments and the comparatively limited amount of research undertaken on non-temperate glaciers as a whole. The recognition of relevant research work is hampered further by the terminological confusion in distinguishing between debris-rich basal ice and ice-rich subglacial sediment (section 2.2.1; Waller *et al.*, 2009a).

Useful insights into the potential spatial and temporal variations in permafrost rheology during subglacial deformation have however been provided by stratigraphic studies of glacitectonites or glacial mélanges that have remained frozen since they were deformed during the Pleistocene (Astakhov *et al.*, 1996). In the Mackenzie Delta region of western Arctic Canada for example, Murton *et al.* (2004) describe Pleistocene permafrost that was glacially deformed beneath the north-western margin of the Laurentide Ice Sheet. The ice-rich permafrost comprises truncated ice blocks and granular and cohesive materials (ice-cemented sand intraclasts) situated within a silt-clay diamicton, which is interpreted as a frozen glacitectonite that has never thawed. The survival of relatively undeformed frozen intraclasts within a more pervasively deformed diamicton and the associated rheological heterogeneity can be explained by subglacial deformation occurring at temperatures below, but close to, the pressure melting point of ice within ice-rich permafrost. As mentioned previously (Section 2.2.2), the presence and quantity of liquid water and therefore the mechanical properties of frozen ground at these relatively warm cryotic temperatures (estimated to be between 0 and −2 °C) are strongly grain-size dependent. Where such 'warm permafrost' develops in sediments of variable particle size, the permafrost is only partially frozen and comprises a complex mixture of sediment, ice and liquid water. The liquid water will be most abundant in clayey sediments, in response to pre-melting at temperatures below the bulk freezing point (Dash *et al.*, 2006), and the pore ice will be most abundant in the sands and gravels. Consequently, when such texturally variable sequences were overridden by Laurentide ice, the shear imposed from the ice sheet resulted in enhanced ductile deformation within the partially frozen, fine-grained matrix, while the frozen intraclasts displayed greater shear strength, allowing them to act as more competent (rigid) blocks (see Section 5.4.2 for more detail).

As permafrozen sediments warm towards the melting point of ice, increasing amounts of liquid water may coexist with dwindling amounts of ice, and the resulting partially frozen mixture of ice, liquid water and sediment will tend to behave in an increasingly ductile manner. Conceptually therefore, the advance and retreat of a glacier or ice sheet across permafrozen sediments can be expected to change the temperature of the permafrost over time and hence its unfrozen water content and its rheology. For example, ice advancing over cold continuous permafrost may initially cause brittle deformation (e.g. proglacial thrusting or faulting) in which shear is most likely to occur preferentially along planes of structural weakness including ice-rich or clay-rich horizons (Astakhov *et al.*, 1996; Fitzsimons *et al.*, 2001). Continued ice advance and an increase in ice thickness (potentially also associated with an increase in velocity and strain heating) is likely to raise the temperature of the subglacial permafrost, promoting a transition to a more pervasive and ductile style of the deformation leading to folding and potentially to diapirism (Astakhov *et al.*, 1996; Murton *et al.*, 2004). Continued warming may eventually result in the complete thaw of the permafrost, generating high pore-water pressures in an unfrozen deforming bed as the remaining ice melts. Detailed examination of such glacially deformed permafrost and unfrozen glacitectonites is consequently required in order to determine whether such sequences contain glaciotectonic structures indicative of polyphase deformation and rheological transitions indicating systematic changes in glacier–permafrost interactions over time (e.g. Waller *et al.*, 2011).

Further detail and related coverage on the observations of basal processes beneath cold-based glaciers and the sedimentological and structural signatures of glacier–permafrost interactions are provided in Sections 3.5 and 5.4, respectively.

2.6 Summary

- The term permafrost encompasses a broad range of Earth surface materials that remain at or below 0 °C for at least two consecutive years. Whilst not being central to its definition, permafrost commonly contains water in both its solid and liquid forms. Liquid water occurs as both porewater and as thin interfacial films, with the latter persisting to temperatures well below the bulk freezing point in fine-grained sediments as pre-melted water. Ground ice is a key component of the global cryosphere that occurs in areas of ice-rich permafrost in a variety of forms depending on the water source and the processes of formation.

- Permafrost most commonly forms in subaerial settings where the mean annual air temperature is generally below −1 °C. The state and thickness of the permafrost in any subaerial environment is significantly influenced by a myriad additional controls that include the nature of the surface boundary layer (e.g. vegetation type, snowpack thickness and water depth), the thermal conductivity and water content of the substrate and the climatic and geological history.

- Permafrost can also grow and persist in subglacial settings and is most commonly associated with cold-based glaciers and ice sheets characterised by basal temperatures below the pressure melting point. Such thermal regimes are generally promoted by cold surface temperatures, the accumulation and downward advection of cold ice, limited ice thicknesses (generally less than the permafrost thickness) and limited basal ice velocities. Alternatively, subaerial permafrost can persist as subglacial permafrost when overridden by glaciers and ice sheets during glaciation. Finally, subglacial permafrost can also aggrade during deglaciation as formerly active ice thins and stagnates.

- Approximately 24% of the land area of the Northern Hemisphere is characterised by permafrost. Occurring primarily in polar and alpine environments, it is commonly juxtaposed with glaciers and ice sheets. Its extent was significantly greater during the LGM when it extended out onto the exposed Arctic continental shelves where some remains as extensive areas of relict permafrost.

- The spatial extent of subglacial permafrost is poorly constrained although borehole observations and geophysical measurements have confirmed its presence beneath both the interiors and the margins of glaciers and ice sheets. Numerical ice-sheet models and geomorphological inverse models both suggest that subglacial permafrost may have been widespread beneath Pleistocene ice sheets, occurring primarily beneath marginal fringes, ice-dispersal centres and in areas of strong flow divergence. Its extent is likely to have been highly variable, however, being more extensive during glaciation, less extensive during ice sheet maxima and the early phases of deglaciation, with the potential for its redevelopment in the latter stages of deglaciation.

- Permafrost in the form of ice–water–sediment mixtures is characterised by a complex range of rheological behaviours that reflect the influence of a diverse range of controls including the relative concentrations of sediment and ice, the particle size of the sediment, the temperature, the solute concentration and the nature of the applied stress. The presence and quantity of pre-melted water plays a particularly influential role by limiting the interaction between sediment particles and the ice matrix. With the amount of pre-melted water in turn being strongly grain-size dependent, permafrost comprising fine-grained sediments can remain ductile down to temperatures well below the bulk freezing point as plastic-frozen permafrost, with field observations within formerly glaciated permafrost regions confirming the ability of such permafrost to act as subglacial deforming layers.

References

Allen, D.M., Michel, F.A. & Judge, A.S., 1988. The permafrost regime in the Mackenzie Delta, Beaufort Sea region, NWT and its significance to the reconstruction of the palaeoclimatic history. *Journal of Quaternary Science*, 3, 3–13.

Andersland, O.B. & Alnouri, I., 1970. Time-dependent strength behavior of frozen soils. *Journal of Soil Mechanics and Foundations Division. ASCE*, 96, 1249–1265.

Arenson, L.U. & Springman, S.M., 2005a. Mathematical descriptions for the behaviour of ice-rich frozen soils at temperatures close to 0°C. *Canadian Geotechnical Journal*, 42, 431–442.

Arenson, L.U. & Springman, S.M., 2005b. Triaxial constant stress and constant strain rate tests on ice-rich permafrost samples. *Canadian Geotechnical Journal*, 42, 412–430.

Arenson, L.U., Johansen, M.M. & Springman, S.M., 2004. Effects of volumetric ice content and strain rate on shear strength under triaxial conditions for frozen soil samples. *Permafrost and Periglacial Processes*, 15, 261–271.

Astakhov, V.I. & Isayeva, L.L., 1988. The 'Ice Hill': an example of 'retarded deglaciation' in Siberia. *Quaternary Science Reviews*, 7, 29–40.

Astakhov, V.I., Kaplyanskaya, F.A. & Tarnogradsky, V.D., 1996. Pleistocene permafrost of West Siberia as a deformable glacier bed. *Permafrost and Periglacial Processes*, 7, 165–191.

Attig, J.W., Mickelson, D.M. & Clayton, L., 1989. Late Wisconsin landform distribution and glacier-bed conditions in Wisconsin. *Sedimentary Geology*, 62, 399–405.

Ballantyne, C.K. & Harris, C., 1994. *The Periglaciation of Great Britain*. Cambridge University Press.

Björnsson, H., Gjessing, Y., Hamran, S.E., *et al.*, 1996. The thermal regime of sub-polar glaciers mapped by multi-frequency radio-echo sounding. *Journal of Glaciology*, 42, 23–32.

Bockheim, J.G., 1995. Permafrost distribution in the southern circumpolar region and its relation to the environment: a review and recommendations for further research. *Permafrost and Periglacial Processes*, 6, 27–45.

Boulton, G.S., 1970b. On the deposition of subglacial and melt-out tills at the margins of certain Svalbard glaciers. *Journal of Glaciology*, 9, 231–245.

Boulton, G.S., 1986. Geophysics: a paradigm shift in glaciology? *Nature*, 322, 18.

Budd, W.F., 1970. Ice flow over bedrock perturbations. *Journal of Glaciology*, 9, 29–48.

Budd, W.F. & Jacka, T.H., 1989. A review of ice rheology for ice sheet modelling. *Cold Regions Science and Technology*, 16, 107–144.

Burn, C.R., 2002. Tundra lakes and permafrost, Richards Island, western Arctic coast, Canada. *Canadian Journal of Earth Sciences*, 39, 1281–1298.

Burn, C.R., 2004. The thermal regime of cryosols. In: Kimble, J.M. (Ed.), *Cryosols: Permafrost-affected Soils*. Springer, Berlin, 391–413.

Burn, C.R., 2007. Permafrost. In: Elias, S.A. (Ed.), *Encyclopedia of Quaternary Sciences*. Elsevier, Amsterdam, 2191–2199.

Clark, P.U., Dyke, A.S., Shakun, J.D., *et al.*, 2009. The Last Glacial Maximum. *Science*, 325, 710–714.

Clarke, G.K., Nitsan, U. & Patrerson, W.S.B., 1977. Strain heating and creep instability in glaciers and ice sheets. *Reviews of Geophysics*, 15, 235–247.

Cuffey, K.M. & Paterson, W.S.B., 2010. *The Physics of Glaciers*. Academic Press.

Cutler, P.M., MacAyeal, D.R., Mickelson, D.M., Parizek, B.R. & Colgan, P.M., 2000. A numerical investigation of ice-lobe–permafrost interaction around the southern Laurentide ice sheet. *Journal of Glaciology*, 46, 311–325.

Dash, J.G., Rempel, A.W. & Wettlaufer, J.S., 2006. The physics of premelted ice and its geophysical consequences. *Reviews of Modern Physics*, 78, 695–741.

Davis, T.N., 2001. *Permafrost: A Guide to Frozen Ground in Transition*. University of Alaska Press.

Dyke, A.S., 1993. Landscapes of cold-centred Late Wisconsinan ice caps, Arctic Canada. *Progress in Physical Geography*, 17, 223–247.

Eckardt, H., 1982. Creep tests with frozen soils under uniaxial tension and uniaxial compression. In: French, H.M. (Ed.), *Proceedings of the 4th Canadian Permafrost Conference*. National Research Council of Canada, Calgary, Alberta, 394–405.

Etzelmüller, B., Hagen, J.O., Vatne, G., Ødegård, R.S. & Sollid, J.L., 1996. Glacier debris accumulation and sediment deformation influenced by permafrost: examples from Svalbard. *Annals of Glaciology*, 22, 53–62.

Fish, A.M. & Zaretsky, Y.K., 1997. Ice strength as a function of hydrostatic pressure and temperature. US Army, Cold Regions Research & Engineering Laboratory, Hanover, New Hampshire, Report 97-6.

Fitzsimons, S.J., McManus, K.J., Sirota, P. & Lorrain, R.D., 2001. Direct shear tests of materials from a cold glacier: implications for landform development. *Quaternary international*, 86, 129–137.

French, H.M., 2017. *The Periglacial Environment* (4th ed.). John Wiley & Sons.

French, H.M. & Harry, D.G., 1990. Observations on buried glacier ice and massive segregated ice, western Arctic coast, Canada. *Permafrost and Periglacial Processes*, 1, 31–43.

French, H. & Shur, Y., 2010. The principles of cryostratigraphy. *Earth-Science Reviews*, 101, 190–206.

Glen, J.W., 1955. The creep of polycrystalline ice. *Proceedings of the Royal Society of London. Series A. Mathematical and Physical Sciences*, 228, 519–538.

Goughnour, R.R. & Andersland, O.B., 1968. Mechanical properties of a sand-ice system. *Journal of the Soil Mechanics and Foundations Division, American Society of Civil Engineers*, 94, 923–950.

Gow, A.J., Epstein, S. & Sheehy, W., 1979. On the origin of stratified debris in ice cores from the bottom of the Antarctic ice sheet. *Journal of Glaciology*, 23, 185–192.

Gregersen, O., Pukan, A. & Johansen, T., 1983. Engineering properties and foundation design alternatives in marine Svea Clay, Svalbard. *4th International Conference on Permafrost*, Fairbanks, AK, 384–388.

Gusmeroli, A., Murray, T., Jansson, P., *et al.*, 2012a. Vertical distribution of water within the polythermal Storglaciaren, Sweden. *Journal of Geophysical Research*, 115, F04002, doi:10.1029/2009JF001539.

Gusmeroli, A., Jansson, P., Pettersson, R. & Murray, T., 2012b. Twenty years of cold surface layer thinning at Storglaciären, sub-Arctic Sweden, 1989-2009. *Journal of Glaciology*, 58, 3–10.

Haldorsen, S. & Heim, M., 1999. An Arctic groundwater system and its dependence upon climatic change: an example from Svalbard. *Permafrost and Periglacial Processes*, 10, 137–149.

Hansen, B.L. & Langway, C.C. Jr., 1966. Deep core drilling in ice and core analysis at Camp Century, Greenland, 1961-1966. *CRREL Special Report*, 126, 207–208.

Harper, J.T., Humphrey, N.F., Meierbachtol, J.W., Graly, J.A. & Fischer, U.H., 2017. Borehole measurements indicated hard bed conditions, Kangerlussuaq sector, western Greenland Ice Sheet. *Journal of Geophysical Research*, 122, 1605–1618.

Herron, S. & Langway, C.C., 1979. The debris-laden ice at the bottom of the Greenland Ice Sheet. *Journal of Glaciology*, 23, 193–207.

Herterich, K., 1988. A three-dimensional model of the Antarctic ice sheet. *Annals of Glaciology*, 11, 32–35.

Holdsworth, G. & Bull, C., 1970. The flow law of cold ice: investigations on Meserve Glacier, Antarctica. *IASH Publication*, 86, 204–216.

Hooke, R.L., 1977. Basal temperatures in polar ice sheets: a qualitative review. *Quaternary Research*, 7, 1–13.

Hughes, T., 1973. Glacial permafrost and Pleistocene ice ages. In: *Permafrost: Second International Conference. North American contribution*. National Academy of Sciences, Washington, DC, 213–223.

Irvine-Fynn, T.D.L., Moorman, B.J., Williams, J.L.M. & Walter, F.S.A., 2006. Seasonal changes in ground-penetrating radar signature observed at a polythermal glacier, Bylot Island, Canada. *Earth Surface Processes & Landforms*, 31, 892–909.

Jones, S.J. & Glen, J.W., 1969. The effect of dissolved impurities on the mechanical properties of ice crystals. *Philosophical Magazine*, 19, 13–24.

Jones, R.L. & Keen, D.H., 1993. *Pleistocene environments in the British Isles*. Chapman & Hall.

Kanevskiy, M., Shur, Y., Jorgenson, M.T., *et al.*, 2013. Ground ice in the upper permafrost of the Beaufort Sea coast of Alaska. *Cold Regions Science and Technology*, 85, 56–70.

Kleman, J. & Glasser, N.F., 2007. The subglacial thermal organisation (STO) of ice sheets. *Quaternary Science Reviews*, 26, 585–597.

Kleman, J. & Hättestrand, C., 1999. Frozen-bed Fennoscandian and Laurentide ice sheets during the Last Glacial Maximum. *Nature*, 402, 63–66.

Kleman, J., Hättestrand, C. & Clarhäll, A., 1999. Zooming in on frozen-bed patches: scale-dependent controls on Fennoscandian ice sheet basal thermal zonation. *Annals of Glaciology*, 28, 189–194.

Koerner, R.M., 1989. Queen Elizabeth Islands glaciers. *Quaternary geology of Canada and Greenland, Geological Survey of Canada, Geology of Canada*, 1, 464–473.

Koerner, R.M. & Fisher, D.A., 1979. Discontinuous flow, ice texture, and dirt content in the basal layers of the Devon Island ice cap. *Journal of Glaciology*, 23, 209–222.

Kupsch, W.O., 1962. Ice-thrust ridges in western Canada. *The Journal of Geology*, 70, 582–594.

Lawson, W., 1996. The relative strengths of debris-laden basal ice and clean glacier ice: some evidence from Taylor Glacier, Antarctica. *Annals of Glaciology*, 23, 270–276.

Lindgren, A., Hugelius, G., Kuhry, P., *et al.*, 2016. GIS-based maps and area estimates of northern hemisphere permafrost extent during the last glacial maximum. *Permafrost and Periglacial Processes*, 27, 6–16.

Lorrain, R.D. & Fitzsimons, S., 2011. Cold-based glaciers. In: Singh, V.P., Singh, P. & Haritashya, U.K. (Eds.), *Encyclopedia of Snow, Ice and Glaciers, 1*. Springer, New York, 157–161.

Lovell, H., Fleming, E.J., Benn, D.I., *et al.*, 2015. Former dynamic behaviour of a cold-based valley glacier on Svalbard revealed by basal ice and structural glaciology investigations. *Journal of Glaciology*, 61, 309–328.

Mackay, J.R., 1972. The world of underground ice. *Annals of the Association of American Geographers*, 62, 1–22.

Mackay, J.R., 1986. The first 7 years (1978–1985) of ice wedge growth, Illisarvik experimental drained lake site, western Arctic coast. *Canadian Journal of Earth Sciences*, 23, 1782–1795.

Mackay, J.R. & Burn, C.R., 2002. The first 20 years (1978-1979 to 1998–1999) of ice-wedge growth at the Illisarvik experimental drained lake site, western Arctic coast, Canada. *Canadian Journal of Earth Sciences*, 39, 95–111.

Mackay, J.R. & Dallimore, S.R., 1992. Massive ice of the Tuktoyaktuk area, western Arctic coast, Canada. *Canadian Journal of Earth Sciences*, 29, 1235–1249.

Marshall, S.J. & Clark, P.U., 2002. Basal temperature evolution of North American ice sheets and implications for the 100-kyr cycle. *Geophysical Research Letters*, 29, 67–61, doi:10.1029/2002GL015192.

Mathews, W.H. & Mackay, J.R., 1960. Deformation of soils by glacier ice and the influence of pore pressures and permafrost. *Transactions of the Royal Society of Canada*, 54, 27–36.

Mooers, H.D., 1990. Ice-marginal thrusting of drift and bedrock: thermal regime, subglacial aquifers, and glacial surges. *Canadian Journal of Earth Sciences*, 27, 849–862.

Moore, P.L., 2014. Deformation of debris-ice mixtures. *Reviews of Geophysics*, 52, 435–467.

Moore, P.L., Iverson, N.R., Brugger, K.A., *et al.*, 2011. Effect of a cold margin on ice flow at the terminus of Storglaciären, Sweden: implications for sediment transport. *Journal of Glaciology*, 57, 77–87.

Muller, S.W., 1943. *Permafrost or Permanently Frozen Ground and Related Engineering Problems*. US Engineering Office, Strategic Engineering Study, Special Report No. 62, 136pp

Murray, T., Stuart, G.W., Miller, P.J., *et al.*, 2000. Glacier surge propagation by thermal evolution at the bed. *Journal of Geophysical Research: Solid Earth*, 105(B6), 13491–13507.

Murton, J.B. & French, H.M., 1994. Cryostructures in permafrost, Tuktoyaktuk coastlands, western arctic Canada. *Canadian Journal of Earth Sciences*, 31, 737–747.

Murton, J.B., Waller, R.I., Hart, J.K., *et al.*, 2004. Stratigraphy and glaciotectonic structures of permafrost deformed beneath the northwest margin of the Laurentide ice sheet, Tuktoyaktuk Coastlands, Canada. *Journal of Glaciology*, 50, 399–412.

Murton, J.B., Whiteman, C.A., Waller, R.I., *et al.*, 2005. Basal ice facies and supraglacial melt-out till of the Laurentide Ice Sheet, Tuktoyaktuk Coastlands, western Arctic Canada. *Quaternary Science Reviews*, 24, 681–708.

Murton, J.B., Peterson, R. & Ozouf, J.C., 2006. Bedrock fracture by ice segregation in cold regions. *Science*, 314(5802), 1127–1129.

Nickling, W.G. & Bennett, L., 1984. The shear strength characteristics of frozen coarse granular debris. *Journal of Glaciology*, 30, 348–357.

Ogata, N., Yasuda, M. & Kataoka, T., 1983. Effects of salt concentration on strength and creep behavior of artificially frozen soils. *Cold Regions Science and Technology*, 8, 139–153.

Paterson, W.S.B., Koerner, R.M., Fisher, D., *et al.*, 1977. An oxygen-isotope climatic record from the Devon Island ice cap, arctic Canada. *Nature*, 266, 508–511.

Reinhardy, B.T.I., Booth, A.D., Hughes, A.L.C., *et al.*, 2019. Pervasive cold ice within a temperate glacier - implications for glacier thermal regimes, sediment transport and foreland geomorphology. *The Cryosphere*, 13, 827–843.

Rempel, A.W., 2007. Formation of ice lenses and frost heave. *Journal of Geophysical Research*, 112(F2), doi:10.1029/2006JF000525.

Rempel, A.W., 2010. Frost heave. *Journal of Glaciology*, 56, 1122–1128.

Rempel, A.W., 2011. Microscopic and environmental controls on the spacing and thickness of segregated ice lenses. *Quaternary Research*, 75, 316–324.

Rempel, A.W., Wettlaufer, J.S. & Worster, M.G., 2004. Premelting dynamics in a continuum model of frost heave. *Journal of Fluid Mechanics*, 498, 227–244.

Robin, G., 1955. Ice movement and temperature distribution in glaciers and ice sheets. *Journal of Glaciology*, 2, 523–532.

Rolandone, F., Mareschal, J.C. & Jaupart, C., 2003. Temperatures at the base of the Laurentide Ice Sheet inferred from borehole temperature data. *Geophysical Research Letters*, 30(18), doi:10.1029/2003GL018046.

Ruskeeniemi, T., Engström, J., Lehtimäki, J., *et al.*, 2018. Subglacial permafrost evidencing re-advance of the Greenland Ice Sheet over frozen ground. *Quaternary Science Reviews*, 199, 174–187.

Shen, P.Y. & Beck, A.E., 1992. Paleoclimate change and heat flow density from temperature data in the Superior Province of the Canadian Shield. *Global Planetary Change*, 98, 143–165.

Smith, M.W., 1975. Microclimatic influences on ground temperatures and permafrost distribution, Mackenzie Delta, Northwest Territories. *Canadian Journal of Earth Sciences*, 12, 1421–1438.

Sollid, J.L. & Sørbel, L., 1994. Distribution of glacial landforms in southern Norway in relation to the thermal regime of the last continental ice sheet. *Geografiska Annaler: Series A, Physical Geography*, 76, 25–35.

Sugden, D.E., 1977. Reconstruction of the morphology, dynamics, and thermal characteristics of the Laurentide ice sheet at its maximum. *Arctic and Alpine Research*, 9, 21–47.

Szewczyk, J. & Nawrocki, J., 2011. Deep-seated relict permafrost in northeastern Poland. *Boreas*, 40, 385–388.

Ting, J.M., 1983. On the nature of the minimum creep rate–time correlation for soil, ice, and frozen soil. *Canadian Geotechnical Journal*, 20, 176–182.

Tsytovich, N.A., 1975. *The Mechanics of Frozen Ground*. McGraw-Hill, New York.

Van Tatenhove, F.G. & Olesen, O.B., 1994. Ground temperature and related permafrost characteristics in West Greenland. *Permafrost and Periglacial Processes*, 5, 199–215.

Vandenberghe, J. & Pissart, A., 1993. Permafrost changes in Europe during the last glacial. *Permafrost and Periglacial Processes*, 4, 121–135.

Vandenberghe, J., French, H.M., Gorbunov, A., *et al.*, 2014. The Last Permafrost Maximum (LPM) map of the Northern Hemisphere: permafrost extent and mean annual air temperatures, 25–17 ka BP. *Boreas*, 43, 652–666.

Vtyurin, B.I. & Glazovskiy, A.F., 1986. Composition and structure of the Ledyanaya Gora tabular ground ice body on the Yenisey. *Polar Geography*, 10, 273–285.

Walder, J. & Hallet, B., 1985. A theoretical model of the fracture of rock during freezing. *Geological Society of America Bulletin*, 96, 336–346.

Waller, R.I., Hart, J.K. & Knight, P.G., 2000. The influence of tectonic deformation on facies variability in stratified debris-rich basal ice. *Quaternary Science Reviews*, 19, 775–786.

Waller, R.I., Murton, J.B. & Knight, P.G., 2009a. Basal glacier ice and massive ground ice: different scientists, same science? In: Knight, J. & Harrison, S. (Eds.), *Periglacial and Paraglacial Processes and Environments*. Geological Society, London, Special Publication, 320, 57–69.

Waller, R.I., Murton, J.B. & Whiteman, C., 2009b. Geological evidence for subglacial deformation of Pleistocene permafrost. *Proceedings of the Geologists' Association*, 120, 155–162.

Waller, R.I., Phillips, E., Murton, J., Lee, J. & Whiteman, C., 2011. Sand intraclasts as evidence of subglacial deformation of Middle Pleistocene permafrost, North Norfolk, UK. *Quaternary Science Reviews*, 30, 3481–3500.

Watanabe, K. & Flury, M., 2008. Capillary bundle model of hydraulic conductivity for frozen soil. *Water Resources Research*, 44, W12402, doi:10.1029/2008WR007012.

Wilch, E. & Hughes, T.J., 2000. Calculating basal thermal zones beneath the Antarctic ice sheet. *Journal of Glaciology*, 46, 297–310.

Williams, P.J. & Smith, M.W., 1989. *The Frozen Earth: Fundamentals of Geocryology*. Cambridge University Press.

Yang, Y., Lai, Y. & Li, J., 2010a. Laboratory investigation on the strength characteristic of frozen sand considering effect of confining pressure. *Cold Regions Science and Technology*, 60, 245–250.

Yang, Y., Lai, Y. & Chang, X., 2010b. Laboratory and theoretical investigations on the deformation and strength behaviours of artificial frozen soil. *Cold Regions Science and Technology*, 64, 39–45.

Yasufuku, N., Springman, S.M., Arenson, L.U. & Ramholt, T., 2003. Stress-dilatancy behaviour of frozen sand in direct shear. In: Phillips, M., Springman, S.M. & Arenson, L.U. (Eds.), *Proceedings of the 8th International Conference on Permafrost*. Lisse, The Netherlands, 1253–1258.

Zhang, T., Heginbottom, J.A., Barry, R.G. & Brown, J., 2000. Further statistics on the distribution of permafrost and ground ice in the Northern Hemisphere. *Polar Geography*, 24, 126–131.

Zhang, T., Barry, R.G., Knowles, K., Heginbottom, J.A. & Brown, J., 2008. Statistics and characteristics of permafrost and ground-ice distribution in the Northern Hemisphere. *Polar Geography*, 31, 47–68.

3

Glaciers and Glacial Processes in Permafrost Environments

3.1 Introduction

Having focused primarily on the characteristics of permafrost within Chapter 2, this chapter considers the distinctive characteristics of glaciers found within permafrost environments and the influence of the resultant glacier–permafrost interactions upon both subglacial and ice-marginal processes. Permafrost environments are generally associated with low mean annual air temperatures; long, cold winters; short, cool summers and limited rates of precipitation. As a consequence, glaciers occurring within permafrost regions typically display distinctive mass balance regimes characterised by low annual accumulation rates, and short ablation seasons limited to the brief summer period during which air temperatures rise above freezing (Bogen & Bønsnes, 2003; Etzelmüller & Hagen, 2005). The refreezing of meltwater within the snowpack during the melt season to create layers of superimposed ice provides an additional form of internal accumulation (Hagen *et al.*, 2003) that constitutes the dominant form of accumulation in some Arctic glaciers (Koerner, 1970).

This climatically driven restriction of the magnitude of the inputs (precipitation) and outputs (meltwater) means that the mass turnover is consequently generally much lower than glaciers occurring in more temperature environments. As a result, glaciers occurring in permafrost areas are generally slower moving (Section 5.3) as the velocity required to maintain a steady ice-surface slope is much reduced. The limited length of the ablation season also means that the volume of meltwater generated is also comparatively low (Section 5.2).

This chapter starts by considering the thermal connections between glaciers and permafrost and the associated thermal regimes of non-temperate glaciers (Section 3.2). Section 3.3 provides an introduction to the basal boundary conditions of glaciers that play a key role in determining their dynamic behaviour. Section 3.4 considers the traditional glaciological paradigm in which critical basal processes are viewed as inactive beneath cold-based glaciers, before Section 3.5 examines recent evidence that has challenged these assumptions and suggested that subglacial processes can remain active at cryotic temperatures. Finally, Section 3.6 considers the distinctive processes that can occur in ice-marginal and proglacial environments as a consequence of glacier–permafrost interactions. In providing detailed insights into the influence of glacier–permafrost interactions on key glaciological

Glacier–Permafrost Interactions, First Edition. Richard I. Waller.
© 2024 John Wiley & Sons Ltd. Published 2024 by John Wiley & Sons Ltd.

characteristics and processes, this chapter provides a precursor to the consideration of their impact on glacier hydrology, glacier dynamics and glacial sediment fluxes in Chapter 4 and their landscape expression in Chapter 5.

3.2 Thermal Regimes of Glaciers and Permafrost

In view of the central importance of temperature to the topic of glacier–permafrost interactions, it is important to revisit the thermal characteristics of glaciers within their broader permafrost environmental settings prior to examining the influence of permafrost on a range of glacial processes in the remainder of the chapter. A theoretical consideration of the key controls on the basal temperatures of glaciers and ice sheets was provided earlier in Section 2.3.2 in the context of the formation and preservation of subglacial permafrost. This section develops this further by considering the terminology used to describe and differentiate different types of glacier from the specific perspective of their thermal regimes. It also considers the more recent conceptual models that have been developed to encourage a more integrated examination of glaciers and permafrost as the key components of a coupled system rather than as discrete elements of the cryosphere. This section also considers the limited number of empirical studies that have provided an insight into the thermal characteristics of glaciers situated in permafrost environments. Finally, the emerging impacts of climate change are introduced, forming a line of discussion revisited in the subsequent chapters.

Glacier–permafrost interactions are generally associated with cold-based glaciers, which can be defined as ice masses with basal temperatures that are below the pressure melting point of ice (Section 2.2.2). Conversely, glaciers with basal temperatures at the pressure melting point (PMP) are referred to as warm-based glaciers. This has resulted in the establishment of a fundamental if somewhat simplistic dichotomy in glaciology between warm-based (temperate) and cold-based (polar) glaciers (Lorrain & Fitzsimons, 2011), the nature and implications of which are explored in Section 3.4 and critically reappraised in Section 3.5. An assumed absence of basal melting beneath cold-based glaciers contrasting with the active meltwater generation beneath warm-based glaciers has for example resulted in the use of the alternative terms 'dry-based' glaciers and 'wet-based' glaciers (e.g. Kleman & Borgström, 1996) that can be considered synonymous.

With many Arctic and high-altitude glaciers simultaneously featuring areas of both warm-based and cold-based ice (see empirical case studies later in this section), the term polythermal (also referred to as sub-polar) is also commonly used to refer to these transitional or hybrid basal thermal regimes. A detailed review of the complex thermal regimes of polythermal glaciers is provided by Irvine-Fynn *et al.* (2011), which highlights their key thermal structures and their implications for glacier hydrology (see Chapter 4). Polythermal glaciers are defined as glaciers with a perennial occurrence of temperate and cold ice (Blatter & Hutter, 1991), with temperate ice existing at its PMP and containing up to c. 9% of interstitial water by volume, whilst cold ice is below the PMP and features negligible interstitial water. The complex interaction of a range of climatic and glaciological controls including the ice thickness, depth of penetration of the winter cold wave, the advection of cold ice, creep and sliding rates and associated heat generation results in polythermal glaciers displaying a wide variety of thermal structures (Figure 3.1). The presence and location of a 'cold-temperate transition zone' is of particular note in view of its relevance to glacier–permafrost interactions when it intersects the glacier bed. These complex thermal structures can in turn vary over space and time as discussed below.

In view of this terminological complexity, where possible the term 'non-temperate' is employed as a singular term that encompasses the range of glaciological settings in which cold-based conditions can occur and in which subglacial permafrost may be present (cf. Weertman, 1967).

Etzelmüller & Hagen (2005) have developed a highly relevant conceptual model that emphasises the wider thermal interactions between glaciers and the permafrost terrains in which they occur through its explicit connection of the glaciological concept of equilibrium line altitude (ELA; the point at which accumulation and ablation are equal) with the concept of mountain permafrost altitude (MPA; the lower limit of extensive permafrost) (Figure 3.2). In situations where the MPA is located well below both the ELA and the terminus, the glacier is likely to be entirely cold-based (Figure 3.2a,b). This is commonly the case where glaciers are thin such that both the insulating effects of the ice and the heat generated by ice flow are limited. Dyke (1993) provides a useful rule of thumb in this regard suggesting that this thermal regime is likely to be maintained if the local permafrost thickness exceeds the ice thickness of the glacier. In these circumstances the glacier can in essence be viewed as forming a constituent part of the permafrost horizon (Hughes, 1973). This scenario is commonly found in High Arctic locations such as the

Figure 3.1 Illustration of the various thermal regimes of polythermal glaciers (Blatter & Hutter, 1991, p262). The shaded areas denote the locations of temperate ice, the dashed lines indicate zones of melting along cold-temperate transitions and the arrows indicate the approximate position of the equilibrium line.

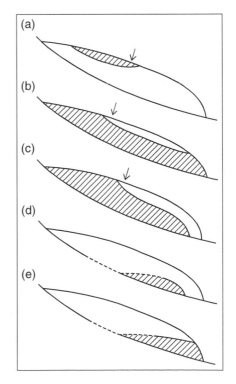

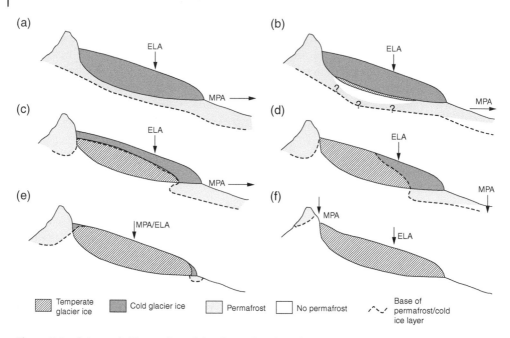

Figure 3.2 Schematic illustration of the thermal regimes in glaciers related to permafrost in mountain areas (Etzelmüller & Hagen, 2005, p12). See text for detailed discussion.

Canadian Arctic Archipelago (e.g. Koerner, 1989) and in High Altitude locations such as Tibet (e.g. Echelmeyer & Zhongxiang, 1987).

Where the thickness of the glacier ice is equal to or greater than the thickness of the local permafrost, glaciers are more likely to be polythermal in nature with some areas being characterised by warm-based conditions and others being associated with cold-based conditions (Figure 3.2c,d). In warmer and wetter environments, the latent heat release associated with the refreezing of meltwater can also promote the development of warm-based conditions in accumulation areas (Etzelmüller & Hagen, 2005). In these circumstances, cold-based conditions and glacier–permafrost interactions are typically limited to marginal areas where the ice thicknesses and velocities are low. This scenario is commonly found in the larger valley glaciers in Svalbard (e.g. Björnsson *et al.*, 1996). However, as discussed at the end of this section, ongoing climate change and a reduction in ice thickness and velocity is counter-intuitively resulting in an expansion of the areas of the bed associated with cold-based conditions.

Finally, in more temperate environments where the MPA is commonly higher than the elevation of the ice margin, the existence of cold-based ice and glacier–permafrost interactions is typically limited to the marginal fringes of glaciers where the ice is at its thinnest (Figure 3.2e). This can allow winter cold and the 0 °C isotherm to penetrate to the bed and generate cryotic conditions that can persist through the following summer. This situation is commonly found in sub-Arctic locations including Iceland, southern Norway and the European Alps (e.g. Matthews *et al.*, 1995; Evans & Hiemstra, 2005; Reinhardy *et al.*, 2019).

Although numerous theoretical studies have considered the thermal characteristics and profiles of non-temperate glaciers, comparatively few studies have provided actual measurements describing the thermal regimes of cold glaciers located within permafrost environments. The studies of the thermal regime of White Glacier undertaken as part of a deep drilling programme run between 1974 and 1981 by members of the McGill Axel Heiberg Research Expedition provide a notable exception (Müller, 1976; Blatter, 1987). In summarising the glacier's thermal characteristics, results reported by Blatter (1987) demonstrate cold-based conditions over most of the bed with basal temperatures of between −10 and −15°C recorded across the entire accumulation area. The lower part of the glacier however featured a layer of basal ice up to 40 m thick with temperatures at or close to the pressure melting point that extended down the centreline of the marginal glacier tongue. Vertical temperature profiles produced along the length of the glacier provide a more detailed insight into the thermal structure and its development (Figure 3.3). The accumulation area is associated with the formation of cold surface ice with a notable and unexpected temperature minimum 100–150 m below the surface thought to relate past colder climatic conditions. This cold ice is advected down glacier but is then removed in the ablation area by the upward vertical motion of the ice to generate a typical ablation profile with a large temperature gradient near the surface as described by Hooke (1977) (Section 2.3.2). Areas of the glacier associated with high surface velocities experience higher levels of frictional heat generation that increases the basal thermal gradient. This is however removed by basal melting resulting in the lowest parts of the glacier featuring isothermal conditions and a layer of temperate ice up to 40 m thick. Finally, the lateral margins and the terminus region are associated with cold ice that is frozen to the bed.

This thermal structure involving the presence of a basal layer of warm-based ice below the central ablation zone of an otherwise cold glacier was emulated during a subsequent study of the Laika Glacier, another small polythermal glacier located on Coburg Island in the Canadian Arctic Archipelago (Blatter & Kappenberger, 1988). This repeat observation encouraged Blatter & Hutter (1991) to attempt to replicate the observed thermal structure within a computational numerical model. The model however invariably generated an entirely cold-based glacier and was incapable of creating the area of warm-based ice observed in the field. The only way in which this could be created was by re-profiling the glacier and increasing the glacier thickness by 50% to correspond with its Little Ice Age (LIA) extent, as illustrated by trimlines observed in the field. This led the authors to conclude that the Laika Glacier was no longer in thermal equilibrium with the contemporary climate conditions and therefore that the warm-based ice represented a thermal remnant inherited from its enlarged LIA geometry. This requires the presence of a positive feedback loop whereby basal sliding, once initiated, can result in the persistence of the warm-based thermal conditions required. A critical point emerging from this study therefore is the time-dependency of a glacier's current thermal behaviour that can reflect both its climate history and related changes in its geometry. The importance of this point is well made by Irvine-Fynn *et al.* (2011) in their aforementioned review when they note that 'the thermal structure of a polar ice mass has a significant memory that is largely inherited' (p3).

Studies in Svalbard have demonstrated the presence of broadly similar two-layered thermal structures, although as indicated by Etzelmüller & Hagen (2005), the warmer and wetter climatic conditions result in warm ice being more extensive than in the Canadian High Arctic. Employing a combination of radio-echo sounding and borehole temperature measurements, Ødegård *et al.* (1992) for example identified a surface layer of cold ice

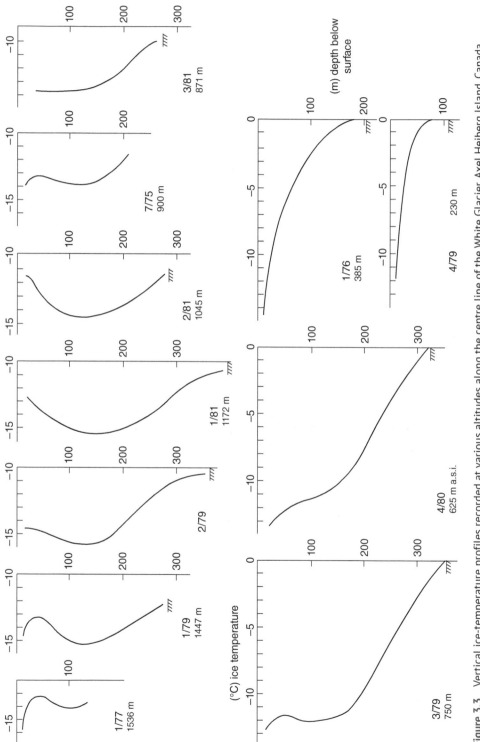

Figure 3.3 Vertical ice-temperature profiles recorded at various altitudes along the centre line of the White Glacier, Axel Heiberg Island, Canada (Blatter, 1987, p205).

20–60 m thick along the centreline of Erikbreen extending from above the equilibrium line to the glacier margin that was underlain by a layer of warm ice. The glacier also displayed a broader heterogeneity in its thermal structure with the marginal areas and parts of the accumulation area featuring cold-based ice, whilst the central part of the accumulation area was composed entirely of warm ice.

More recent studies focusing on the geometrical and thermal changes of glaciers in Svalbard caused by ongoing climatic change provide an additional and increasingly topical insight into the time-dependent nature of glaciers and glacier–permafrost interactions. A comparative study of polythermal and cold-based glaciers in northwest Spitsbergen by Glasser & Hambrey (2001) revealed a reduction in the extent of crevassing and of the formation of subglacial bedforms indicative of active basal melting following recession from their neoglacial maxima, consistent with a reduction in velocity and a progressive change from predominantly wet-based, through partly frozen to completely frozen basal conditions. More recently, Lovell *et al.* (2015) have considered the potential impacts of such thermal changes in relation to basal ice formation beneath Tellbreen, a glacier in Central Spitsbergen that in common with many valley glaciers in Svalbard has experienced significant thinning and recession during the twentieth century. The authors argue that a 2- to 4-m thick debris-rich basal ice layer represents the end product of a thermal transition from a polythermal to a cold-based basal thermal regime. This is thought to have occurred in response to significant thinning and recession that led to the basal adfreezing of a former water-saturated traction till. In describing a basal ice layer comprising frozen diamict, the authors refer to 'clean ice lenses' that are particularly extensive close to the ice-bed interface. These may represent segregated ice lenses that are commonly described within frost-susceptible permafrost sequences and that form in response to the migration of a frozen fringe and the cryosuction of porewater from the subjacent unfrozen materials (Section 3.5.4).

Geophysical surveys at Midtdalsbreen, an outlet glacier in south-central Norway, suggest that cold ice may be more extensive within some temperate glaciers than previously assumed and that as in Svalbard, its extent may be increasing in response to ongoing glacier recession (Reinhardy *et al.*, 2019). Systematic GPR surveys undertaken in April revealed a 40–50 m wide zone of cold-based ice beneath the majority of the glacier margin promoted both by the thin ice (<10 m) and the presence of ice-marginal snowbanks that persist into September and limit summer warming. The cold ice zones were also found in the accumulation areas around two nunataks, leading the authors to suggest that the distribution of permafrost within these rock masses has a potentially significant impact on the thermal regime of the surrounding glacier. In considering the potential impacts of future climate change, they hypothesise that a continued thinning of the parent Hardangerjøkulen ice field could result in an increase in the extent of cold ice and frozen-bed conditions in the accumulation area, whilst the earlier ablation of seasonal snow patches may conversely result in a decrease in the ablation area.

If correct, these inferences illustrate the ways in which changes in glacier mass balance and geometry can influence the thermal regime, in turn leading to a renewed phase of subglacial permafrost aggradation. These complex connections between glacier dynamics, basal thermal regime, sediment entrainment and basal ice formation are explored in further detail in Section 3.5.4.

3.3 Basal Boundary Conditions and Processes

3.3.1 Introduction

The basal boundary conditions and processes that occur at the base of a glacier can exert a profound influence on a range of key glaciological characteristics including morphology, dynamic behaviour and geomorphological and geological impacts (e.g. Weertman, 1957; Kamb & LaChapelle, 1964; Boulton & Jones, 1979; Boulton & Hindmarsh, 1987; Hart & Boulton, 1991; Piotrowski *et al.*, 2004). Recognition of the influential role played by basal processes extends back over a century (e.g. Reid, 1882; Russell, 1893; Lamplugh, 1911) with considerable subsequent research effort focusing on developing our understanding of the role of basal processes in determining patterns and rates of glacier motion (Raymond, 1971; Engelhardt *et al.*, 1978; Vivian, 1980; Iken & Bindschadler, 1986; Iverson *et al.*, 1995; Kavanaugh & Clarke, 2006). Particular attention has been paid to identifying the processes involved in instigating, maintaining and terminating states of fast ice-flow (Clarke, 1987) associated with surge-type glaciers (Clarke *et al.*, 1984; Kamb *et al.*, 1985; Kamb, 1987; Porter *et al.*, 1997; Truffer *et al.*, 2001), ice-streams (Alley *et al.*, 1986; Engelhardt *et al.*, 1990; Engelhardt & Kamb, 1997) and tidewater glaciers (Humphrey *et al.*, 1993; Hart & Smith, 1997). The broader significance of basal processes to these ongoing research questions within glaciology is neatly summarised by Paterson (1994) who commented that,

> *...an understanding of basal processes is the key to such problems as explaining the rapid decay of the ice-age ice sheets that began some 14,000 years ago or predicting whether global warming will lead to the disintegration of the west Antarctic ice sheet.* (p132)

Incorporating the influence of basal processes and boundary conditions into numerical ice-sheet models in order to enable them to address these important glaciological research questions has posed a significant challenge to modellers. Ice-sheet models therefore typically acknowledge the importance of basal boundary conditions and ice-bed interactions through the incorporation of some sort of 'basal slipperiness' parameter that determines the dynamic response to a given driving stress. Boulton & Hagdorn (2006) for example incorporate a 'décollement parameter' into their model of the former British Ice Sheet to integrate the combined influence of basal sliding and subglacial sediment deformation. This was adjusted along with climatic parameters in order to tune the model and identify model set-ups that provided the best fit with the field evidence. This for example involved prescribing the location of 'slippery patches' where a given shear stress would drive a more rapid horizontal basal velocity that in turn promoted the development of fast-flowing ice steams.

The basal boundary conditions of any part of a glacier can be subdivided according to the type of bed present (rigid or soft) and the basal thermal regime (warm-based or cold-based). This provides a simple fourfold classification of warm/rigid, warm/soft, cold/rigid and cold/soft that can be used to assess the relative extent of our knowledge and degree of understanding of these different types of basal boundary condition (Figure 3.4). Both types of warm-based glacier (rigid and soft bed) have been widely investigated within the literature and our knowledge of their dynamic behaviour and the contribution of basal processes is quite comprehensive, although the relative importance of basal sliding and

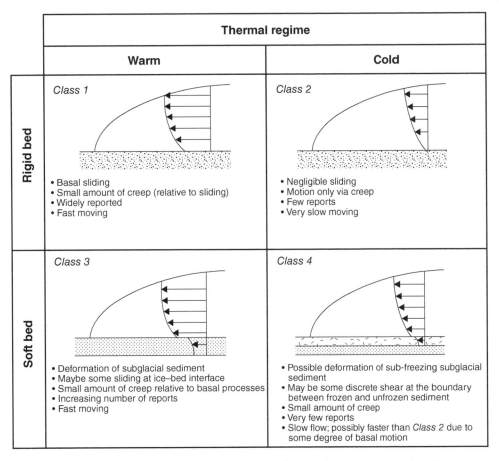

Figure 3.4 Diagrammatic summary of the generalised flow profiles, processes and attributes associated with different glaciological boundary conditions (Waller, 2001, p124).

subglacial sediment deformation in soft-bedded glaciers remains the subject of ongoing debate (Boulton, 1986; Robin, 1986; Clark, 1995; Iverson *et al.*, 1995; Section 3.3.3). A small number of studies of cold-based, rigid-bedded glaciers have been undertaken, mostly on small valley glaciers and ice-caps situated in the High Arctic (e.g. Koerner & Fisher, 1979) and Antarctica (e.g. Holdsworth, 1974) (Section 4.3.2). Glaciers and ice-sheets character-ised by cold-based ice and cryotic, unconsolidated sediments however remain the least well investigated and the most poorly understood (e.g. Menzies, 1981).

In view of their central significance to glaciology, the remainder of this section provides an introduction to the basal processes traditionally ascribed to rigid-bed (Section 3.3.2) and soft-bed glaciers (Section 3.3.2). The subsequent sections focus more specifically on the influence of basal thermal regime on the operation of the basal processes that are of most relevance to glacier–permafrost interactions. Section 3.4 considers the traditional ortho-doxy most concisely summarised in the frequently used statement that cold-based glaciers resting on permafrost are 'frozen to their beds'. Section 3.5 then considers the research that has challenged this orthodoxy and led to a reappraisal of the role played by basal processes beneath cold-based glaciers.

3.3.2 Rigid-Bed Glaciers and Basal Sliding

With glaciers traditionally being conceptualised as resting on rigid beds composed of undeformable bedrock, research into the nature and influence of basal processes initially focused on the ice–rock interface and the process of basal sliding with the bed itself being considered a passive substrate. Early empirical observations of glacier flow indicated that the creep of ice alone was unable to account for observed surface velocities. Gerrard *et al.* (1952) for example used borehole inclinometry to develop vertical flow profiles that led to the conclusion that approximately half of the surface flow of ice in a basin below the Jungfraujoch was accommodated by sliding at the glacier bed.

In developing a theory capable of estimating the contribution of basal sliding to overall glacier flow, Weertman (1957) suggested that the process was in fact associated with two distinct mechanisms, both of which are related to the occurrence of bedrock irregularities. The first mechanism is associated with the phenomena of pressure melting whereby ice flow against a protuberance results in an increase in pressure, a reduction in the pressure melting point and basal melting. The meltwater generated flows to the lee-side of the obstacle where lower pressures result in refreezing through a process referred to as regelation. This in turn liberates latent heat energy that establishes a heat flux through the obstacle promoting further melting on its up-glacier side. In this way, pressure melting and regelation lead to the mass transfer of ice past the obstacle. Weertman suggested that this process is likely to be most effective where obstacles are small (<1 m in size) such that the latent heat generated by refreezing can be conducted back through the obstacle to promote further melting. The second mechanism is associated with the tendency of bedrock obstacles to generate stress concentrations that in turn enhance the creep rate of the ice. In contrast to the first mechanism, this is most effective where the protuberances are large (>1 m in size). Weertman suggested that the relative importance of these two mechanisms is therefore dependent on the roughness of the glacier bed and the size of the obstacles. With regelation facilitating fast movement around small obstacles and creep rate enhancement facilitating flow around larger obstacles, obstacles of intermediate size are considered to provide the greatest resistance to basal sliding. Weertman (1964) subsequently considered the influence of a thin basal water film on the rate of sliding, concluding that a film of only a few millimetres in thickness could increase the sliding velocity by 40–100% by lifting the glacier sole and reducing the contact area and level of friction with the bed. This progressive decoupling of the glacier with its bed results in increased stress concentrations around the protuberances that remain in contact with the glacier sole and a concomitant increase in creep rate.

As part of an audacious research project entailing the mechanical excavation of tunnels, Kamb & LaChapelle (1964) were able to measure basal sliding beneath the Blue Glacier in Washington. In testing Weertman's model, this seminal work confirmed the general mechanisms but suggested that the contribution of enhanced creep relative to pressure melting had been overemphasised. The observation of a debris-rich 'regelation layer' also demonstrated the significance of the process in terms of debris entrainment and transport (see Section 3.5.4). The authors also provided empirical evidence for the existence of thin water films at the ice–rock interface associated with pressure melting, noting that rapidly excavated blocks of basal ice were easily removed, whereas if the surrounding ice was removed and the overburden pressure released, the ice would freeze to the bed.

With the rate of sliding being dependent upon the thickness of the water film, subsequent research has revealed mechanisms whereby an increase in the supply of meltwater to the glacier bed can result in a significant increase in velocity. Iken & Bindschadler (1986) for example recorded both the subglacial water pressures and surface velocities of Findelengletscher in Switzerland and discovered that events associated with very high water pressures (relating for example to lake drainage) where associated with both a vertical displacement and an abrupt increase in surface velocity. More recently, Zwally *et al.* (2002) related variations in surface meltwater production to periods of accelerated flow around the equilibrium zone of the west-central Greenland. In this case, they suggest that the rapid migration of supraglacial meltwater to the glacier bed via crevasses and moulins drives the aforementioned reduction in basal friction and increase in basal sliding. This provides a potential mechanism whereby climate change and an increased in surface meltwater production can directly contribute to an increase in glacier flow and potentially deglaciation over longer time periods (see Section 4.3.3 for more detail).

3.3.3 Soft-Bed Glaciers and Subglacial Sediment Deformation

Over the last three decades, glaciologists have perhaps belatedly recognised that not all glaciers are associated with hard rock beds and that many rest on unconsolidated and potentially deformable substrates referred to as 'soft beds'. The ability of glaciers to deform the materials over which they flow has been recognised for over a century, with Reid (1882) for example describing the 'disturbed chalk' and 'contorted drifts' of northeast Norfolk and likening the associated glacier-induced deformation to the movement of a hand over a tablecloth (cited in Lee *et al.*, 2011). However, it was not until much more recently that research into basal processes shifted from an exclusive focus on rigid-bed glaciers to one that increasingly considers the distinctive basal processes and active ice–substrate interactions associated with unconsolidated beds. Boulton (1986) suggested that this represents an important paradigm shift in glaciology, noting that,

> ... *the fundamental conceptual difference between analyses of fast low-stress glacier movement over rigid or deformable surfaces is that whereas in the former the surface can be regarded as passive, the strong interaction in the latter requires a coupled analysis of both ice and sediment deformation.* (p18)

Soft-bed glaciology is therefore characterised by an increased emphasis on the nature of the bed and in particular on the dynamic and hydrological connections between the glacier and its substrate. High rates of meltwater input to the bed combined with the presence of subglacial sediments with low permeabilities (typically fine-grained tills) in particular can lead to the development of high porewater pressures. These counterbalance the ice overburden pressure resulting in the development of very low effective normal stresses (Boulton & Jones, 1979; Boulton & Hindmarsh, 1987). As discussed previously (Section 2.5.2) in the context of the Mohr–Coloumb equation, this substantially reduces the frictional strength component and the yield stress, in turn facilitating shear failure of the subglacial material even under relatively low basal driving stresses. The associated process of subglacial sediment deformation can lead to the development of a subglacial deforming layer in which the majority of the glacier motion can be accommodated through simple shear.

Field studies undertaken in the 1970s and 1980s provided empirical evidence of ice-bed coupling and the influential role played by subglacial sediment deformation in controlling the dynamic behaviour of glaciers or ice sheets with soft beds. Field investigations at Breiðamerkurjökull, Iceland involving the burial and subsequent excavation of a marker array indicated that the pervasive deformation of a layer of saturated, subglacial sediment c. 0.6 m thick accommodated 88% of the total glacier flow with the remaining 12% occurring through slip at the ice-bed interface (Boulton & Jones, 1979; Boulton & Hindmarsh, 1987) (Figure 3.5). In addition, geophysical surveys beneath one of the Siple Coast Ice Streams in Antarctica (Whillans Ice Stream, formerly known as Ice Stream B) used seismic velocity

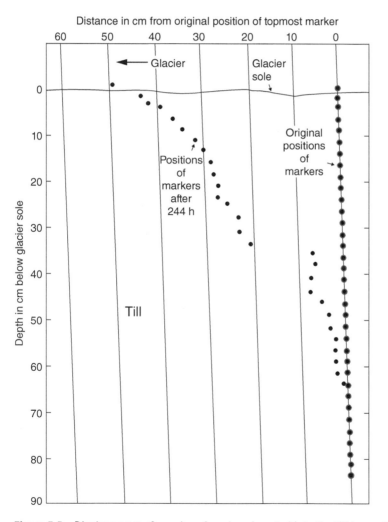

Figure 3.5 Displacement of a series of markers inserted into the till beneath Breiðamerkurjökull over a 244-h period (Boulton & Jones, 1979, p30). This indicated that 88% of the surface motion was caused by deformation of the substratum.

data to infer the existence of a dilatant and water saturated till up to 12 m in thickness. This led the authors to conclude that the rapid ice flow of the ice stream was caused primarily by active subglacial sediment deformation (Alley *et al.*, 1986; Blankenship *et al.*, 1987).

Subsequent field and laboratory investigations have led to ongoing debates concerning the nature and broader glaciological significance of subglacial sediment deformation and deforming layers. Borehole observations beneath the Whillans Ice Stream have for example suggested that basal motion is restricted to a subglacial shear zone 3–25 cm thick that is more appropriately conceptualised as basal sliding over a soft bed (Engelhardt & Kamb, 1998). With other studies reaching similar conclusions, this has led to the suggestion that as with rigid-bed glaciers, states of fast ice-flow in soft-bed glaciers are associated with the intense shear of a thin subglacial layer and a decoupling of the ice from the bed, rather than the pervasive deformation of a thick layer of sediment (e.g. Iverson *et al.* 1995). In this scenario, the ice-bed interface acts as a shear zone that in turn is likely to limit the transmission of shear stresses into the underlying bed and prevent significant subglacial sediment deformation. A detailed overview of this ongoing debate is beyond the scope of this book however and the reader is referred to the discussion included within Piotrowski *et al.* (2001) and Boulton *et al.* (2001) as an illustration of the key points of contention.

3.4 'Frozen Beds' and Traditional Glaciological Models

3.4.1 Frozen Beds and the Operation of Basal Processes

Glaciologists have traditionally assumed that the basal processes introduced in the previous section only occur beneath warm-based glaciers or ice sheets where the basal ice is at the pressure melting point (e.g. Weertman, 1957), with the production of significant amounts of basal meltwater being central to the operation of both basal sliding and subglacial sediment deformation. In driving more rapid rates of ice-flow, this in turn increases the strain heating of the basal ice, further enhancing the rate of basal melting and the supply of meltwater. Conversely, these selfsame processes are considered inoperative beneath cold-based glaciers where the prevalence of temperatures below the bulk freezing point severely limits the supply and availability of liquid water. With basal processes being central to a wide range of additional glaciological processes such as erosion and entrainment (Sections 3.4.2 and 3.5.3), cold-based glaciers resting on permafrost have consequently been viewed as slow moving and geomorphologically ineffectual.

A long-standing and extensively encountered assumption within glaciological literature is that cold-based glaciers are 'frozen to their beds' and that as a direct consequence basal processes are by definition inactive. This conceptualisation is perhaps unsurprising when one considers the traditional paradigm that envisaged basal boundaries as universally comprising ice–rock interfaces. Lliboutry (1966) for example argued that even if the base of a glacier or ice sheet is at the pressure melting point, sliding will not occur if protuberances penetrate into the cold ice above,

> *If a thick bottom layer of temperate ice does not exist, the hillocks protrude into ice that remains cold, whatever the roughness of their surface may be. The ice then sticks on this surface and no sliding occurs. (Lliboutry, 1966, pp525–526).*

In explicitly considering the occurrence of sliding beneath non-temperate glaciers, Weertman (1967) made the noteworthy point that the enhanced creep flow mechanism will continue to operate at temperatures below the pressure melting point, although he argued that the operation of this process alone was unlikely to result in significant sliding speeds if the bed is characterised by a range of obstacle sizes. In contrast to Lliboutry, he also suggested that basal sliding through regelation could occur beneath non-temperate glaciers so long as the glacier sole is at the pressure melting point (i.e. it is warm-based), arguing that the resulting increased hydrostatic pressure around obstacles projecting into the 'cold ice' would result in a reduction in the pressure melting point such that melting and regelation occur. Nonetheless, in order for significant sliding rates to occur, Weertman's sliding model requires the basal ice in contact with the bedrock obstacles to be at the pressure melting point with subsequent papers pointing to the dependency of significant rates of basal sliding upon the development and maintenance of thin water films at the ice-bed interface (Weertman, 1964).

Cold-based glaciers have consequently been viewed as being incapable of sliding over rigid beds for two primary reasons (Sugden & John, 1976). First, the bond between the ice and the bedrock exceeds the shear strength of the ice such that shear deformation will tend to occur within the basal ice. Second, the absence of regelation limits the ability of the cold-based ice to circumvent small-scale obstacles present at the bed. Glacier flow is therefore restricted to the creep deformation of the ice. Even when one considers this sole remaining mechanism, with ice tending to harden as temperatures decrease (see Section 2.5.1), cold ice temperatures will also tend to limit the rate at which cold ice masses creep.

Early empirical observations used to constrain sliding theory were thought to provide direct evidence for this view with the use of a variety of techniques beneath Meserve Glacier, a small cirque glacier in Antarctica with a mean basal temperature of $-18.8\,^\circ$C for example, failing to detect any movement across the ice-bed contact (Holdsworth, 1974). While subsequent research has shown this conclusion to relate in part to limitations in the contemporary equipment employed (Sections 3.5.2 and 4.3.2), such early initial findings led Weertman (1967) to assert that 'field observations have shown that temperate glaciers slide over their beds and cold glaciers do not' (p521).

If one considers the more recent recognition of soft beds, the associated process of subglacial sediment deformation is dependent on the production of significant quantities of basal meltwater that can raise the porewater pressure in the substrate and dramatically reduce the effective normal stress and the substrate's yield stress (see Section 2.5.2). It has again been assumed that a reduction in basal temperatures below the pressure melting point will result in the formation of a pore ice cement that increases sediment cohesion and prevents the dilation required for significant deformation to occur (Williams & Smith, 1989). In his early work on the influence of thermal regime on sedimentation, this led Boulton (1974) for example to state that, 'deformation of frozen subglacial sediments does not occur' (p1; see Section 5.4.3 from a subsequent revision of this viewpoint within Hart & Boulton, 1991). These assumptions have subsequently been widely employed as important boundary conditions within numerical ice-sheet models that typically prescribe a basal velocity of zero if the basal temperatures are calculated to be below the pressure melting point and the ice is considered to be frozen to its bed (e.g. Pattyn *et al.*, 2012).

Consequently, with states of fast and unstable ice flow being associated with active basal processes and 'slippery beds', cold-based glaciers have in contrast been viewed as being slow moving, geomorphologically ineffective and therefore of limited glaciological interest. Of most interest here has been the perpetuation of a binary conceptualisation of glacier dynamics whereby warm-based glaciers are fast flowing due to the operation of basal processes and cold-based glaciers are slow flowing due to their inactivity, with the pressure melting point acting as the clearly defined threshold between these two contrasting states of behaviour. This can be seen for example in the set-up of the ice-sheet model used by Boulton & Hagdorn (2006) to examine the dynamic behaviour of the former British Ice Sheet (Section 3.3). In relation to the basal boundary conditions, the authors state that, 'if the ice/bed contact is frozen, no decollement occurs, if unfrozen, decollement is assumed to be switched on, either by sliding, subglacial sediment or both' (p3367).

A consideration and accompanying critique of the wider glaciological implications resulting from what is now considered by some to be an overly simplistic conceptualisation for glacier dynamics and the geomorphological impact of glaciers are discussed in more detail in Chapters 4 (Section 4.3) and 5, respectively.

3.4.2 Thermal Controls on the Processes of Subglacial Erosion

Active basal sliding is commonly considered a pre-requisite for the geomorphic process of abrasion whereby clasts entrained within debris-rich basal ice are dragged over a bedrock surface to create classic erosional features such as striations. With basal sliding being widely regarded as inoperative beneath cold-based glaciers, it follows that abrasion should also be limited, leading to the assumption that cold-based glaciers resting on permafrost are geomorphologically ineffective as well as slow moving. Early experimental work again provided evidence that was seemingly entirely consistent with this prediction. An experimental study by Hope *et al.* (1972) for example, involving the rotation of sandstone cylinders loaded onto an ice surface at $-12\,°C$ demonstrated little evidence of wear and a tendency for shear to occur within the ice rather than at the ice–rock interface. Early observations of localities covered by cold-based ice caps similarly described landscapes characterised by an *absence* of subglacial landforms and deposits and the preservation of pre-glacial features such as tors, with the presence of scattered erratics providing the only clear evidence of glaciation (Dyke, 1976; section 5.2). The ongoing prevalence of the assumption of subglacial erosion being negligible beneath cold-based ice is illustrated by its incorporation within numerical ice-sheet models that have been used more recently to estimate spatial variations in the rate of erosion beneath the Antarctic Ice Sheet. In describing their erosion model, Jamieson *et al.* (2010) for example state that 'a pre-requisite to the occurrence of erosion at the bed is the presence of water which allows sliding to occur' (p4). This is in turn related to the basal thermal regime with an absence of water beneath cold or 'dry-based' ice resulting in an erosion rate of zero.

Whilst erosion rates beneath cold-based ice are likely to be considerably lower than those beneath temperate ice, numerous studies have argued that they are unlikely to be zero and therefore that geomorphic modifications may occur given sufficient time.

Following on from the earlier work of Röthlisberger (1968), Drewry (1986) notes that even if the ice is frozen to the bed and no sliding takes place, the strong shear occurring with the basal ice and the consequent rotation of an embedded clast may result in a limited amount of net forward motion that can result in some interaction between entrained debris and the bed.

If one takes a broader view of the processes of erosion that can take place beneath glaciers, a number of authors have suggested that processes other than abrasion may remain active and may indeed prove to be more effective beneath cold-based ice. Boulton (1972) for example define four thermal boundary conditions that can be used to elucidate the links between basal thermal regime, basal processes and the processes of erosion and sedimentation (see Section 5.4.2). The four zones are defined by contrasting basal boundary conditions: net basal melting (zone A), a balance between melting and freezing (zone B), refreezing of meltwater is sufficient to maintain the glacier sole at the pressure melting point (zone C), and the temperature at the sole is below the pressure melting point (zone D). Active basal processes in zones A to C promote abrasion and crushing, with the highest rates occurring in zone A where the production of meltwater and sliding rates are greatest. In contrast, basal sliding is considered to be inactive in zone D, where glacier–permafrost interactions promote a high level of adhesion at the ice–bed interface. Whilst this severely limits abrasion, Boulton however suggested that erosion can still result from two alternative processes. First, as mentioned above, debris transported englacially due to internal creep could come into contact with the bed if the flow lines intersect protuberances. Second, adhesion of the glacier sole to the bed could quarry or pluck large lithified or unlithified erratics if the bed is characterised by thin permafrost and favourably orientated discontinuities that in combination with high porewater pressures facilitate block removal (Figure 3.6). In a later allied paper, Boulton (1974) subsequently suggested that 'the existence of a significant adhesive force between the glacier and its bed should enhance the

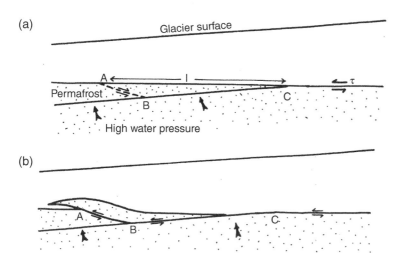

Figure 3.6 Illustration of the subglacial plucking of a large erratic (ABC) in zone D (with frozen subglacial materials) associated with thin subglacial permafrost (the base of which is illustrated by the arrows) (Boulton, 1972, p9). Fracture and displacement (illustrated in part b) will occur if the shear stress ($\tau \cdot l$) exceeds the strength of the frozen sediments between A & B and the frictional resistance along BC.

plucking process considerably' (p84). In this way, the same ice-bed adhesion that prevents abrasion can facilitate the displacement of large bedrock fragments. For this to occur, however, the shear resistance of the underlying materials still needs to be overcome and traditional conceptualisations might suggest this is difficult to achieve if the bed as a whole is frozen and undeformable as a result.

In developing a process-form model capable of accounting for the development of the diverse glacial geomorphology of North America, Clayton & Moran (1974) distinguish between abrasion zones and quarrying zones whose location is determined primarily by the basal thermal regime (see Section 5.5.1). They relate the abrasion zone to the central or sub-marginal part of a glacier characterised by a thawed bed (i.e. warm-based ice), excess subglacial porewater pressures and parallel or convergent flowlines, all of which act to promote basal sliding (Figure 3.7). Conversely, the quarrying zone is associated with the marginal zone of the glacier where frozen-bed conditions prevail, porewater pressures are lower and glacier flow is divergent. This promotes the quarrying of blocks of subglacial sediment that are then entrained into the glacier due to the upwards flow trajectory promoted by divergent flow. The extension of a subglacial aquifer from the warm-based interior to the cold-based margin can result in the generation of very high porewater pressures and the entrainment of large masses of sediment, particularly if drainage is impeded. Perhaps most surprisingly, they contend that the majority of the glacial erosion takes place in the quarrying zone that is reflective of glacier-permafrost conditions. In this respect, they in essence present an antithesis to the traditional argument that cold-based glaciers are geomorphologically ineffectual.

Following on from previous attempts to simulate the basal thermal regime of the entire Laurentide Ice Sheet at its Last Glacial Maximum extent (Sugden, 1977; see Section 2.4.2), Sugden (1978) compared the distribution of different erosional landscapes with the model's predictions in order to ground truth the results and assess the influence of three key controls: basal thermal regime, bed topography and geology. In spatially correlating the simulated glaciological conditions with the resultant landscapes, a number of interesting insights emerged. As expected, areas with little or no erosion coincided with areas of cold-based ice. However, landscapes of *areal scouring* were found to be associated not only with areas of basal melting and converging flow but also with areas of cold-based ice located down glacier of areas of basal freezing and debris entrainment, thereby highlighting the significance of basal debris content as well as the thermal regime. In using lake basin density as a proxy for the intensity of glacial erosion, the zone of maximum erosion was found to form a ring located between the centre and periphery of the ice sheet coincident with a zone of basal freezing, with the formation of a debris-rich ice layer estimated to be 20–50 m thick providing an effective means of debris evacuation (see Section 3.5.4). Sugden (1978) concluded with a cautionary note emphasising the need for critical testing of a range of hypotheses, with those of most relevance to glacier–permafrost interactions stating that (quoted from pp389–390):

- Landscapes of glacial erosion are related primarily to the basal thermal regime of the ice sheet.
- Mechanisms allowing evacuation of debris rather than those of abrasion or fracture may be the most important in influencing the amount of glacial erosion achieved by an ice sheet.

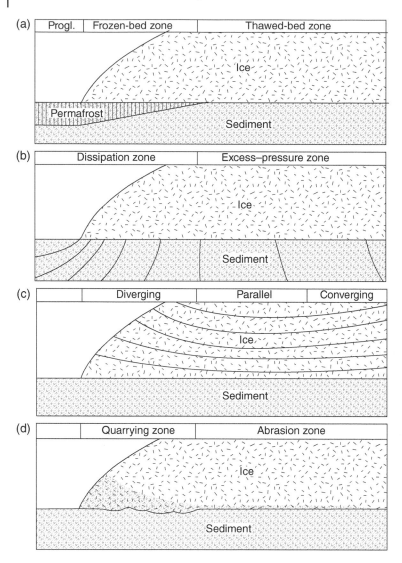

Figure 3.7 Schematic illustration of the thermal zones of an ice sheet (a) and the associated groundwater-pressure zones with lines of equipotential (b), glacial-flow zones (c) and erosion zones (d) (Clayton & Moran, 1974, p92).

- Cold-based ice may accomplish erosion if it has a debris load, inherited for example from an up-stream zone of regelation.

In summary, whilst the absence of basal sliding is widely believed to preclude significant abrasion where glaciers and permafrost interact, it does not necessarily follow that glacial erosion *in toto* is also negligible, with 'frozen bed' conditions potentially facilitating the quarrying and entrainment of sediment. These issues and their wider implications are discussed further in the following section on active subglacial processes, whilst the distinctive products are explored further in Chapter 5.

3.5 Active Subglacial Processes at Sub-Freezing Temperatures – An Emerging Paradigm?

3.5.1 Early Experimental Studies

In marked contrast to the discussion in the preceding section, a range of theoretical, laboratory and field studies have argued that in certain circumstances, active subglacial processes can remain active at temperatures below the bulk freezing point. Whilst these studies have grown in volume and persuasiveness over the last 20 or 30 years to represent what could be argued to be an incipient paradigm shift, the origins of this idea can be extended as far back as the nineteeth century.

At the core of this reconsideration of the potential role played by glacier–permafrost interactions is a recognition of the complex rheology of ice–sediment mixtures that change in response to temperature, ice content, stress and strain (Section 2.5.1) and in particular, the ability of premelted water to remain present in significant quantities as a result of the presence of impurities or the influence of interfacial effects (Section 2.2.2). A critical consequence of this is a direct challenge to the assumption that sediments beneath glaciers can exist in one of only two thermal and rheological states – unfrozen and deformable or frozen and rigid – that are separated by a clearly defined thermal boundary, which is typically considered to be the pressure melting point. The associated process-oriented assumption commonly utilised within numerical ice-flow models (Section 3.4.1) and geomorphological inverse models (Section 5.5.1) that basal processes switch on and off either side of a clearly defined temperature is increasingly regarded as being overly simplistic and certainly not universally applicable to all subglacial environments. This section will review the growing body of research that has highlighted the situations in which basal processes can remain active at temperatures well below the pressure melting/bulk-freezing point and thereby potentially contribute both to a glacier's dynamic behaviour (Chapter 4) and its geomorphic impact (Chapter 5).

Experimental studies extending back for over a hundred years have illustrated the unusual and distinctive properties of ice close to its melting point. Forbes (1858) for example demonstrated the phenomenon of regelation whereby two thawing ice masses bought into contact have a tendency to adhere and conjoin. In reply, Thomson (1860) explained this behaviour as a consequence of the existence of an intervening film of water maintained by capillary tension which in turn causes two ice masses to unite when these capillary films are bought into contact. In undertaking a range of laboratory experiments, Faraday (1860) went on to demonstrate the persistence of regelation both in air and underwater. He also demonstrated that the phenomena was peculiar to water and failed to occur when the experiments were repeated with metals such as tin or lead.

Over a hundred years later, Telford and Turner (1963) demonstrated the continued operation of pressure melting and regelation at temperatures glaciologists considered too low for these processes to occur. In pulling 0.45 mm steel wires through ice under a driving stress of c. 3 MPa at various temperatures below the bulk freezing point, they recorded speeds ranging from $160 \, \text{mm} \, \text{a}^{-1}$ at $-3.5\,°C$ to $800 \, \text{mm} \, \text{a}^{-1}$ at $-0.7\,°C$. Gilpin (1979) theorised that regelation at temperatures below the bulk freezing point was enabled by the existence of a nanometre thick liquid-like layer of water between the ice and the foreign object, a phenomenon more recently referred to as interfacial premelting (Dash *et al.*, 2006; Section 2.2.2). Gilpin (1980) subsequently undertook his own experiments of a similar design to Telford & Turner (1963) that confirmed the ability of regelation to operate at very low temperatures, recording a speed of $1.3\,mm\,a^{-1}$ at $-35\,°C$ for a $0.0127\,mm$ diameter tungsten wire under a driving stress of $0.68\,MPa$. These experimental data were used to validate the model originally developed in 1979 that in turn predicted liquid water contents consistent with measurements made by Barer *et al.* (1977) who used nuclear magnetic resonance to determine the amount of unfrozen water in ice containing dispersed silica particles.

In integrating Gilpin's model into Nye's theory of glacier sliding (Nye, 1970), Shreve (1984) predicted non-zero sliding rates at subfreezing temperatures that are controlled by the average thickness of the interfacial liquid layer and in turn the ambient temperature. Whilst sliding could still occur at sub-freezing temperatures, the predicted rates were so slow (e.g. $<10^{-1}\,m\,a^{-1}$ at $-1\,°C$) that their influence on glacier dynamic behaviour was considered to be very limited and of a magnitude too low to be detected by contemporary field observations (e.g. Holdsworth, 1974, see Section 4.3.1). Shreve did however point out that these very slow rates of basal sliding could become significant if viewed from a geomorphological perspective over the potential lifetime of the glacier. The total sliding distance over a period of 10,000 years under an effective basal stress of $100\,kPa$ was for example calculated to range from c. $35\,m$ at $-20\,°C$ to c. $350\,m$ at $-5\,°C$. This could allow the development of specific geomorphological features such as striations and subglacial precipitates that are commonly considered to be restricted to and diagnostic of warm-based glaciers (Section 5.3.3).

3.5.2 Direct Observations of Basal Processes Beneath Cold-Based Glaciers

More recent field investigations have confirmed both that basal processes can remain active at temperatures below the pressure melting point and that their potential contribution to glacier flow has been underestimated. In directly observing the operation of basal processes beneath a predominantly cold-based glacier in China, Echelmeyer & Zhongxiang (1987) published a seminal paper in this field that has forced glaciologists to reappraise their views on the dynamics of glaciers resting on permafrost. Following the excavation of a $70\,m$ long tunnel close to the terminus of the Urumqi Glacier Number 1, the authors observed both sliding over bedrock and active deformation of a layer of 'ice-laden drift' even though the recorded temperatures at the ice-bed interface never exceeded $-1.75\,°C$, were typically below $-4\,°C$ and were therefore well below the pressure melting point. Measurements of basal sliding over an embedded boulder recorded rates of c. $0.5\,mm\,d^{-1}$, corresponding to c. 5% of the total surface motion. More significantly, a marker array spanning the interface between the basal ice and the ice-laden subglacial drift indicated a cumulative deformation rate of $>9\,mm\,d^{-1}$ that was accommodated both via pervasive deformation of the ice-rich sediment and as slip along discrete ice-rich shear planes, a figure that corresponds to c. 80% of the total surface motion. This led the authors to conclude that,

> *These results indicate that the assumption of a zero basal velocity at sub-freezing temperatures is not valid, and that, on the contrary, the majority of the overall motion of a glacier or ice sheet can be accommodated at the bed* (p83).

It is interesting to note that in terms of a percentage contribution to ice surface velocity, the subglacial deformation of an ice-rich sediment layer of no more than 0.35 m in thickness under cold-based ice is no less significant than the deformation of water-saturated sediment beneath the warm-based ice of Breiðamerkurjökull recorded by Boulton & Jones (1979) (Section 3.3.3). In line with the theoretical work described in the preceding section, Echelmeyer & Zhongxiang (1987) attributed the active basal processes to the retention of liquid water adsorbed to particles at sub-freezing temperatures (i.e. interfacial films). This facilitated ice-particle and particle-particle slip and resulted in a substantial reduction in the viscosity of the ice-rich substrates that as a consequence deformed more than 120 times faster than the overlying debris-poor englacial ice.

Follow up research by Maohuan (1992) associated with the excavation of a new tunnel beneath the Urumqi Glacier No. 1 provides additional detail on the potential mechanisms associated the continuation of bed deformation at sub-freezing temperatures. Uniaxial compression testing of samples of debris-rich ice (with debris contents of <10% by weight) at temperatures of $-1\pm0.3\,°C$ failed to provide clear evidence of enhanced creep rates relative to debris-free glacier ice at the same temperature. With previous Chinese researchers at the site having observed shear planes within the subglacial ice-laden till, the authors consequently suggest that discrete shear (i.e. brittle failure) is the primary mechanism of deformation and that such shear is promoted by the presence of ice-rich layers or alternatively by layers of solute-rich liquid water.

More recent work at Storglaciaren, a glacier considered by Maohuan (1992) to have a comparative polythermal thermal regime to the Urumqi Glacier No. 1, provides additional corroborating evidence for the continued occurrence of basal processes beneath the glacier's cold-based margin (Moore *et al.*, 2011) (see Section 4.3.2). A change in basal boundary conditions from warm-based interior ice to the cold-based marginal ice was predicted to result in a reduction in ice velocity. Field measurements of the surface velocities however provided no evidence for the predicted significant change across this boundary. In addition, results obtained from borehole inclinometry and the installation of 'slidometers' at the ice-bed interface indicated that the ice motion relates primarily to basal processes that continue to operate down-glacier of the basal thermal transition to the glacier margin. Related modelling work suggested that '…cold ice must be able to move over the bed to account for the observed surface velocities' (p85) and therefore that the transition from warm to cold-based ice does not represent a slip/no-slip transition. As in the case of the Urumqi No. 1 glacier in China, the authors concluded that the continued basal motion can be explained by displacement either within a frozen subglacial debris layer or at the interface between this frozen debris layer and the underlying unfrozen sediments. In addition, a discrepancy between the estimates of basal slip derived from the two techniques employed provided potential evidence for movement deeper within the substrate and the entrainment of frozen subglacial sediment as predicted by the concept of effective beds (Section 3.5.5).

Field observations at Hagafellsjökull-Eystri, a surge-type outlet glacier in Iceland, suggest that active subglacial deformation at temperatures below the pressure melting point may in certain circumstances also occur beneath the margins of predominantly temperate glaciers (Bennett *et al.*, 2003, 2005). In examining a part of the foreland overridden by a lateral piedmont lobe during a winter surge lasting from Autumn 1998 to April 1999, the authors discovered that the subglacial sediments contained a series of sub-rounded, imbricated ice blocks hosted in a frozen diamicton characterised by frequent ice-rich fractures (Figure 3.8). With the foreland also being associated with an extensive network of crevasse-fill ridges, a feature commonly found within surge-affected areas and indicative of subglacial sediment mobility, the authors suggest that the combination of active subglacial deformation and sub-freezing conditions provides evidence for two distinct phases of development and a progressive cooling of a subglacial deforming layer. The first stage of the surge was associated with unfrozen conditions and ductile deformation that resulted in the formation of the crevasse-fill ridges and the partial melting of the rigid ice blocks that were entrained following serac collapse at the advancing margin. The second stage was associated with a transition to partially frozen conditions and a change in the rheological behaviour to a more brittle style of deformation that resulted in the formation of the ice-rich layers and the attenuation of the ice blocks into augen-like structures (Figure 3.8). In addition to providing additional evidence for active basal processes at sub-freezing temperatures, this study suggests that the basal thermal regime changed during the duration of the winter surge potentially as a result of the overriding of seasonally frozen ground and the penetration of cold air into the heavily crevassed margin.

Shifting the focus to the opposite end of the climatic spectrum, work in Antarctica has demonstrated that basal processes can remain active beneath some of the coldest glaciers on Earth. Successive studies of the basal zone of Meserve Glacier, a small glacier in the largely ice-free Wright Valley, have reached contrasting conclusions (see Section 4.3.2), thereby providing a microcosm of the broader shifts in debates regarding the operation of basal processes beneath cold-based ice. Early investigations by Holdsworth (1974) involved the excavation of a tunnel into the glacier margin which was characterised by a basal temperature of −18 °C and a debris-rich basal ice layer c. 0.6 m thick. The use of dial gauges fixed to rock-bolts in the bed and wooden pegs set into the tunnel walls seemingly confirmed an absence of sliding across the boulder pavement on which the glacier margin rested. The deformation of peg arrays within the basal ice did however indicate relatively rapid creep of the debris-rich basal ice that was related the presence of impurities and the upward

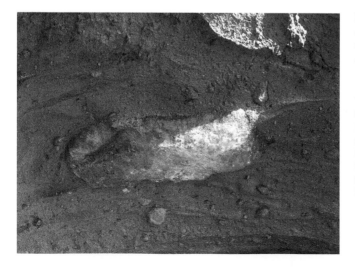

Figure 3.8 Ice block situated within frozen diamicton at the surge-affected margin of Hagafellsjökull-Eystri, Iceland. Note the ice-rich fractures extending from (and surrounding) the block to produce an augen-like structure. Ice flow was from left to right and the main body of the ice block is c. 60 cm in length (Image: Richard Waller).

diffusion of salts from the bed. The author also describes the presence of 'smears' of fine-grained debris within the basal ice that are thought to be associated with the rotation of clasts within the basal ice and their interaction with static boulders. This raises the possibility of ice-bed interactions even though the glacier was viewed as being frozen to the bed, providing field evidence consistent with the theoretical models of cold-based erosion advocated by Boulton (1972) and Drewry (1986) (Section 3.4.2).

On revisiting the site in 1995–1996, Cuffey *et al.* (1999) reached the opposite conclusion regarding the operation of basal processes, providing multiple lines of evidence for basal sliding remaining active beneath the Meserve Glacier in spite of a recorded temperature at the ice-bed interface of −17 °C. Following the excavation of another tunnel at the glacier bed, the authors observed 'striated' cavity roofs with wavelengths equivalent to the roughness of the bed composed of boulders. They also installed displacement markers and displacement transducers capable of measuring much smaller velocities than the dial micrometres previously used by Holdsworth. The displacement markers were observed to move down glacier over a period of up to 10 months at a rate more than an order of magnitude faster than the rate predicted for creep deformation at the recorded basal temperature. The transducers demonstrated sliding rates of between 2 and 8 mm a^{-1}, rates at least an order of magnitude greater than the 0.2 mm a^{-1} predicted by the Gilpin/Shreve model (see Section 4.3.2 for further detail). Active sliding at rates much higher than predicted by theory were again related to the presence of interfacial films at ice–rock interface, field evidence for which was provided in the form of clean segregated ice lenses within the debris-rich basal ice. In order to explain the observed velocities, the films were estimated to be between 20 and 40 nm in thickness, which is far greater than expected for the ambient temperatures. This was in turn related to the presence of very high solute concentrations that are characteristic of grain boundaries and veins in polycrystalline ice and further promoted by the solute rejection caused by freezing that can result in solute concentrations 10^4 times higher than the bulk ice. With this in mind, the authors concluded that, 'the relatively rapid sliding of this polar glacier . . . is made possible by a brine layer at the ice-rock interface' (p753).

3.5.3 Active Erosion and Debris Entrainment at Sub-Freezing Temperatures

The identification of theoretical and field evidence for the continued operation of basal processes at temperatures below the bulk-freezing point has led in turn to a re-consideration of the erosive impacts of cold-based ice. Field-based studies on cold-based glaciers in Antarctica in particular have challenged traditional conceptualisations of cold-based glaciers as being geomorphologically ineffective and have provided further insights into the potential processes of erosion that can remain active beneath cold-based ice as predicted by earlier theoretical and experimental studies (Section 3.5.1). As early as the 1970s for example, Mercer (1971) reported the occurrence of striated englacial debris within cold-based glaciers at a variety of sites in Antarctica suggestive of the continued occurrence of subglacial erosion. This section examines some of the process-based studies that have helped elucidate the mechanisms that can lead to active erosion beneath contemporary cold-based glaciers, prior to a more detailed consideration of the distinctive landforms, sediments and landsystems indicative of glacier–permafrost interactions in Chapter 5.

The excavation of a tunnel beneath the Suess Glacier, a small 'dry-based' Alpine glacier in Antarctica, has provided evidence of active substrate deformation and the erosion of the glacier bed in spite of a recorded basal temperature of −17 °C (Fitzsimons *et al.*, 1999). The tunnel walls revealed that the basal zone of the glacier comprised a debris-rich basal ice layer 3.8 m thick, the lower 2 m of which contained large blocks of frozen sand and fine-gravel (Figure 3.9). The preservation of primary sedimentary structures and the limited disaggregation indicate that these blocks had been eroded from the bed and entrained into the basal ice *en masse* with little if any modification. With processes of sediment entrainment associated with the bulk freezing of liquid water being assumed to be of limited relevance to entirely cold-based glaciers (see next section), debris entrainment is considered more likely to occur through a mechanical coupling between the glacier and the subjacent permafrost as discussed previously in Section 3.4.2. In considering the formation of debris bands in Svalbard comprising frozen sediment cemented by interstitial ice, Boulton (1970a) for example noted that the 0 °C isotherm is likely to occur in the subglacial sediments even when the glacier sole is at the pressure melting point. He suggested that this would encourage the freezing on of these sediments with slip subsequently occurring at the subjacent boundary between frozen and unfrozen sediments such that, 'subglacial accumulations of sediment could be incorporated within the ice without undue disturbance of their structure' (p223) (see Sections 3.5.5 and 5.4.2).

A potential problem with this mechanism is that ice-sediment mixtures with high concentrations of coarse sediment are commonly considered to be stronger and more resistant to deformation than pure ice at very low temperatures, due to the high levels of interparticle friction and limited quantities of premelted water (Section 2.5.2). As such, one would expect any deformation to occur within the clean basal ice rather than within an ice-rich permafrost substrate. In an attempt to constrain the mechanical properties of the basal materials and thereby identify the mechanisms responsible for the entrainment of the frozen sediment blocks at the Suess Glacier, Fitzsimons *et al.* (2001) consequently undertook direct shear tests of the glacier ice and the frozen substrate in the field. These indicated that the average peak shear strength of the substrate samples (2.53 MPa) was

Figure 3.9 Intraclast composed of frozen sand located within the base of Suess Glacier, Antarctica. Note the well preserved bedding. Pocket knife blade for scale (Image Courtesy: Sean Fitzsimons).

indeed approximately double that of the basal ice (1.28 MPa) and the glacier ice (1.39 MPa). Earlier field observations had however revealed the presence of physical discontinuities within the frozen subglacial sediment in the form of cavities, layers and lenses of segregated ice and thin mud layers (Fitzsimons *et al.*, 1999). These features therefore provide localised planes of weakness within the otherwise high strength frozen material along which deformation may preferentially occur. If a segregated ice layer within the frozen substrate occurs close to the glacier–bed interface, the rate of deformation is likely to be higher within this discontinuity resulting in the migration of the 'effective bed' into the substrate (see Section 3.5.5) and the mobilization and entrainment of the overlying high strength material (Figure 3.10). Subsequent observations of frozen subglacial sediment within the basal ice layer of the Upper Victoria Glacier led Fitzsimons *et al.* (2008) to suggest that,

> the nature of the glacier substrate acts as an independent control on erosion and entrainment processes. Strong beds such as crystalline bedrock or frozen coarse sediments may be characterized by low rates of erosion, whereas sedimentary beds that contain substantial volumes of ice (e.g., overridden permafrost) may be susceptible to deformation and entrainment by cold-based glaciers (p9).

In addition to the *en masse* entrainment of large blocks of frozen material, Cuffey *et al.* (2000) argued that the processes of abrasion and entrainment can also continue to operate at temperatures well below the bulk freezing point. As part of a wider reappraisal of the activity of basal processes at the Meserve Glacier reported in the previous section, Cuffey *et al.* (2000) asserted that basal sliding at the very low temperature of −17 °C had resulted in abrasion, striation formation and the generation of fine-grained material comprised primarily of silt and fine sand. They suggested that the presence of solute-rich interfacial films had also enabled entrainment through regelation at these very low temperatures resulting in the formation of a basal layer of comparatively debris-rich 'amber ice' approximately 1 m thick.

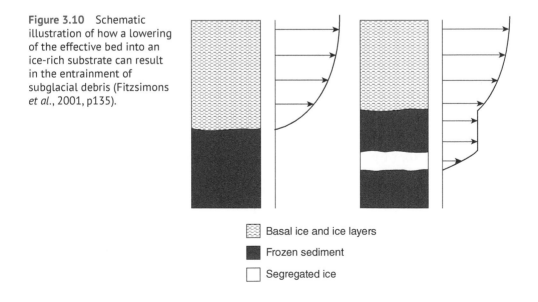

Figure 3.10 Schematic illustration of how a lowering of the effective bed into an ice-rich substrate can result in the entrainment of subglacial debris (Fitzsimons *et al.*, 2001, p135).

Basal ice and ice layers

Frozen sediment

Segregated ice

The combination of abrasion, sediment entrainment and glacier flow thereby provides a mechanism through which sediment can be evacuated by cold-based glaciers, with measured sediment fluxes of $10–30\,m^3\,a^{-1}$ corresponding to average denudation rates of between 9×10^{-7} and $3\times10^{-6}\,m\,a^{-1}$. Although this rate is very low in comparison to warm-based glaciers, given sufficient time to operate, it has the potential to generate significant landforms. In this case, net erosion totals over a period of 10 million years (the estimated time the Dry Valleys have experienced a cold and dry climatic regime) amount to $10–30\,m$, which can account for the generation of the bedrock depression in which the Meserve Glacier is situated. Consequently, the authors concluded that, 'the common assumption that cold-based glaciers are "protective rather than erosional" is not true in the absolute sense' (Cuffey *et al.*, 200, p354) (see Section 5.2).

3.5.4 Formation of Debris-Rich Basal Ice

Sediment that has been actively generated through the processes of subglacial erosion, or that is pre-existing and has been overridden by a glacier, can be entrained by a variety of processes to produce debris-rich basal ice. Basal ice differs from the overlying englacial glacier ice as a consequence of its formation at and interaction with the glacier bed. Basal ice layers can reach tens of metres in thickness and can typically be distinguished from firnified glacier ice on the basis of their relatively high debris content, anisotropic structure and distinctive isotopic and chemical composition (Hubbard & Sharp, 1989). It has long been recognised that the basal ice layers of non-temperate or polythermal glaciers tend to be thicker and more extensive than those associated with temperate glaciers (e.g. Boulton, 1970a). As a result, Boulton (1970a, 1972) identified thermal regime as a primary control on the entrainment of basal sediment with thick basal ice layers being promoted by situations in which there is net transfer of meltwater from a warm-based interior or from the glacier surface to a cold-based margin (potentially underlain by permafrost) where basal freezing can take place. Boulton (1970a) noted a clear coincidence between the entrainment of debris at the margins of Makarovbreen and Aavatsmarkbreen in Svalbard and their marginal zones of cold ice leading him to conclude that,

> This apparent contrast between the englacial debris load of temperate and polar glaciers supports the suggestion that temperature regime is the controlling factor in the englacial incorporation of large volumes of basally derived debris, and that only those glaciers in which basal freezing is an important process show large amounts of such debris (p228).

A range of distinctive processes have been advocated within the literature to explain the formation of debris-rich basal ice. With a number of these processes being related either directly or indirectly to changes in basal thermal regime, they are potentially related to glacier–permafrost interactions and are therefore worthy of explicit consideration. However, rather than attempting to provide a comprehensive review (which can be found in Hubbard & Sharp, 1989 and Knight, 1997), this section will focus on the processes of basal ice formation in which glacier–permafrost interactions have the potential to play a prominent role.

Early research focused on the hypothesised role of shearing that was commonly thought to occur around the cold-based margins of ice masses in response to a deceleration of the ice and associated longitudinal compression. In documenting field observations made at the East Branch glacier in north-west Greenland, Chamberlin (1895) for example noted that,

> *Here tongues of debris, having their origin in the bowlder-clay (sic.) below the glacier, were seen to reach out into the basal portion of the ice as though they were being introduced into it by the differential movement of the layers upon each other.* (p207)

More focused work on the process was undertaken by Goldthwait (1951) in his investigations of the sources of the debris responsible for the formation of a zone of marginal 'dirty ice' and associated end moraines (see Section 5.3.2) at the southern margin of the Barnes Ice Cap on Baffin Island. He described the ice margin as featuring 'myriads of dirt-filled thin fractures' containing small pebbles and occasional large boulders that are thought to represent shear planes responsible for the transfer of material from the glacier bed (Figure 3.11). The density of these shear planes, each up to 1 inch thick, was in combination responsible for the creation of a debris-rich ice margin with a total thickness of 100–200 ft. Whilst field measurements failed to provide direct evidence for any differential motion across these shear planes, the presence of augen structures and the preferential orientation of clasts along these shear planes coupled with an up-glacier dip of 10°–36° was considered consistent with an origin through ice-marginal thrusting. Swinzow (1962) reached similar conclusions following his detailed examination of debris-rich shear planes exposed in the walls of a series of tunnels excavated into the margin of the Greenland Ice

Figure 3.11 Margin of Fountain Glacier on Bylot Island featuring a thick basal ice layer and distinct debris bands similar to those described on nearby Baffin Island by Goldthwait (1951). The vertical ice margin is approximately 10 m high (Image: Richard Waller).

Sheet near Thule. As part of these studies, he explicitly acknowledged what he believed to be the influential role played by glacier–permafrost interaction with the shear planes considered responsible for debris entrainment forming at the boundary between active ice and immobile marginal ice 'fused' to the underlying permafrost.

Subsequent work undertaken by Weertman (1961) however argued that the shear hypothesis of debris entrainment was theoretically unsound, suggesting for example that it is incapable of explaining the density of the shear planes observed at non-temperate glacier margins. Boulton (1970a) similarly suggested that although shearing may redistribute debris held within a glacier, it is incapable of entraining new material. Basal ice research has consequently focused largely on a number of thermally regulated processes of debris entrainment that result from melting and refreezing (regelation) operating at various spatial scales. It is worth noting here that whilst the classic process of regelation associated with the sliding of a temperate glacier over an irregular rigid bed (Section 3.3.2) may continue to operate beneath some cold-based glaciers (see previous section), it is considered unable to explain the formation of the thick basal ice layers commonly observed at the margins of polythermal glaciers that can reach tens of metres in thickness (Boulton, 1970a; Sugden *et al.*, 1987). This is because ice formed in the lee of one obstacle tends to be destroyed by melting against obstacles further down-glacier with the thickness of the basal ice layer therefore being limited to the amplitude of the bedrock obstacles (Hubbard & Sharp, 1993). Iverson & Semmens (1995) have suggested that regelation may instead operate downwards into porous sediments in response to a normal driving stress resulting in debris entrainment beneath soft-bedded glaciers. As with rigid-bed regelation however, this mechanism is thought to be most effective beneath temperate glaciers rather than those underlain by permafrost.

Following the rejection of the shear hypothesis in his consideration of the formation of debris bands at the margin of the Barnes Ice Cap and the north-western Greenland Ice Sheet, Weertman (1961) proposed an alternative 'freezing model'. This is of direct relevance to glacier–permafrost interactions in that its operation is associated with polythermal basal thermal regimes in which the marginal ice is 'frozen to its bed' (p970). Weertman argued that under the thick ice of the ice-sheet interior, the temperature gradient would be insufficient to remove the heat generated at the bed. This excess heat would therefore lead to basal melting, producing water that flows along the local pressure gradient towards the margin. As the ice thins nearer the margin, the temperature gradient steepens, removing all the heat generated at the bed and returning the base to freezing conditions, where the water refreezes. As a result, there is a large-scale net transfer of mass (water) from the ice-sheet interior to locations closer to the margin allowing the accretion of thick basal ice sequences as a consequence. Aerogeophysical surveys of the East Antarctic Ice Sheet undertaken as part of the International Polar Year (2007–2008) have provided an insight into the potential extent and glaciological significance of basal freeze-on in terms of mass inputs (Bell *et al.*, 2011). They estimate that 24% of the basal area of Dome A is experiencing basal freeze-on which in places has contributed half of the ice sheet's observed thickness. Radar imagery indicates that the freeze on occurs primarily around valley heads and valley walls in the subglacial Gamburtsev Mountains. The authors estimate that that a total of 8600 km^3 of ice has been frozen on to the base of the ice sheet, a thickness sufficient to disturb the internal layering of the ice sheet and potentially to influence its surface morphology.

A common misconception of this mechanism is that it simply involves the migration of water from the ice-sheet interior to the margin driven by a hydrostatic pressure gradient. Weertman (1961) however cautioned that this mechanism is unable to entrain debris if steady-state conditions prevail, arguing that debris can only be entrained if there is a cyclical shift in the basal heat balance, resulting for example from a change in the supply of water to the ice-bed interface (Figure 3.12a). Such a reduction in the water supply to the bed would result in a reduction in the release of latent heat causing the glacier to 'freeze to its bed' and the 0 °C isotherm to migrate into the substrate (Figure 3.12b). A subsequent increase in water supply would increase the release of latent heat, reverting to a situation where the 0 °C isotherm returns to the ice-bed interface such that ice is again accreted to the bed (Figure 3.12c). In this way, cyclical changes in water supply and basal heat balance can result in the incorporation of the discrete debris layers commonly observed at polythermal glacier margins (Figure 3.12d).

Since its initial proposition by Weertman (1961), the process of basal adfreezing has been widely accepted as the dominant mechanism of large-scale basal ice formation capable of accounting for the formation of basal ice layers up to tens of metres in thickness. Boulton (1970a) was one of the first to advocate its operation in a field-based context. In considering the origin of a number of thick basal debris bands near the margin of Makarovbreen in Spitzbergen, he rejected the shear hypothesis and instead suggested debris was frozen onto the base and then 'locked' into the glacier by the freezing of water beneath this debris. However, one important difference to the model proposed by Weertman is that Boulton suggested that this water was derived not from basal melting further up-glacier, but from surface melting and its subsequent transport to the bed through a heavily crevassed area just up-glacier of the margin.

Whatever the water source, the adfreezing of basal meltwater down-glacier of a thermal transition has the potential to elevate entrained material into the body of the glacier, thereby promoting its long distance transport (Hooke *et al.*, 2013). Convincing field evidence for basal adfreezing was provided via a study of the basal ice layer of Aavatsmarkbreen in which Boulton (1970a) tested the prediction that the lowermost material should be locally derived whilst the uppermost material should be the farthest travelled. In traversing a series of distinctive lithologies, he was able to demonstrate a sequence that reflected the successive entrainment of the lithologies that the glacier had traversed. A similar pattern has been replicated by Zdanowicz *et al.* (1996) on Bylot Island in the Eastern Canadian Arctic that used clay mineralogy data to demonstrate sequential basal ice accretion from the upper and the marginal zones of the glaciers studied.

Research in western Greenland has however challenged the universal applicability of the freezing model as an explanation of the banded ice sequences commonly found at the margins of polythermal glaciers and ice sheets. In comparing the co-isotopic characteristics of the debris bands and the intercalated clean ice, Knight (1989) tested and falsified a model of simple freeze-on. Field observations, structural and sedimentological data led Knight (1994) to suggest that the debris bands represent a structural extension of the stratified basal ice created by longitudinal compression. Therefore, whilst the role of shearing in debris entrainment remains a matter of debate, deformation can clearly result in the elevation and modification of pre-existing basal ice. Post-formational, flow-related deformation within the basal ice itself may in addition result in the formation of distinctive

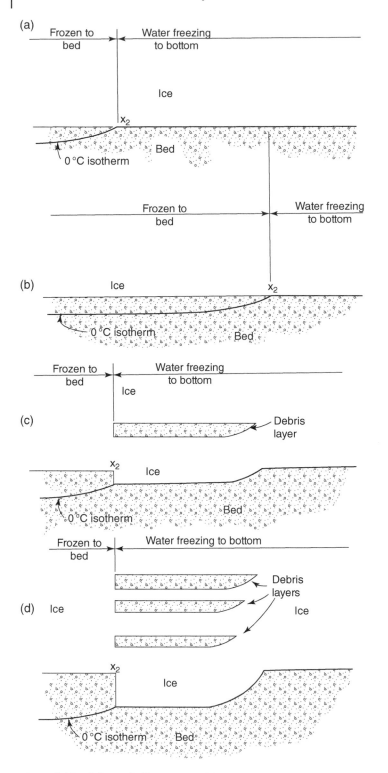

Figure 3.12 Schematic illustration of the entrainment of debris bands via basal adfreezing under non-steady-state conditions (Weertman, 1961, p972). See text for details.

tectonically-derived sub-facies within pre-existing basal ice layers irrespective of the original mechanism of debris entrainment (Waller *et al.*, 2000).

Propagation of the 0 °C isotherm (or more accurately the freezing front) into subglacial sediments beneath cold-based ice can be considered to be conceptually similar to the aggradation of permafrost that occurs in subaerial environments in response to an increasingly negative surface heat balance and a reduction in the mean annual ground temperature (Section 2.3.1). Basal ice formation can therefore in certain circumstances be likened to subglacial permafrost aggradation. This provides an alternative perspective on the processes of glacial sediment entrainment and deposition, if the subsequent thaw of this subglacial permafrost is taken into consideration. With this in mind, Menzies (1981) argues that the temperatures within subglacial sediments are key to understanding their entrainment, transport and subsequent deposition. In the absence of any direct measurements (considered to be a major gap in our current understanding), he presents four theoretical models that illustrate the potential impacts of oscillating freezing fronts on materials of contrasting textures and permeabilities (Figure 3.13). As with Weertman (1961), he suggests that these oscillating freezing fronts are most likely to occur at subglacial thermal boundaries between warm and cold-based ice that can migrate up and down glacier in response to changes in velocity, ice thickness and water supply. Menzies however develops the model further by considering the potential impacts of these oscillations on porewater migration within the subglacial sediment and its resultant geotechnical characteristics. If the subglacial materials are permeable, he suggests that porewater will migrate towards the advancing freezing front resulting in the downwards migration of the zone of maximum shear, the entrainment of debris and an increase in the shear strength in the sediment from which the porewater has been withdrawn (Figure 3.13b). Conversely, if the subglacial materials are fine-grained, porewater is expelled from the advancing freezing front. If the subjacent materials are coarse grained and permeable, this will potentially lead to the development of a series of shear horizons as the freezing front continues to propagate (Figure 3.13c). Alternatively, if they are more impermeable, this will slow the migration of the freezing front, promoting the formation of a smaller number of more persistent shear planes (Figure 3.13d). Finally, porewater expulsion into a lower stratum of impermeable material some distance below the freezing front can lead to the generation of very high porewater pressures (Figure 3.13e). This may in turn generate a flow instability and surge type basal condition in which the sediment liquifies.

The work of Menzies (1981) introduces important theoretical connections that might occur between the basal thermal regime, basal processes and boundary conditions, and the dynamic behaviour of glaciers and ice sheets. Focusing here on their implications for debris entrainment and basal ice formation, research by Christoffersen & Tulaczyk (2003a,b) has developed this research thread in relation to the potential mechanisms associated with the initiation and termination of states of fast ice flow (see Section 4.3.3 for further detail on the implications of this work for ice stream dynamics). In comparing borehole observations from two of the Siple Coast ice streams in Antarctica - the active Whillans Ice Stream (formerly called Ice Steam B) and the slow-moving Kamb Ice Stream (formerly called Ice Stream C) – Christoffersen & Tulaczyk (2003a) observed that the latter was characterised by: (i) comparatively steep basal temperature gradients, (ii) comparatively thick layers of debris-rich basal ice ~12 to 25 m thick, (iii) unfrozen subglacial till displaying supercooled

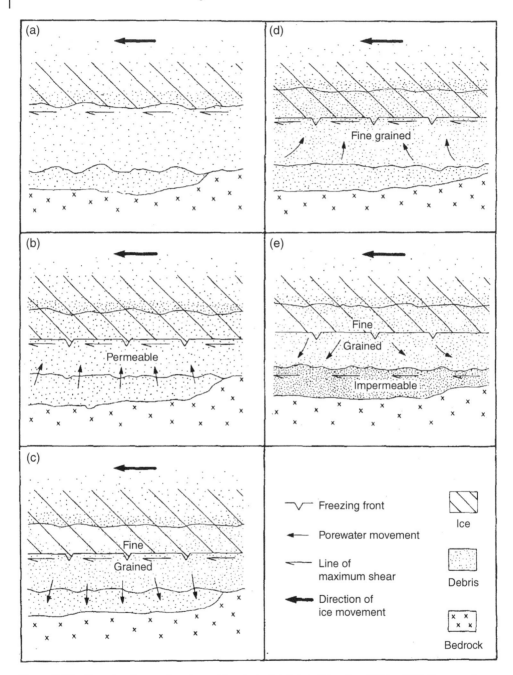

Figure 3.13 Freezing front migration with subglacial debris (Menzies, 1981, p272). See text for details.

temperatures (up to −0.35 °C) and (iv) a decrease in till porosity close to the ice–sediment interface. They subsequently hypothesised that whilst the continued rapid motion of the Whillans Ice Stream is promoted by active basal melting and a reduction in effective pressure and shear strength within the subjacent till, the stagnation of the Kamb Ice Stream is caused by basal freeze-on and an associated consolidation and strengthening of the subglacial till.

In developing a numerical, vertical-column model of the freezing process to represent these contrasting glaciological scenarios, Christoffersen & Tulaczyk (2003b) noted that, 'there is a paucity of theoretical and empirical investigations of heat, water, and solute flow during basal freeze-on' (p2). Recognising the general similarity of ice formed by basal freezing and by near-surface frost heave, the authors utilised theoretical treatments of frost heave originally developed by permafrost engineers, thereby providing a specific illustration of the beneficial applications of permafrost theory to glaciological research questions. More specifically, they employed the Clapeyron equation that provides a means for coupling pressure and temperature in a freezing porous medium and the fundamental basis for frost-heave models (e.g. Rempel, 2007). The model's prediction of the development of intercalated layers of debris-rich ice (frozen till) and debris-poor ice (segregation ice) is consistent with subsequent borehole observations of a debris-rich basal ice layer 10–14 m thick at the base of the Kamb Ice Stream as described by Vogel *et al.* (2005) and Engelhardt & Kamb (2013) (see Section 4.3.3).

Christoffersen *et al.* (2006) subsequently used the same model to provide quantitative definitions for four distinct types of basal ice beneath soft-bedded glaciers whose formation is associated with subglacial frost-heave processes, with the facies variability in turn being determined by the delivery of water to the freezing front. Clear ice (type I) forms when the influx of water matches or exceeds the freezing rate, such that the heat budget is satisfied with no extraction of water from the underlying till. In the other three ice types, a progressive reduction in the water supply results in enhanced ice-bed coupling and debris entrainment through regelation. Laminated ice (type II) is produced by periodic regelation events associated with an influx of groundwater slightly lower than the freezing rate. Brief periods of soft-bed regelation entrain thin layers of debris-rich ice whilst intervening longer periods of congelation entrain thicker layers of clear ice. Massive regelation ice (type III) is created when the influx of water is roughly half that required to satisfy the freezing rate. Regelation events are still episodic, but the increase in effective pressure results in an increase in the regelation rate and the thickness of the regelation ice layers, ultimately generating a facies of c. 50% debris by volume. Finally, solid dirty regelation ice (type IV) is associated with a closed system whereby all the water is drawn from the till pore spaces. In this case, very little congelation ice is generated and the effective stress increases rapidly, producing a rapidly thickening regelation layer with a sediment content of >60% by volume (i.e. lacking excess ice).

In concluding this section, it is worth noting that the shearing hypothesis originally proposed by Goldthwait (1951) has experienced something of a renaissance as a potential explanation for basal-ice formation beneath entirely cold-based glaciers where mechanisms involving the mass transfer and subsequent refreezing of water are considered untenable (e.g. Shaw, 1977a,b). Following the analysis of the isotopic characteristics and petrography of the ice as well as the nature of the entrained sediment, Tison *et al.* (1993) for example concluded that the overriding of active ice over stagnant ice and subsequent shearing provided the only tenable hypothesis of debris entrainment at a site at Terre Adélie in Eastern Antarctica associated with basal temperatures of −7 to −8 °C. This type of entrainment through the mechanical coupling of cold-based glaciers and their frozen beds is considered in the following section, whilst the role played by the overriding of ice-marginal aprons is discussed in more detail in Section 3.6.2.

3.5.5 Effective Beds and Dynamic Soles – Insights from Formerly Glaciated Environments

Complementary research in formerly glaciated environments has provided additional insights into the mechanisms whereby cold-based glaciers can dynamically couple with pre-existing permafrost, resulting in the mobilisation and entrainment of significant volumes of pre-existing material *en masse* (e.g. Mathews & Mackay, 1960; Kupsch, 1962; Hughes, 1973; Boulton, 1972; Moran *et al.*, 1980; Mooers, 1990; Astakhov *et al.*, 1996). In this way, glacier–permafrost interactions can result in erosion, entrainment and geomorphic change even in situations where the permafrost substrate is 'hard frozen' and highly resistant to deformation (Tsytovich 1975).

Thinking more broadly about the basal boundaries of glaciers, in temperate glacial settings, the bed or sole of a glacier is traditionally delineated on a textural basis and is typically considered to represent a distinct ice–rock interface, or following the recognition of 'soft beds', an ice–sediment interface. It is readily apparent in the preceding discussions that this basal zone is glaciologically significant in that it constitutes a location where important processes such as basal sliding occur. It is also widely viewed as defining the boundary between the glacier and the substrate or subglacial zone in which a distinctive suite of processes can occur.

In circumstances where cold-based glaciers overly permafrost, such a texturally defined interface can however be difficult if not impossible to identify, particularly where the basal zone of the glacier is debris rich and the underlying permafrost is unconsolidated and ice-rich (Hughes, 1973; Clayton & Moran, 1974; Fitzsimons, 2006; Waller *et al.*, 2009a). The associated tendency of cold-based ice to bond with subglacial materials can in addition result in a mechanical coupling of the glacier and the permafrost and the effective transmission of basal shear stress into the bed such that the two elements behave as a combined mechanical unit (Figure 3.14). In the absence of a clear cut textural or dynamic boundary between the glacier and the bed, Fitzsimons (2006) argues therefore that the base of the glacier is more appropriately conceptualised as a complex zone characterised by a 'continuum of material properties and deformation processes that is variable in time and space' (p335).

This results in a rather more dynamic conceptual view of the glacier bed in which it comprises the boundary between glacially mobilised and static material and is referred to as the 'effective bed' or 'dynamic sole'. In considering the basal boundary conditions beneath frozen-bed zones associated with quarrying (Sections 3.4.2 and 5.5.1), Clayton & Moran (1974) explicitly acknowledge this concept suggesting that, '...the shear plane marking the "base of the glacier" may move up and down with time' (p97). The base of the glacier is therefore defined as the 'point in the sediment-ice column where the movement

Figure 3.14 Ice-rich subglacial sediment adhering to the margin of the Russell Glacier (western Greenland) and being ripped-up by continued ice motion. See metre rule for scale (Image: Richard Waller).

stops' (p97) and is something which can migrate within a basal zone associated with gradational variations in the ice–sediment–water content.

Research in formerly glaciated permafrost environments has suggested that the dynamic sole is commonly located at the base of the underlying permafrost and the lower limit of ice-cemented materials (e.g. Kupsch, 1962). The generation of high porewater pressures at the boundary between subglacial permafrost and the underlying unfrozen sediment further increases the likelihood of this interface providing an obvious plane of décollement (e.g. Mathews & Mackay, 1960; Rutten, 1960; Menzies, 1981; Mooers, 1990; Boulton et al., 1999). Mathews & Mackay (1960) and Menzies (1981) describe how such high porewater pressures might result from the aggradation of permafrost down into the substrate and the associated expulsion of porewater ahead of the freezing front. Alternatively, Mooers (1990) and Boulton et al. (1993) have demonstrated how elevated porewater pressures are promoted at the base of pre-existing permafrost as subglacial drainage is focused along this interface, field evidence for which has been has been provided beneath cold-based zones of the Trapridge Glacier in Canada (Clarke et al., 1984). Consequently, the boundary between frozen and unfrozen material, rather than the ice-bed interface can represent the most important zone, both in terms of subglacial drainage (Section 4.2.5) and basal motion (Section 4.3.2).

With permafrost in High Latitude environments commonly being several tens or even hundreds of metres in thickness, some authors have argued that this scenario can result in the displacement of huge volumes of material. Astakhov et al. (1996) for example suggest that interaction between the Pleistocene, Barents Sea ice-sheet and pre-existing permafrost, resulted in the entrainment and deformation of up to 300 m of the stratigraphy, as the effective bed was driven down to the base of the pre-existing permafrost (see Section 5.4.3) (Figure 3.15).

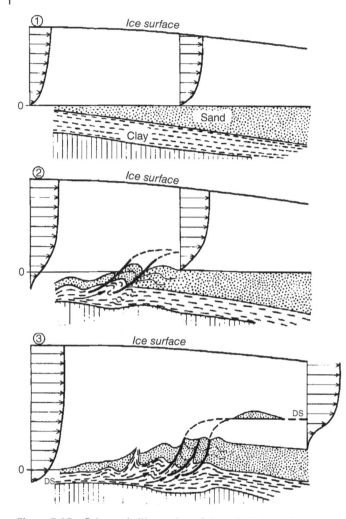

Figure 3.15 Schematic illustration of the subglacial deformation of permafrost in western Siberia resulting from the advance and thickening of the Barents Sea Ice Sheet (Astakhov *et al.*, 1996, p186). Parallel arrows illustrate the inferred velocity profiles. Note in particular how the 'dynamic sole' (DS) results in the deformation and entrainment of the subglacial permafrost.

In summary, when interacting with permafrost characterised by comparative rigidity (*hard-frozen permafrost*; Tsytovich, 1975), it is suggested that cold-based glaciers are able to mobilise and entrain the permafrozen, subglacial materials *en masse*, due to the increased coupling between the cold-based ice and the underlying permafrost (particularly if ice-rich), the more effective transmission of basal shear stress across the ice-bed interface and the generation of a décollement at the permafrost table. It is this process that is thought to result in the formation of some of the most spectacular features potentially indicative of glacier–permafrost interactions including the displacement of large bedrock rafts (Section 5.4.2) and the formation of large thrust-block or push moraines (Section 5.3.2).

3.6 Ice-Marginal and Proglacial Processes

3.6.1 Introduction

Ice-marginal and proglacial environments constitute a dynamic interface between glacial and permafrost process environments and are therefore a key setting in which the processes and products of glacier–permafrost interactions can be particularly evident. Preceding coverage has highlighted the thermal and physical connections that are both characteristic of these environments and fundamental to the distinctive processes that operate. Subaerial permafrost almost invariably extends beneath the peripheries of the glacier or ice sheet (Section 2.4.2) such that 'glaciers ending in a stable terrestrial permafrost environment will always be polythermal, as at least marginal areas of the glaciers will be cold-based' (Etzelmüller & Hagen, 2005, p13). The strong ice–bed coupling between cold-based ice and the permafrozen substrate (Figure 3.14) in turn establishes a mechanical connection that in addition to promoting subglacial erosion (Section 3.5.3), can also facilitate the propagation of glacier-induced stresses into the proglacial area (Section 5.3.2).

Glacier margins are also much easier to access than subglacial environments and therefore lend themselves more readily to the field research and empirical data collection that can enhance our understanding and appreciation of the distinctive processes and products. Extensive research undertaken around the terrestrial margins of glaciers in Svalbard for example, provides a useful introduction to the distinctive processes and products that can occur when polythermal glaciers terminate in regions of continuous permafrost. Etzelmüller (2000) provides a conceptual model (Figure 3.16) that illustrates the diverse range of interconnected processes associated with the advance and subsequent recession of a glacier over both rigid and deformable permafrost. It is these *processes* of debris entrainment, tectonic deformation of both the glacier margin and the foreland, and the deposition and long-term preservation of stagnant glacier ice, that form the focus of this section. More detailed discussion of the key landscape *products* is provided in Sections 5.3 and 5.5.

3.6.2 Apron Entrainment

As mentioned earlier in relation to the processes of debris entrainment and basal ice formation (Section 3.5.4), processes involving phase changes and the mass transfer of water either over short distances in terms of regelation, or longer distances in relation to net basal adfreezing, have been argued to be of little relevance to glaciers that are entirely cold based. At the same time, many glacier margins situated in some of the coldest parts of the planet have been observed to feature thick, debris-rich basal ice layers, thereby necessitating a reconsideration of the processes whereby debris can be entrained (e.g. Tison *et al.*, 1993).

In order to account for a basal ice layer several metres in thickness observed around the margin of Taylor Glacier in the arid polar Dry Valleys of Antarctica, Shaw (1977a,b) contended that the overriding and entrainment of frontal aprons provided the primary mechanism whereby sediment can be entrained (Figure 3.17). These aprons comprise a heterogeneous mix of sediment, snow and ice produced by dry calving at the cliff-like ice margins that can be overridden and entrained during ice advance to produce a debris-rich basal ice layer. This entrained debris is then subject to shear deformation and flow metamorphism, leading to the

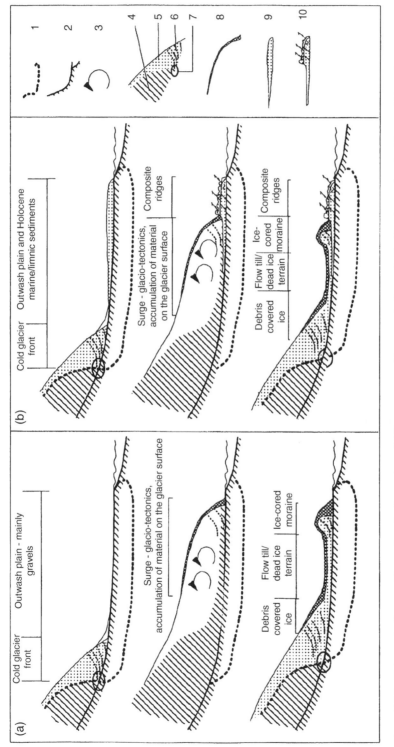

Figure 3.16 Schematic diagram illustrating the formation of different types of ice-marginal area in Svalbard associated with a rapid ice advance or surge over both rigid bedrock or undeformable sediments (a) and a deformable ice-rich sediment layer (b) (Etzelmüller, 2000, p344). Key: (1) base of permafrost, (2) bedrock, (3) glacitectonic deformation of the marginal ice, (4) temperate ice, (5) cold ice, (6) shear zone and/or englacial debris bands, (7) transition zone between warm and cold-based ice, (8) debris covered ice, (9) frozen marine/limnic sediments, (10) deformed sediment layers and composite ridges. Note how in both cases, the processes and products are associated with a zone of ice-marginal permafrost.

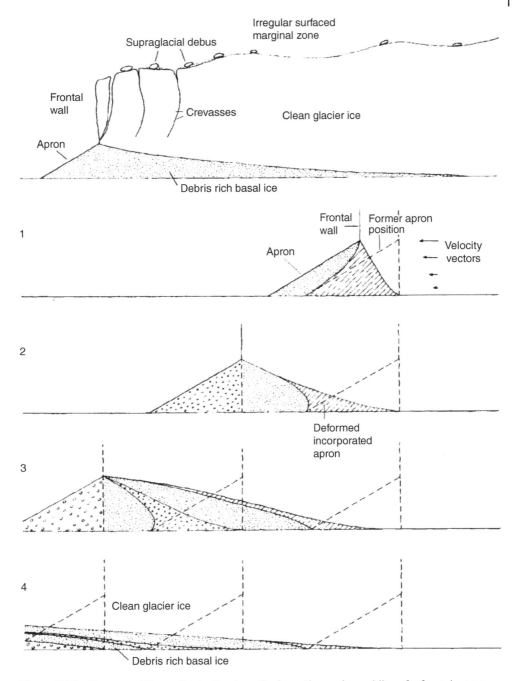

Figure 3.17 Conceptual figure illustrating how the formation and overriding of a frontal apron leads to debris entrainment and the progressive deformation of the apron sediments (stages 1–4) (Shaw, 1977a,b, p193).

attenuation of the debris and the formation of a debris foliation comprising alternating layers of clean and debris-rich ice of varying thicknesses. This foliation is most prominently developed in the upper part of the basal ice, highlighting the potential role played by flow-related deformation within the basal zone (see Waller *et al.*, 2000). Shaw (1977a) also argued that the lack of water in these arid environments and subsequent removal of ice primarily through sublimation increased the chances of these features being preserved within the associated till deposits in the form of a massive lower and foliated upper unit (see Section 5.4.1).

The operation of this mechanism of debris entrainment has subsequently been extended to the more arid regions of the High Arctic, with Evans (1989) invoking apron entrainment as the principal mechanism responsible for the entrainment of debris and the formation of debris-rich basal ice beneath a series of sub-polar glaciers on north-west Ellesmere Island. In this case, ice-marginal fluctuations resulted in the entrainment of ice-marginal fluvial and deltaic sediments in combination with dry-calved ice blocks to produce debris-poor and debris-rich folia of varying thickness as well as distinctive augens. In some instances, large blocks of alluvium and glacimarine sediment were also entrained into the basal ice. The predominance of alluvium within the aprons reflects an increase in meltwater availability and activity at the ice margin and a distinct difference to the aprons described within the hyperarid Dry Valleys of Antarctica. The associated accumulation of material is also important in that it can lead to the burial and preservation of buried ice (Section 3.6.4) that is then available for re-entrainment during subsequent re-advances.

More recent field research in the Dry Valleys of Antarctica by Fitzsimons *et al.* (2008) examining the morphology, structure, composition and deformation of both the ice-marginal apron and basal ice layer of the Victoria Upper Glacier was specifically designed to critically test apron entrainment as a mechanism of basal ice formation in polar glaciers. The research findings revealed that the apron itself comprised two distinct structural elements: an inner element with strongly foliated ice and a steep up-glacier dip and an outer element with no clear foliation and a down-glacier dip parallel to the surface of the apron. In combination with differences in the sedimentology, solute content and isotopic characteristics, this led the authors to conclude that the basal ice and the apron ice are genetically different, with the basal ice layer formed through subglacial processes overriding the apron ice formed through ice-marginal processes. Whilst the authors are keen to stress that this does not refute the ability of the apron entrainment mechanism to create basal ice layers elsewhere, the results contradict any suggestion that debris entrained into the basal zones of polar glaciers can form solely through this process. An important and noteworthy geomorphological implication of this conclusion is the provision of additional indirect evidence for the persistence of subglacial erosion and sediment entrainment beneath cold ice as discussed previously in Section 3.5.

3.6.3 Proglacial Thrusting

The ability of cold-based glaciers to physically connect with permafrost substrates such that the two components can act as a coupled mechanical unit has led numerous authors to identify proglacial deformation in general and proglacial thrusting in particular as important ice-marginal processes related to glacier–permafrost interactions. Whilst more extensive coverage considering the role of permafrost in the development of push moraines (which is still subject to debate) is provided in Section 5.3.2, outline coverage of the formative process is included here as it introduces the complex connections between a glacier's thermal regime, its dynamic behaviour, its rheological interaction with its foreland and its hydrological characteristics that are addressed in Chapters 4 and 5.

Pioneering work undertaken by Mackay (1956, 1959) described a series of exposures within the western Canadian Arctic featuring tilted, folded and thrust frozen sediments occurring within distinctive landsystem settings characterised by arcuate ridges and elongate lakes. In one of the earlier descriptions of glacitectonic deformation, Mackay argued that they reflected the occurrence of 'glacier ice thrust' at the margin of the Mackenzie Valley lobe of the Laurentide Ice Sheet, aided by compression against a topographic obstruction. Numerous field studies undertaken primarily within Svalbard (e.g. Boulton *et al.*, 1999) and the Canadian Arctic (e.g. Kälin, 1971) have subsequently provided additional detail on the processes of push moraine formation, the mechanical interactions between glaciers and permafrozen forelands and the potential role played by permafrost in facilitating the formation of large glacitectonic landforms. In summary, permafrost is argued to promote their formation by enabling: (i) the transmission of shear stresses into subglacial and ice-marginal sediments through an increase in sediment cohesion and the development of strong ice-bed coupling, and (ii) by facilitating the deformation of the sediments through the inclusion of ice-rich cements or the development of a sub-permafrost décollement associated with high porewater pressures (Haeberli, 1979).

It is the indirect hydrogeological implications of ice-marginal permafrost however that are considered particularly influential in promoting proglacial thrusting and push moraine formation. In considering the causes of the aforementioned glacitectonic deformation observed in the western Canadian Arctic, Mackay & Mathews (1964) argued that in preventing the upward flow of groundwater at the margin of the Laurentide Ice Sheet, a permafrost layer a few hundred metres thick would have promoted the generation of high porewater pressures at the base of the permafrost layer that could then act as a décollement along which thrusting could occur. More recent investigations in modern-day glacial environments have provided clear empirical evidence for this model with a study by Boulton *et al.* (1999) focusing on the formation of a large multi-crested push moraine at Holmstrømbreen identifying a single detachment surface at a depth of 30–50 m along which the majority of the deformation occurred. This was coincident with the expected depth of the permafrost table at which the generation of artesian porewater pressures enabled the mobilisation of the overlying materials (see Section 5.3.2 for further detail).

A number of studies have noted clear spatial associations between thrust moraines and buried aquifers, further emphasising the hydrological and hydrogeological importance of ice-marginal permafrost. Moran *et al.* (1980) suggested that the down-glacier pinching out of an aquifer, its capping by glacier ice or most relevantly, the presence of thick ice-marginal permafrost would result in a down-glacier reduction in hydraulic transmissivity and therefore an elevation in subglacial porewater pressures. This would then facilitate the shear failure of the substrate, the displacement of large blocks of pre-existing rock and sediment and the formation of thrust moraines. Mooers (1990) has similarly identified the influence of permafrost on subglacial aquifers as one the key controls on the development of glacial thrust systems in Minnesota. A steady-state ice-sheet model was used to demonstrate that the production of meltwater from warm-based ice up glacier was likely to have been at least two orders of magnitude greater than the amount that could be refrozen onto a cold-based margin initially calculated to be 1.5–2 km wide. This excess water therefore resulted in the recharge of a thick subglacial aquifer, the drainage from which was in turn limited by ice-marginal permafrost, generating elevated fluid pressures sufficient to cause failure within the subglacial sediment. Mooers (1990) concluded by suggesting that the thrusting and associated incorporation of sediment blocks was most likely to occur during ice-marginal fluctuations that may in turn generate flow instabilities (i.e. surge behaviour).

Boulton & Caban (1995) suggest that the ideal hydrogeological setting for thrusting, the displacement of large masses of material and the development of large push moraines involves an ice-marginal permafrost zone with a narrow zone of unfrozen ground beyond it. This situation commonly occurs in Svalbard and the Canadian High Arctic where glaciers terminate in permafrozen coastal forelands, the distal end of which is unfrozen as it has only recently been exposed to subaerial conditions by isostatic uplift. These conditions have also been described in a Pleistocene context where Saalian end moraines are associated with large river valleys that would have provided proglacial taliks. In these circumstances, the extension of the permafrost promotes the development of artesian porewater pressures within the groundwater system adjacent to the glacier. This in turn reduces the drag along the permafrost table that as noted earlier acts as a décollement with the rigidity of the permafrost enabling the propagation of stress into the foreland. Finally, the presence of an unfrozen zone allows the distal toe of the permafrost layer to glide over the foreland when pushed during a glacier advance. In addition to push moraines, the authors identify a broader range of geological features associated with fluid overpressures and ice-marginal permafrost including sediment dykes, diapiric structures and rampart-rimmed hollows, the formation of which is examined in Section 5.3.5.

3.6.4 The Deposition and Preservation of Ice Within Proglacial Environments

If apron entrainment and proglacial thrusting are considered two of the main ice-marginal processes related to ice advance, the deposition and subsequent preservation of ice within proglacial environments can be considered one of the most important processes relating to glacier–permafrost interactions that occurs during periods of glacier recession. As mentioned previously, glacier recession in cold climate regions is commonly associated with the expansion and aggradation of permafrost (Section 2.4.3), which limits the rate of ice loss (or 'deglaciation' in its literal sense) and thereby promotes the preservation of any buried ice present. This section provides a short introduction to the key processes involved in the deposition and preservation of ice prior to their more detailed consideration in relation to the products in the form of both ice-cored moraines (Section 5.3.2) and massive ground ice (Section 5.3.4), the thermokarst processes associated with their degradation and finally the potential impacts of climate change (Section 6.5).

Glaciers and ice sheets terminating within areas of permafrost are commonly associated with the development of extensive areas of stagnant or 'dead ice'. This type of 'dissipation' is associated with the detachment of large tracts of ice from the main body of the glacier or ice sheet such that it ceases to move and downwastes *in situ,* in contrast to the more classic models envisaging the retreat of active ice with a clearly delineated ice front (e.g. Flint, 1929). As discussed earlier in this section, polythermal basal thermal regimes promote the formation of thick layers of basal ice as basal meltwater is accreted to cold-based marginal ice

(Section 3.5.4). The melt-out of ice-marginal exposures of debris-rich ice can subsequently result in the accumulation of supraglacial sediment that reduces the rate of ablation, thereby promoting the detachment of this ice from the debris-poor and dynamically active ice. The subsequent formation of ice-cored landforms is consequently considered one of the most distinctive features of forelands associated with glacier–permafrost interactions (Etzelmüller & Hagen, 2005; section 5.3.2). The ice buried within the (partially) deglaciated forelands can similarly be regarded as one of the primary depositional products (Hambrey & Glasser, 2012; section 5.4.1) with Sollid *et al.* (1994) concluding that it is essentially ubiquitous beneath the forelands of glaciers in northern Spitsbergen for example.

It is important to recognise however that the deposition and widespread occurrence of stagnant ice in proglacial areas is neither exclusive to permafrozen forelands nor diagnostic of glacier–permafrost interactions. Extensive research in Iceland (e.g. Kruger & Kjær, 2000; Everest & Bradwell, 2003; Schomacker & Kjær, 2007) and south-west Alaska (e.g. Russell, 1893; Clayton, 1964; Mickelson & Ham, 1995) has provided numerous examples of extensive tracts of stagnant ice occurring within the forelands of temperate glaciers where permafrost is absent. Most if not all of these cases relate to specific instances in which glacier surges have delivered significant volumes of ice to the terminus that experiences widespread stagnation during the subsequent quiescent phase (Figure 3.18). With climate exerting a primary control on the rates of ice loss (Schomacker, 2008), any ice deposited

Figure 3.18 Area of stagnant buried ice associated with the 1991 surge of Skeiðarárjökull in southern Iceland. Note the exposures of stagnant debris-rich ice in the mid ground (Image: Richard Waller).

with temperate humid environments will typically be removed within a few decades (Krüger & Kjær, 2000). If the stagnant ice is debris-rich however, the accumulation of a sediment cover that accompanies its removal can provide a negative feedback by reducing the ablation rate. This has resulted in the preservation of buried ice for over a century in climatic settings normally considered inimical to its survival (Everest & Bradwell, 2003); sufficient time in some instances in coastal Alaska (most notably the Malaspina Glacier) for forest ecosystems to develop on these ice-cored terrains (Stephens, 1969).

With climate providing the fundamental control on ground thermal regimes and the rates of ice loss, it is the association of permafrost aggradation with the removal of glaciers and ice sheets from landscapes that is ultimately required to preserve relict glacier ice over longer and potentially geological timescales. Permafrost researchers have long recognised the existence of significant bodies of 'massive ground ice' within areas of continuous permafrost including the Canadian Arctic (e.g. Dyke & Savelle, 2000) and western Siberia (e.g. Astakhov & Isayeva, 1988). Whilst its origin has been disputed, there is an emerging consensus that much of this buried ice is glacial in origin. It has consequently been referred to as the product of 'retarded' or 'incomplete deglaciation' (Kaplyanskaya & Tarnogradskiy, 1986) in which the continued presence of permafrost has enabled significant amounts of relict ground ice to persist over millennial timescales (Section 5.3.4). This provides a palaeoenvironmental and palaeoglaciological archive, the potential value of which has only recently been explored (e.g. Lacelle *et al.*, 2018). In addition, it is the presence of large amounts of relict glacier ice – particularly within former ice-cored moraine complexes – that has resulted in some of the most dramatic landscape impacts of Arctic climate change (Section 5.6.2).

3.7 Summary

- Glacier–permafrost interactions are generally if not exclusively associated with non-temperate glaciers (featuring areas of cold-based ice) – a classification that includes both polythermal glaciers and those that are entirely cold-based. The thermal structure and extent of cold-based ice is controlled by a complex combination of glaciological characteristics as well as the permafrost regime, with the most significant glacier–permafrost interactions occurring in situations where the lower limit of the permafrost intersects with the glacier. Climate change and history can also lead to dynamic changes in the nature and extent of glacier–permafrost interactions through its influence on glacier geometry, flow and hydrology.
- The conditions and processes occurring at the base of glacier or ice sheet exert a profound influence on its morphology, dynamic behaviour and geomorphological impact. In addition to the basal thermal regime, the nature of the substrate is also fundamentally important with a key distinction being drawn between rigid-bed glaciers and soft-bed glaciers. Rigid bed glaciers resting on bedrock are most commonly associated with the processes of basal sliding and regelation. Resting on unconsolidated substrates, soft-bed glaciers are capable of actively coupling with and deforming their beds such that an understanding of their behaviour requires an understanding of both the glacier and the subjacent materials.

- Glaciologists have traditionally assumed that basal processes including basal sliding and subglacial deformation are inactive beneath cold-based ice. Glaciers resting on permafrost are therefore commonly referred to as being 'frozen to their beds' with the substrate being rigid and undeformable regardless of its composition. Whilst the amount of abrasion is therefore considered to be very limited, the strong ice–bed coupling associated with cold-based ice can facilitate plucking and block removal resulting in significant erosion. Nonetheless, such glaciers have more generally been viewed as being slow moving, geomorphologically ineffectual and therefore of limited intrinsic research interest.

- A growing body of more recent research has demonstrated the ability of basal processes to remain active at temperatures below the bulk freezing point and to make significant contributions to ice–surface motion, thereby disproving the assumption that basal processes are universally inactive beneath cold-based ice. This also challenges the conceptual view of two binary states of glacier flow (warm, fast and erosive vs cold, slow and protective) separated by the pressure melting point as a clearly defined thermal threshold.

- In addition to erosion through block removal, the continued operation of basal sliding suggests that abrasion rates beneath cold-based ice are limited rather than zero and are therefore capable of creating distinctive landforms given sufficient periods of time. Polythermal basal thermal regimes associated with cold-based margins also promote debris entrainment through the process of basal adreezing, resulting in the formation of thick layers of debris-rich basal ice that enhance glacial sediment fluxes and abrasion rates.

- With the base of polythermal glaciers comprising similar ice–sediment mixtures to ice-rich permafrost, their basal boundary is often texturally indistinct. The interface between the glacier and underlying permafrost is therefore commonly conceptualised as a 'dynamic sole' or 'effective bed' that can constitute a distinct décollement surface. As the point below which glacier-induced motion ceases, this boundary can migrate up and down over time in response to various glaciological controls resulting in the entrainment, mobilisation and deposition of significant amounts of material.

- The interactions of glaciers with ice-marginal and proglacial permafrost can lead to the operation of a range of distinctive processes that vary according to the glacier fluctuations. Periods of glaciation and ice advance can lead to the incorporation of ice and sediment via the processes of apron entrainment. In enhancing the mechanical rigidity of proglacial materials, permafrost can also facilitate the propagation of stresses into the glacier foreland with its hydrogeological impacts and the generation of high porewater pressures further promoting its mobilisation and subsequent deformation. Glacier recession and dissipation meanwhile can result in the stagnation of large tracts of ice that can be regarded as one of the key depositional products. Subsequent permafrost aggradation can promote the long-term preservation of the resultant ice-rich landscapes that are the result of 'incomplete deglaciation'.

References

Alley, R.B., Blankenship, D.D., Bentley, C.R. & Rooney, S.T., 1986. Deformation of till beneath ice stream B, West Antarctica. *Nature*, 322, 57–59.

Astakhov, V.I. & Isayeva, L.L., 1988. The 'Ice Hill': an example of 'retarded deglaciation' in Siberia. *Quaternary Science Reviews*, 7, 29–40.

Astakhov, V.I., Kaplyanskaya, F.A. & Tarnogradsky, V.D., 1996. Pleistocene permafrost of West Siberia as a deformable glacier bed. *Permafrost and Periglacial Processes*, 7, 165–191.

Barer, S.S., Kvlividze, V.I., Sobolev, V.D., Churaev, N.V. & Kurzaev, A.B., 1977. Thickness and viscosity of thin non-freezing interlayer between ice and quartz surfaces. *Doklady Akademii Nauk USSR*, 235, 601–603.

Bell, R.E., Ferraccioli, F., Creyts, T.T., *et al.*, 2011. Widespread persistent thickening of the East Antarctic Ice Sheet by freezing from the base. *Science*, 331, 1592–1595.

Bennett, M.R., Waller, R.I., Midgley, N.G., *et al.*, 2003. Subglacial deformation at sub-freezing temperatures? Evidence from Hagafellsjökull-Eystri, Iceland. *Quaternary Science Reviews*, 22, 915–923.

Bennett, M.R., Huddart, D. & Waller, R.I., 2005. The interaction of a surging glacier with a seasonally frozen foreland: Hagafellsjökull-Eystri, Iceland. *Geological Society, London, Special Publications*, 242, 51–62.

Björnsson, H., Gjessing, Y., Hamran, S.E., *et al.*, 1996. The thermal regime of sub-polar glaciers mapped by multi-frequency radio-echo sounding. *Journal of Glaciology*, 42, 23–32.

Blankenship, D.D., Bentley, C.R., Rooney, S.T. & Alley, R.B., 1987. Till beneath Ice Stream B: 1. Properties derived from seismic travel times. *Journal of Geophysical Research: Solid Earth*, 92(B9), 8903–8911.

Blatter, H., 1987. On the thermal regime of an Arctic valley glacier: a study of White Glacier, Axel Heiberg Island, NWT, Canada. *Journal of Glaciology*, 33, 200–211.

Blatter, H. & Hutter, K., 1991. Polythermal conditions in Arctic glaciers. *Journal of Glaciology*, 37, 261–269.

Blatter, H. & Kappenberger, G., 1988. Mass balance and thermal regime of Laika ice cap, Coburg Island, NWT, Canada. *Journal of Glaciology*, 34, 102–110.

Bogen, J. & Bønsnes, T.E., 2003. Erosion and sediment transport in High Arctic rivers, Svalbard. *Polar Research*, 22, 175–189.

Boulton, G.S., 1970a. On the origin and transport of englacial debris in Svalbard glaciers. *Journal of Glaciology*, 9, 213–229.

Boulton, G.S., 1972. The role of thermal regime in glacial sedimentation. In: Price, R.J. & Sugden, D.E. (Eds.), *Polar Geomorphology*. Institute of British Geographers Special Publication No. 4, 1–19.

Boulton, G.S., 1974. Processes and patterns of glacial erosion. In: Coates, D.R. (Ed.), *Glacial Geomorphology*. A proceedings volume of the Fifth Annual Geomorphology Symposia series. Springer, New York, 41–87.

Boulton, G.S., 1986. Geophysics: a paradigm shift in glaciology? *Nature*, 322, 18.

Boulton, G.S. & Caban, P., 1995. Groundwater flow beneath ice sheets: part II—its impact on glacier tectonic structures and moraine formation. *Quaternary Science Reviews*, 14, 563–587.

Boulton, G.S. & Hagdorn, M., 2006. Glaciology of the British Isles Ice Sheet during the last glacial cycle: form, flow, streams and lobes. *Quaternary Science Reviews*, 25, 3359–3390.

Boulton, G.S. & Hindmarsh, R.C.A., 1987. Sediment deformation beneath glaciers: rheology and geological consequences. *Journal of Geophysical Research: Solid Earth*, 92(B9), 9059–9082.

Boulton, G.S. & Jones, A.S., 1979. Stability of temperate ice caps and ice sheets resting on beds of deformable sediment. *Journal of Glaciology*, 24, 29–43.

Boulton, G.S., Slot, T., Blessing, K., *et al.*, 1993. Deep circulation of groundwater in overpressured subglacial aquifers and its geological consequences. *Quaternary Science Reviews*, 12(9), 739–745.

Boulton, G.S., Van der Meer, J.J.M., Beets, D.J., Hart, J.K. & Ruegg, G.H.J., 1999. The sedimentary and structural evolution of a recent push moraine complex: Holmstrømbreen, Spitsbergen. *Quaternary Science Reviews*, 18, 339–371.

Boulton, G.S., Dobbie, K.E. & Zatsepin, S., 2001. Sediment deformation beneath glaciers and its coupling to the subglacial hydraulic system. *Quaternary International*, 86, 3–28.

Chamberlin, T.C., 1895. Recent glacial studies in Greenland. *Bulletin of the Geological Society of America*, 6, 199–220.

Christoffersen, P. & Tulaczyk, S., 2003a. Thermodynamics of basal freeze-on: predicting basal and subglacial signatures of stopped ice streams and interstream ridges. *Annals of Glaciology*, 36, 233–243.

Christoffersen, P. & Tulaczyk, S., 2003b. Response of subglacial sediments to basal freeze-on 1. Theory and comparison to observations from beneath the West Antarctic Ice Sheet. *Journal of Geophysical Research: Solid Earth*, 108(B4), doi:10.1029/2002JB001935.

Christoffersen, P., Tulaczyk, S., Carsey, F.D. & Behar, A.E., 2006. A quantitative framework for interpretation of basal ice facies formed by ice accretion over subglacial sediment. *Journal of Geophysical Research: Earth Surface*, 111(F1), doi:10.1029/2005JF000363.

Clark, P.U., 1995. Fast glacier flow over soft beds. *Science*, 267, 43–44.

Clarke, G.K., 1987. Fast glacier flow: ice streams, surging, and tidewater glaciers. *Journal of Geophysical Research: Solid Earth*, 92(B9), 8835–8841.

Clarke, G.K., Collins, S.G. & Thompson, D.E., 1984. Flow, thermal structure, and subglacial conditions of a surge-type glacier. *Canadian Journal of Earth Sciences*, 21, 232–240.

Clayton, L., 1964. Karst topography on stagnant glaciers. *Journal of Glaciology*, 5, 107–112.

Clayton, L. & Moran, S.R., 1974. A glacial process-form model. In: Coates, D.R. (Ed.), *Glacial Geomorphology*. A proceedings volume of the Fifth Annual Geomorphology Symposia series. Springer, Dordrecht, Netherlands, 89–119.

Cuffey, K.M., Conway, H., Hallet, B., Gades, A.M. & Raymond, C.F., 1999. Interfacial water in polar glaciers and glacier sliding at −17°C. *Geophysical Research Letters*, 26, 751–754.

Cuffey, K.M., Conway, H., Gades, A.M., *et al.*, 2000. Entrainment at cold glacier beds. *Geology*, 28, 351–354.

Dash, J.G., Rempel, A.W. & Wettlaufer, J.S., 2006. The physics of premelted ice and its geophysical consequences. *Reviews of Modern Physics*, 78, 695–741.

Drewry, D.J., 1986. *Glacial Geologic Processes*. Edward Arnold.

Dyke, A.S., 1976. Tors and associated weathering phenomena. Somerset Island, District of Franklin. *Canada Geological Survey, Paper, 76-IB*, 209–216.

Dyke, A.S., 1993. Landscapes of cold-centred Late Wisconsinan ice caps, Arctic Canada. *Progress in Physical Geography*, 17, 223–247.

Dyke, A.S. & Savelle, J.M., 2000. Major end moraines of Younger Dryas age on Wollaston Peninsula, Victoria Island, Canadian Arctic: implications for paleoclimate and for formation of hummocky moraine. *Canadian Journal of Earth Sciences*, 37, 601–619.

Echelmeyer, K. & Zhongxiang, W., 1987. Direct observation of basal sliding and deformation of basal drift at sub-freezing temperatures. *Journal of Glaciology*, 33, 83–98.

Engelhardt, H. & Kamb, B., 1997. Basal hydraulic system of a West Antarctic ice stream: constraints from borehole observations. *Journal of Glaciology*, 43, 207–230.

Engelhardt, H. & Kamb, B., 1998. Basal sliding of ice stream B, West Antarctica. *Journal of Glaciology*, 44, 223–230.

Engelhardt, H. & Kamb, B., 2013. Kamb Ice Stream flow history and surge potential. *Annals of Glaciology*, 54, 287–298.

Engelhardt, H.F., Harrison, W.D. & Kamb, B., 1978. Basal sliding and conditions at the glacier bed as revealed by bore-hole photography. *Journal of Glaciology*, 20, 469–508.

Engelhardt, H., Humphrey, N., Kamb, B. & Fahnestock, M., 1990. Physical conditions at the base of a fast moving Antarctic ice stream. *Science*, 248, 57–59.

Etzelmüller, B., 2000. Quantification of thermo-erosion in pro-glacial areas - examples from Svalbard. *Zeitschrift für Geomorphologie*, 44(3), 343–361.

Etzelmüller, B. & Hagen, J.O., 2005. Glacier-permafrost interaction in Arctic and alpine mountain environments with examples from southern Norway and Svalbard. In: Harris, C. & Murton, J.B. (Eds.), *Cryospheric Systems: Glaciers & Permafrost*. Geological Society, London, Special Publications, 242, 11–27.

Evans, D.J.A., 1989. Apron entrainment at the margins of sub-polar glaciers, north-west Ellesmere Island, Canadian High Arctic. *Journal of Glaciology*, 35, 317–324.

Evans, D.J.A. & Hiemstra, J.F., 2005. Till deposition by glacier submarginal, incremental thickening. *Earth Surface Processes & Landforms*, 30, 1633–1662.

Everest, J. & Bradwell, T., 2003. Buried glacier ice in southern Iceland its wider significance. *Geomorphology*, 52, 347–358.

Faraday, M., 1860. Note on regelation. *Proceedings of the Royal Society of London*, 10, 440–450.

Fitzsimons, S., 2006. Mechanical behaviour and structure of the debris-rich basal ice layer. In: Knight, P.G. (Ed.), *Glacier Science and Environmental Change*. Blackwell Publishing, Oxford, 329–335.

Fitzsimons, S.J., McManus, K.J. & Lorrain, R.D., 1999. Structure and strength of basal ice and substrate of a dry-based glacier: evidence for substrate deformation at sub-freezing temperatures. *Annals of Glaciology*, 28, 236–240.

Fitzsimons, S.J., McManus, K.J., Sirota, P. & Lorrain, R.D., 2001. Direct shear tests of materials from a cold glacier: implications for landform development. *Quaternary international*, 86, 129–137.

Fitzsimons, S., Webb, N., Mager, S., *et al.*, 2008. Mechanisms of basal ice formation in polar glaciers: an evaluation of the apron entrainment model. *Journal of Geophysical Research: Earth Surface*, 113(F2), doi:10.1029/2006JF000698.

Flint, R.F., 1929. The stagnation and dissipation of the last ice sheet. *Geographical Review*, 19, 256–289.

Forbes, J.D., 1858. On some properties of ice near its melting point. *Proceedings of the Royal Society of Edinburgh*.

Gerrard, J.A.F., Perutz, M.F., Roch, A. & Taylor, G.I., 1952. Measurement of the velocity distribution along a vertical line through a glacier. *Proceedings of the Royal Society of London. Series A. Mathematical and Physical Sciences*, 213, 546–558.

Gilpin, R.R., 1979. A model of the "liquid-like" layer between ice and a substrate with applications to wire regelation and particle migration. *Journal of Colloid and Interface Science*, 68, 235–251.

Gilpin, R.R., 1980. Wire regelation at low temperatures. *Journal of Colloid & Interface Science*, 77, 435–448.

Glasser, N.F. & Hambrey, M.J., 2001. Styles of sedimentation beneath Svalbard valley glaciers under changing dynamic and thermal regimes. *Journal of the Geological Society*, 158, 697–707.

Goldthwait, R.P., 1951. Development of end moraines in east-central Baffin Island. *The Journal of Geology*, 59, 567–577.

Haeberli, W., 1979. Holocene push-moraines in alpine permafrost. *Geografiska Annaler: Series A*, 61, 43–48.

Hagen, J.O., Kohler, J., Melvold, K. & Winther, J.G., 2003. Glaciers in Svalbard: mass balance, runoff and freshwater flux. *Polar Research*, 22, 145–159.

Hambrey, M.J. & Glasser, N.F., 2012. Discriminating glacier thermal and dynamic regimes in the sedimentary record. *Sedimentary Geology*, 251, 1–33.

Hart, J.K. & Boulton, G.S., 1991. The interrelation of glaciotectonic and glaciodepositional processes within the glacial environment. *Quaternary Science Reviews*, 10, 335–350.

Hart, J.K. & Smith, B., 1997. Subglacial deformation associated with fast ice flow, from the Columbia Glacier, Alaska. *Sedimentary Geology*, 111, 177–197.

Holdsworth, G., 1974. *Meserve Glacier Wright Valley, Antarctica: Part I. Basal Processes*. Research Foundation and the Institute of Polar Studies, Report No. 37, The Ohio State University Research Foundation.

Hooke, R.L., 1977. Basal temperatures in polar ice sheets: a qualitative review. *Quaternary Research*, 7, 1–13.

Hooke, R.L., Cummings, D.I., Lesemann, J.E. & Sharpe, D.R., 2013. Genesis of dispersal plumes in till. *Canadian Journal of Earth Sciences*, 50, 847–855.

Hope, R., Lister, H. & Whitehouse, R., 1972. The wear of sandstone by cold, sliding ice. *Institute of British Geographers Special Publication*, 5, 21–31.

Hubbard, B. & Sharp, M., 1989. Basal ice formation and deformation: a review. *Progress in Physical Geography*, 13, 529–558.

Hubbard, B. & Sharp, M., 1993. Weertman regelation, multiple refreezing events and the isotopic evolution of the basal ice layer. *Journal of Glaciology*, 39, 275–291.

Hughes, T., 1973. Glacial permafrost and Pleistocene ice ages. In: *Permafrost: Second International Conference. North American contribution*. National Academy of Sciences, Washington, DC, 213–223.

Humphrey, N., Kamb, B., Fahnestock, M. & Engelhardt, H., 1993. Characteristics of the bed of the lower Columbia Glacier, Alaska. *Journal of Geophysical Research: Solid Earth*, 98(B1), 837–846.

Iken, A. & Bindschadler, R.A., 1986. Combined measurements of subglacial water pressure and surface velocity of Findelengletscher, Switzerland: conclusions about drainage system and sliding mechanism. *Journal of Glaciology*, 32, 101–119.

Irvine-Fynn, T.D., Hodson, A.J., Moorman, B.J., Vatne, G. & Hubbard, A.L., 2011. Polythermal glacier hydrology: a review. *Reviews of Geophysics*, 49, RG4002.

Iverson, N.R. & Semmens, D.J., 1995. Intrusion of ice into porous media by regelation: a mechanism of sediment entrainment by glaciers. *Journal of Geophysical Research: Solid Earth*, 100, 10219–10230.

Iverson, N.R., Hanson, B., Hooke, R.L. & Jansson, P., 1995. Flow mechanism of glaciers on soft beds. *Science*, 267, 80–81.

Jamieson, S.S., Sugden, D.E. & Hulton, N.R., 2010. The evolution of the subglacial landscape of Antarctica. *Earth and Planetary Science Letters*, 293, 1–27.

Kälin, M., 1971. *The active push moraine of the Thompson Glacier, Axel Heiberg Island, Canadian Arctic Archipelago, Canada*. McGill University, Axel Heiberg Island Research Reports, Glaciology 4

Kamb, B., 1987. Glacier surge mechanism based on linked cavity configuration of the basal water conduit system. *Journal of Geophysical Research: Solid Earth*, 92(B9), 9083–9100.

Kamb, B. & LaChapelle, E., 1964. Direct observation of the mechanism of glacier sliding over bedrock. *Journal of Glaciology*, 5, 159–172.

Kamb, B., Raymond, C.F., Harrison, W.D., *et al.*, 1985. Glacier surge mechanism: 1982-1983 surge of Variegated Glacier, Alaska. *Science*, 227, 469–479.

Kaplyanskaya, F.A. & Tarnogradskiy, V.D., 1986. Remnants of the Pleistocene ice sheets in the permafrost zone as an object for paleoglaciological research. *Polar Geography*, 10, 257–266.

Kavanaugh, J.L. & Clarke, G.K., 2006. Discrimination of the flow law for subglacial sediment using in situ measurements and an interpretation model. *Journal of Geophysical Research: Earth Surface*, 111(F1), doi:10.1029/2005JF000346.

Kleman, J. & Borgström, I., 1996. Reconstruction of palaeo-ice sheets: the use of geomorphological data. *Earth Surface Processes and Landforms*, 21, 893–909.

Knight, P.G., 1989. Stacking of basal debris layers without bulk freezing-on: isotopic evidence from West Greenland. *Journal of Glaciology*, 35, 214–216.

Knight, P.G., 1994. Two-facies interpretation of the basal layer of the Greenland ice sheet contributes to a unified model of basal ice formation. *Geology*, 22, 971–974.

Knight, P.G., 1997. The basal ice layer of glaciers and ice sheets. *Quaternary Science Reviews*, 16, 975–993.

Koerner, R.M., 1970. The mass balance of the Devon Island ice cap, Northwest Territories, Canada, 1961-66. *Journal of Glaciology*, 9, 325–336.

Koerner, R.M., 1989. Queen Elizabeth Islands glaciers. *Quaternary geology of Canada and Greenland, Geological Survey of Canada, Geology of Canada*, 1, 464–473.

Koerner, R.M. & Fisher, D.A., 1979. Discontinuous flow, ice texture, and dirt content in the basal layers of the Devon Island ice cap. *Journal of Glaciology*, 23, 209–222.

Krüger, J. & Kjær, K.H., 2000. De-icing progression of ice-cored moraines in a humid, subpolar climate, Kötlujökull, Iceland. *The Holocene*, 10(6), 737–747.

Kupsch, W.O., 1962. Ice-thrust ridges in western Canada. *The Journal of Geology*, 70, 582–594.

Lacelle, D., Fisher, D.A., Coulombe, S., Fortier, D. & Frappier, R., 2018. Buried remnants of the Laurentide Ice Sheet and connections to its surface elevation. *Nature Scientific Reports*, doi:10.1038/s41598-018-31166-2.

Lamplugh, C.W., 1911. On the shelly moraine of the Sefström Glacier and other Spitsbergen phenomena illustrative of British glacial conditions. *Proceedings of the Yorkshire Geological Society*, 17, 216–241.

Lee, J.R., Phillips, E., Evans, H.M. & Vaughan-Hirsch, D., 2011. An introduction to the glacial geology and history of glacitectonic research in Northeast Norfolk. In: Phillips, E., Lee, J.R. & Evans, H.M. (Eds.), *Glacitectonics: Field Guide*. Quaternary Research Association, London, 101–115.

Lliboutry, L., 1966. Bottom temperatures and basal low-velocity layer in an ice sheet. *Journal of Geophysical Research*, 71, 2535–2543.

Lorrain, R.D. & Fitzsimons, S., 2011. Cold-based glaciers. In: Singh, V.P., Singh, P. & Haritashya, U.K. (Eds.), *Encyclopedia of Snow, Ice and Glaciers, 1*. Springer, New York, 157–161.

Lovell, H., Fleming, E.J., Benn, D.I., *et al.*, 2015. Former dynamic behaviour of a cold-based valley glacier on Svalbard revealed by basal ice and structural glaciology investigations. *Journal of Glaciology*, 61, 309–328.

Mackay, J.R., 1956. Deformation by glacier-ice at Nicholson Peninsula, N.W.T., Canada. *Arctic*, 9, 218–228.

Mackay, J.R., 1959. Glacier ice-thrust features of the Yukon Coast. *Geographical Bulletin*, 13, 5–21.

Mackay, J.R. & Mathews, W.H., 1964. The role of permafrost in ice-thrusting. *The Journal of Geology*, 72, 378–380.

Maohuan, H., 1992. The movement mechanisms of Ürümqi Glacier No. 1, Tien Shan mountains, China. *Annals of Glaciology*, 16, 39–44.

Mathews, W.H. & Mackay, J.R., 1960. Deformation of soils by glacier ice and the influence of pore pressures and permafrost. *Transactions of the Royal Society of Canada*, 54, 27–36.

Matthews, J.A., McCarroll, D. & Shakesby, R.A., 1995. Contemporary terminal-moraine ridge formation at a temperate glacier: Styggedalsbreen, Jotunheimen, southern Norway. *Boreas*, 24, 129–139.

Menzies, J., 1981. Freezing fronts and their possible influence upon processes of subglacial erosion and deposition. *Annals of Glaciology*, 2, 52–56.

Mercer, J.H., 1971. Correspondence - Cold glaciers in the central Transantarctic Mountains, Antarctica: Dry ablation areas and subglacial erosion. *Journal of Glaciology*, 10, 319–321.

Mickelson, D.M. & Ham, N.R., 1995. Burroughs Glacier, Wachusett Inlet, Glacier Bay, Alaska. In: Engstrom, D.R. (Ed.), *Proceedings of the Third Glacier Bay Science Symposium, 1993*. US Department of the Interior, Anchorage, Alaska, 55–65.

Mooers, H.D., 1990. Ice-marginal thrusting of drift and bedrock: thermal regime, subglacial aquifers, and glacial surges. *Canadian Journal of Earth Sciences*, 27, 849–862.

Moore, P.L., Iverson, N.R., Brugger, K.A., *et al.*, 2011. Effect of a cold margin on ice flow at the terminus of Storglaciären, Sweden: implications for sediment transport. *Journal of Glaciology*, 57, 77–87.

Moran, S.R., Clayton, L., Hooke, R.L., Fenton, M.M. & Andriashek, L.D., 1980. Glacier-bed landforms of the prairie region of North America. *Journal of Glaciology*, 25, 457–476.

Müller, F., 1976. On the thermal regime of a high-Arctic valley glacier. *Journal of Glaciology*, 16, 119–133.

Nye, J.F., 1970. Glacier sliding without cavitation in a linear viscous approximation. *Proceedings of the Royal Society of London. A. Mathematical and Physical Sciences*, 315, 381–403.

Ødegård, R.S., Hamran, S.E., Bø, P.H., *et al.*, 1992. Thermal regime of a valley glacier, Erikbreen, northern Spitsbergen. *Polar Research*, 11, 69–79.

Paterson, W.S.B., 1994. *Physics of Glaciers*. Butterworth-Heinemann.

Pattyn, F., Schoof, C., Perichon, L., *et al.*, 2012. Results of the marine ice sheet model intercomparison project, MISMIP. *The Cryosphere*, 6, 573–588.

Piotrowski, J.A., Mickelson, D.M., Tulaczyk, S., Krzyszkowski, D. & Junge, F.W., 2001. Were deforming subglacial beds beneath past ice sheets really widespread? *Quaternary International*, 86, 139–150.

Piotrowski, J.A., Larsen, N.K. & Junge, F.W., 2004. Reflections on soft subglacial beds as a mosaic of deforming and stable spots. *Quaternary Science Reviews*, 23, 993–1000.

Porter, P.R., Murray, T. & Dowdeswell, J.A., 1997. Sediment deformation and basal dynamics beneath a glacier surge front: Bakaninbreen, Svalbard. *Annals of Glaciology*, 24, 21–26.

Raymond, C.F., 1971. Flow in a transverse section of Athabasca Glacier, Alberta, Canada. *Journal of Glaciology*, 10, 55–84.

Reid, C., 1882. *The Geology of the Country Around Cromer:(Explanation of Sheet 68 E.)* (Vol. 68). HM Stationery Office.

Reinhardy, B.T.I., Booth, A.D., Hughes, A.L.C., *et al.*, 2019. Pervasive cold ice within a temperate glacier - implications for glacier thermal regimes, sediment transport and foreland geomorphology. *The Cryosphere*, 13, 827–843.

Rempel, A.W., 2007. Formation of ice lenses and frost heave. *Journal of Geophysical Research*, 112(F2), doi:10.1029/2006JF000525.

Robin, G., 1986. Fast glacier flow: a soft bed is not the whole answer. *Nature*, 323, 490–491.

Röthlisberger, H., 1968. Erosive processes which are likely to accentuate or reduce the bottom relief of valley glaciers. *International Association of Hydrological Sciences Publication*, 79, 87–97.

Russell, I.C., 1893. Malaspina glacier. *The Journal of Geology*, 1, 219–245.

Rutten, M.G., 1960. Ice-pushed ridges, permafrost and drainage. *American Journal of Science*, 258, 293–297.

Schomacker, A., 2008. What controls dead-ice melting under different climate conditions? A discussion. *Earth-Science Reviews*, 90, 103–113.

Schomacker, A. & Kjær, K.H., 2007. Origin & de-icing of multiple generations of ice-cored moraines at Brúarjökull, Iceland. *Boreas*, 36, 411–425.

Shaw, J., 1977a. Tills deposited in arid polar environments. *Canadian Journal of Earth Sciences*, 14, 1239–1245.

Shaw, J., 1977b. Till body morphology and structure related to glacier flow. *Boreas*, 6, 189–201.

Shreve, R.L., 1984. Glacier sliding at subfreezing temperatures. *Journal of Glaciology*, 30, 341–347.

Sollid, J.L., Etzelmüller, B., Vatne, G. & Ødegard, R.S., 1994. Glacial dynamics, material transfer and sedimentation of Erikbreen and Hannabreen, Liefdefjorden, northern Spitsbergen. *Zeitschrift für Geomorphologie, Supplementband*, 97, 123–144.

Stephens, F.R., 1969. A forest ecosystem on a glacier in Alaska. *Arctic*, 22, 441–444.

Sugden, D.E., 1977. Reconstruction of the morphology, dynamics, and thermal characteristics of the Laurentide ice sheet at its maximum. *Arctic and Alpine Research*, 9, 21–47.

Sugden, D.E., 1978. Glacial erosion by the Laurentide ice sheet. *Journal of Glaciology*, 20, 367–391.

Sugden, D.E. & John, B.S., 1976. *Glaciers and Landscape: A Geomorphological Approach*. Edward Arnold, London.

Sugden, D.E., Knight, P.G., Livesey, N., *et al.*, 1987. Evidence for two zones of debris entrainment beneath the Greenland ice sheet. *Nature*, 328, 238–241.

Swinzow, G.K., 1962. Investigation of shear zones in the ice sheet margin, Thule area, Greenland. *Journal of Glaciology*, 4, 215–229.

Telford, J.W. & Turner, J.S., 1963. The motion of a wire through ice. *Philosophical Magazine*, 8, 527–531.

Thomson, J., 1860. I. On recent theories and experiments regarding ice at or near its melting-point. *Proceedings of the Royal Society of London*, 10, 151–160.

Tison, J.L., Petit, J.R., Barnola, J.M. & Mahaney, W.C., 1993. Debris entrainment at the ice-bedrock interface in sub-freezing temperature conditions (Terre Adélie, Antarctica). *Journal of Glaciology*, 39, 303–315.

Truffer, M., Echelmeyer, K.A. & Harrison, W.D., 2001. Implications of till deformation on glacier dynamics. *Journal of Glaciology*, 47, 123–134.

Tsytovich, N.A., 1975. *The Mechanics of Frozen Ground*. McGraw-Hill, New York.

Vivian, R., 1980. The nature of the ice-rock interface: the results of investigation on 20000 m^2 of the Rock Bed of Temperate Glaciers. *Journal of Glaciology*, 25, 267–277.

Vogel, S.W., Tulaczyk, S., Kamb, B., *et al.*, 2005. Subglacial conditions during and after stoppage of an Antarctic Ice Stream: is reactivation imminent? *Geophysical Research Letters*, 32(14), doi:10.1029/2005GL022563.

Waller, R.I., 2001. The influence of basal processes on the dynamic behaviour of cold-based glaciers. *Quaternary International*, 86, 117–128.

Waller, R.I., Hart, J.K. & Knight, P.G., 2000. The influence of tectonic deformation on facies variability in stratified debris-rich basal ice. *Quaternary Science Reviews*, 19, 775–786.

Waller, R.I., Murton, J.B. & Knight, P.G., 2009a. Basal glacier ice and massive ground ice: different scientists, same science? In: Knight, J. & Harrison, S. (Eds.), *Periglacial and Paraglacial Processes and Environments*. Geological Society, London, Special Publication, 320, 57–69.

Weertman, J., 1957. On the sliding of glaciers. *Journal of Glaciology*, 3, 33–38.

Weertman, J., 1961. Mechanism for the formation of inner moraines found near the edge of cold ice caps and ice sheets. *Journal of Glaciology*, 3, 965–978.

Weertman, J., 1964. The theory of glacier sliding. *Journal of Glaciology*, 5, 287–303.

Weertman, J., 1967. Sliding of Nontemperate Glaciers. *Journal of Geophysical Research*, 72, 521–523.

Williams, P.J. & Smith, M.W., 1989. *The Frozen Earth: Fundamentals of Geocryology*. Cambridge University Press.

Zdanowicz, C.M., Michel, F.A. & Shilts, W.W., 1996. Basal debris entrainment and transport in glaciers of southwestern Bylot Island, Canadian Arctic. *Annals of Glaciology*, 22, 107–113.

Zwally, H.J., Abdalati, W., Herring, T., *et al.*, 2002. Surface melt-induced acceleration of Greenland ice-sheet flow. *Science*, 297, 218–222.

4

Hydrology, Dynamics and Sediment Fluxes

4.1 Introduction

As with glaciological studies more generally, the majority of work on both the hydrology and dynamic behaviour of glaciers has been undertaken on temperate glaciers located in areas where permafrost is absent. In contrast, both the amount of research work undertaken on non-temperate glaciers occurring in permafrost environments and consequently our associated level of understanding of these systems remains comparatively limited. Nonetheless, there have been a series of seminal papers, recent studies and reviews that have helped to highlight a number of key differences the examination of which forms the focus for this chapter.

Within a recent review paper, Irvine-Fynn *et al.* (2011) emphasised the importance of considering glacier hydrology and ice dynamics from an integrated perspective, with recent research on non-temperate glaciers in the Arctic indicating that 'glacier hydrology and ice motion are intricately linked as a result of the role of meltwater accessing the glacier sole and reducing basal traction' (p2). This coupling of a glacier's hydrological systems with its dynamic behaviour helps for example to identify the causes of glacier surges observed in high-latitude environments such as Svalbard and also to address some of the most important research questions such as the potential future impacts of climate change on the rates of ice loss from polar ice masses (e.g. Zwally *et al.*, 2002). In view of their fundamental connection, this chapter provides an integrated consideration of both the hydrological and dynamic behaviour of glaciers that interact with permafrost. With the ice dynamics and hydrological characteristics also determining the sediment transfer through glacial systems, this chapter culminates in a consideration of their implications for glacial sediment transfer prior to a systematic examination of their impact on landscape expression in Chapter 5.

4.2 The Hydrology of Non-Temperate Glaciers

4.2.1 Introduction

Non-temperate glaciers play an important role within any catchment in which they occur in that they constitute a significant hydrological reservoir with the potential to augment streamflow within the catchment. Their presence can influence a range of hydrometeorological characteristics such that permafrost catchments containing glaciers differ markedly from those in which they are absent (e.g. Bogen & Bønsnes, 2003). The occurrence of polythermal glaciers with areas of warm-based ice for example is associated with the production of meltwater fluxes all year round that can deliver a persistent base flow to the wider catchment leading to the development of thick icings within the proglacial environment (Section 5.3.5).

As with the other glaciological processes and environments covered, the majority of the work and the most significant research progress have focused on temperate glaciers, with comparatively little work having been undertaken on non-temperate glaciers located in catchments containing permafrost. As noted by Hodgkins (1997) in his review of glacier hydrology in Svalbard, the resultant limitations in our knowledge and understanding of the hydrology of non-temperate glaciers are distinctly problematic when one recognises that large areas of the Pleistocene ice sheets are likely to have experienced these conditions for the majority of the glacial cycles (Section 2.4).

In a situation that mirrors the preceding discussion on basal processes (Section 3.4), this paucity of research has led to the development of a series of long-standing, widely held but sometimes misguided or overly simplistic assumptions regarding the hydrology of non-temperate glaciers. Glacier ice occurring at temperatures below the pressure melting point has for example traditionally been considered to behave as an aquiclude due to the combined influence of (i) the closure of the vein networks that can transport water along crystal boundaries in temperate ice, (ii) the slow rates of flow-related deformation that limit the formation of crevasses and therefore the development of large-scale permeability, and (iii) the formation and persistence of superimposed ice created by the refreezing of meltwater at the base of the snowpack (Hodgkins, 1997). As a consequence, the hydrology of non-temperate glaciers is commonly conceptualised as being limited to the generation of surface runoff and the development of supraglacial drainage networks. With surface drainage having limited access to sediment supplies, both solute and suspended sediment fluxes have been assumed to be very limited as a consequence.

Although limited in number, there are however a growing number of investigations that are providing a clearer insight into the hydrology of non-temperate glaciers, the primary differences with their temperate counterparts and consequently the influence of thermal regimes and of permafrost. In contrasting the water and sediment discharge from comparable alpine and Arctic catchments – the Haut Glacier d'Arolla (Switzerland) and Austre Brøggerbreen (Svalbard) respectively – Gurnell *et al.* (1994) identified three main differences with the Arctic example displaying (i) more subdued seasonal discharge fluctuations, (ii) an absence of diurnal solute fluctuations and (iii) an increase in the supply of suspended sediment supply during the melt season in marked contrast to the progressive 'exhaustion' seen in the alpine glacier. The significance of glacier–permafrost interaction and the potential sources of sediment are alluded to in relation to the latter point with the authors acknowledging that,

> . . .*in Arctic basins sediment availability is only partly governed by meltwater-sediment contact. The temperature regime of the sediment is also important and so increasing sediment availability and decreasing sediment lead times could also indicate the significance of spatially and temporally variable ground thaw processes* (p333).

This section considers our current knowledge and understanding of the hydrology of non-temperate glaciers, taking advantage of salient reviews undertaken by Hodgkins (1997) and more recently by Irvine-Fynn (2011). In addition to considering the drainage systems that form within the glacier itself, this section will also explicitly consider glacial groundwater fluxes (Section 4.2.5), the significance of which have only recently been appreciated by glaciologists (e.g. Boulton *et al.*, 1995; Piotrowski, 1997). In so doing, this provides an opportunity to appraise the interactions between glaciers and permafrost within their broader catchment setting.

4.2.2 Seasonal Variations

When glaciers are considered to represent a component part of the wider cryosphere, it is unsurprising that their general seasonal flow variations are considered broadly similar to the nival flow regimes characteristic of permafrost river systems. Annual melting typically commences in the late Spring or early Summer with the meltwater generated initially percolating into the snowpack and refreezing to form layers of superimposed ice. This process continues until the snowpack becomes saturated, after which the water can start to flow laterally, commonly accumulating in surface depressions to create supraglacial lakes (Figure 4.1). Observations of supraglacial lake development in Svalbard by Liestøl *et al.* (1980) indicate that they can persist as transient hydrological stores for a few weeks to a few months at the start of each melt season.

This early season supraglacial storage constitutes one of the distinctive hydrological characteristics of non-temperate glaciers with such storage being comparatively limited on

Figure 4.1 Supraglacial lake located at the margin of the accumulation area of Aktineq Glacier, Bylot Island. Note the presence of superimposed ice, the supraglacial channels feeding the lake, and the prominent trimline (Image: Richard Waller).

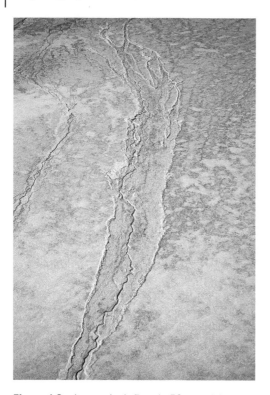

Figure 4.2 Large slush flow (c. 30 m wide) located in the accumulation area of Aktineq Glacier, Bylot Island. Note the prominent marginal levees and formation of an active supraglacial channel within the area affected by the flow (Image: Richard Waller).

temperate glaciers. In the case of non-temperate glaciers, this surface storage is facilitated both by the limited crevassing and by the formation of superimposed ice within the snowpack. In addition to the storage in lakes, water can also be stored over the summer months in areas devoid of snow cover within both a surficial weathering crust and the cryoconite holes that develop as a result of the enhanced ablation associated with accumulations of surface debris (Irvine-Fynn *et al.*, 2011).

Eventually a threshold is reached whereby this stored water can be rapidly released. Saturation of the snowpack can for example lead to the development of deep layers of slush that can mobilise to form slush flows (Figure 4.2). These slush flows can in turn contribute to the re-activation of perennial supraglacial drainage networks that have been blocked by snow and ice during the preceding winter (see Section 4.2.3). Drainage from these supraglacial streams is then commonly routed over the surface to the glacier margins where it contributes to flow within ice-marginal, lateral meltwater channels.

This time period during which the stored water is released can be likened to the spring snowmelt discharge events that dominate the annual discharge regimes of permafrost river systems and that can strongly influence the hydrometeorological response of the wider catchment. Measurements of meltwater discharge at Scott Turnerbreen in Svalbard for example revealed two distinct modes of flow during the course of the melt season (Hodgkins, 1997, 2001). During the first part of the melt season (mid-July to early August) when the supraglacial meltwater stores were still active, meltwater discharges were high and variable, displaying little evidence of the classic diurnal signal commonly associated with variations in the surface energy balance. It was only during the later stages of the melt season when the supraglacial meltwater stores had been exhausted that a second mode of flow was established, characterised by much lower and less variable discharges and the establishment of a clearer diurnal signal. Episodic periods of increased discharge were also observed to occur during storm events and periods of heavy precipitation.

This partitioning of the active drainage season into an initial phase of high and fluctuating discharges associated with active snowmelt and a subsequent phase of relatively low discharges punctuated by periodic flood events is reminiscent of the nival flow regimes characteristic of the High Arctic. Whilst they are broadly similar, when considered in more detail, the hydrological influence of glaciers within permafrost catchments typically results in the development of distinctive proglacial flow regimes with notable differences to classic nival flow regimes (Church, 1974; Woo, 1986) (Figure 4.3). First, whilst the peak discharge typically occurs close to the start of the melt season within a nival flow regime with the majority of the total annual runoff commonly occurring within a few days, a proglacial flow regime is associated with a more gradual increase in runoff with peak discharge typically occurring in mid-summer. Second, whilst the removal of the winter snow cover in nival flow regimes results in a loss of the diurnal signal in the latter part of the melt season, the replacement of snowmelt with glacier melt in proglacial flow regimes commonly results in diurnal fluctuations that are most prominent during the later stages of the summer melt season (Hodgkins, 1997, 2001).

Although these two models provide convenient conceptual end members that highlight the theoretical hydrological differences between glaciated and non-glaciated catchments, a rare study of streamflow regimes on Devon Island by Marsh & Woo (1981) demonstrated the occurrence of significant hydrological variations over small areas as a consequence of the varied contributions made by a range of water sources including rainfall, spring snowmelt and the ablation of both semi-permanent snowbanks as well as the glaciers within the study catchments. One stream for example displayed what resembled a proglacial flow regime in spite of an absence of glaciers within its catchment, with the persistent discharge fluctuations later in the summer relating instead to the ablation of perennial snowbanks. In addition, interannual variations in weather conditions can further complicate the picture with delayed runoff sometimes relating to cool summers rather than the influence of a glacier within the catchment. As a consequence, the authors concluded that Arctic hydrological systems are in reality likely to display a broad spectrum of flow regimes that vary according to the relative contributions of several water sources within each catchment.

Another potential hydrological distinction associated with glaciated permafrost catchments relates to the periodic release of water throughout the winter months that can sustain active meltwater discharge into the wider catchment at a time of year when most subaerial drainage systems are completely frozen. A range of studies have demonstrated the ability of non-temperate glaciers to store water during the winter months with the blockage of ice-marginal drainage routes resulting in the retention of water within large voids within the glacier (e.g. englacial channels and moulins). In observing the flooding of an englacial drainage system at Hansbreen, Svalbard, in April 2007, Benn *et al.* (2009) estimated that at least 1.3 million m^3 of water had been stored within the englacial drainage system over the preceding winter period. As noted earlier, water may also be stored in a more diffuse form within the small interstices that occur between individual ice crystals to define what Irvine-Fynn *et al.* (2006) have referred to as the specific retention capacity of the glacier. This water that can be created by strain heating in the basal parts of the glacier will gradually drain in response to the influence of gravity and can therefore impose increasing hydrostatic pressures on any blocked ice-marginal conduits.

Evidence for periodic or continuous water flow from glaciers terminating in regions of permafrost is graphically illustrated by the development of conspicuous proglacial icings (also known as aufeis or naled) that can reach several metres in thickness. A study of icing formation in Svalbard using aerial photos acquired in the summer of 1995 revealed their

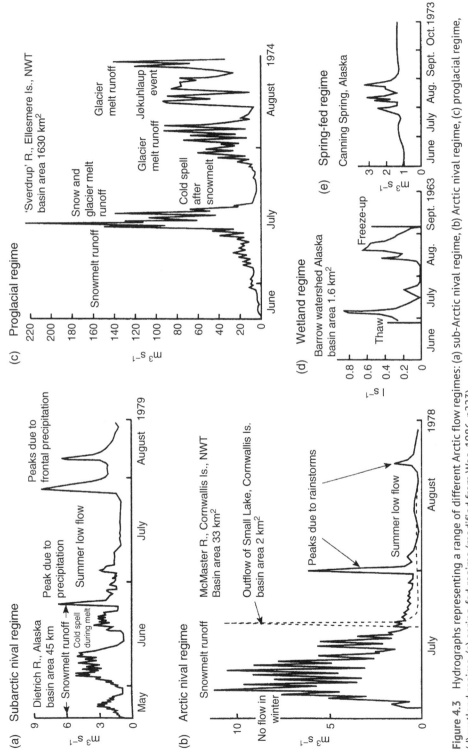

Figure 4.3 Hydrographs representing a range of different Arctic flow regimes: (a) sub-Arctic nival regime, (b) Arctic nival regime, (c) proglacial regime, (d) wetland regime, (e) spring-fed regime (modified from Woo, 1986, p223).

presence in front of 217 glaciers, combining to produce a total areal coverage of 12.3 km^2 in spite of the unusually warm summer conditions experienced that year (Bukowska-Jania & Szafraniec, 2005). Analysis of aggregate data suggested a positive correlation between the mean size of the glacier and the mean area of the icing, a prediction consistent with voids within the glacier storing water over the winter months. The topography of the glacier foreland was also seen as being important with flat proglacial zones located in wide valleys tending to favour the formation of larger icings. This is certainly the case at Fountain Glacier on Bylot Island, where the combination of a wide proglacial braidplain with a valley constriction at its lower end promotes the development of a perennial icing over 1 km in length and up to 6 m in thickness (Figure 4.4). The presence of the constriction in particular facilitates the proglacial accumulation of water, thereby maximising the potential for ice accretion during the winter months with the water supply being provided by the nearby polythermal glacier (Wainstein *et al.*, 2014) (see Section 5.3.5 for further details).

The requirement for proglacial discharge to persist during the winter months in order for aufeis to form has led to the assertion that icings are indicative of polythermal basal regimes, with the warm-based ice providing a winter water source. The observation of icing formation at Scott Turnerbreen, an entirely cold-based glacier, led Hodgkins *et al.* (2004) however to argue that their presence cannot be used as a reliable guide to the parent glacier's thermal regime. In this case, the authors argued that aufeis formation was associated instead with the drainage of water stored within the glacier over the winter months by percolation through a thawed sub-channel sediment matrix (i.e. groundwater flow).

Figure 4.4 Large proglacial icing c. 1 km in length situated in front of the margin of the Fountain Glacier, Bylot Island. Note how the icing occupies the broad outwash plain situated between the glacier margin and a prominent down valley constriction (Image: Richard Waller).

4.2.3 Drainage Pathways

Non-temperate glaciers are well known for the development of supraglacial drainage networks that are typically more extensive, more deeply incised and longer lasting than those associated with temperate glaciers. The formation of these extensive surface drainage networks is promoted by surface crevassing that is generally more limited than that typically associated with temperate glaciers that are generally faster flowing and more commonly found within high relief mountain environments. With surface ablation rates on non-temperate glaciers also being more limited, supraglacial streams are capable of more rapid incision into the glacier surface, with Müller (2007, cited in Irvine-Fynn *et al.*, 2011) recording supraglacial incision rates of c. $0.07\,\mathrm{m\,d^{-1}}$ in comparison to surface ablation rates of $0.01\,\mathrm{m\,d^{-1}}$ on the Longyearbreen glacier in Svalbard. This enables the progressive development of more deeply incised channels that commonly display meandering profiles and step-pool sequences. Acting in combination with the reduced rates of ice deformation, this can result in the establishment of perennial supraglacial drainage networks that are reoccupied each year during the commencement of the melt season (Figure 4.5).

The cold surface ice with limited surface crevassing and associated secondary permeability has traditionally been assumed to preclude the development of englacial drainage systems within non-temperate glaciers. This is consequently thought to promote surface drainage to the glacier margins where it can flow as marginal or sub-marginal drainage channels, leading to the creation of lateral meltwater channels that are thought to constitute one of the most distinctive landscape features associated with cold-based glaciers (Section 5.3.1).

Figure 4.5 Perennial supraglacial channel close to the margin of Fountain Glacier, Bylot Island. The depth of incision is c. 6 m below the glacier surface and the channel is reoccupied by supraglacial drainage each year. Note the pronounced meaders and high level of channel sinuosity (Image: Richard Waller).

A number of studies have however revealed the presence of extensive englacial drainage networks and have consequently demonstrated a variety of mechanisms through which these channels might form and persist within areas of cold ice. Gulley *et al.* (2009) have shown that the rapid incision of supraglacial channels coupled with the tendency for snow bridges and ice deformation to cover them can result in a process of 'cut and closure' and the progressive transformation of deeply incised supraglacial channels into enclosed englacial channels. Once established, these englacial channels have been observed to develop large steps or knickpoints of between 5 and 20 m in height that can migrate up-glacier at rates equivalent to the overall surface motion of the glacier, resulting in the formation of large and static englacial moulins. Vadose englacial channels can also progressively incise downwards to the glacier bed. Once established, these processes have the potential to create stable subglacial drainage systems, although their drainage capacity is likely to be limited in comparison to temperate glaciers as a consequence of the shorter melt season and the reduced basal frictional heat generation.

Supraglacial and englacial drainage has been observed to be strongly controlled by the structure of the glacier. Water flows commonly exploit structural discontinuities leading to the establishment of preferential drainage patterns that highlight important process associations between glacier flow, crevasse formation and glacier hydrology. Where drainage is flowing along the base of open crevasse systems for example, their subsequent closure provides an additional mechanism whereby they can become isolated to form englacial conduits. A remarkable study by Benn *et al.* (2009) involving the use of speleological and mountaineering techniques to explore and survey three englacial drainage systems in three glaciers has provided a direct insight into the morphology and evolution of englacial drainage systems and the ways in which the hydrofracturing of existing discontinuities can allow these systems to reach the glacier bed. Exploration of 'Crystal Cave' situated near the confluence of Hansbreen and Tuvbreen in Svalbard for example revealed a sub-horizontal passage c. 10–20 m below the glacier surface orientated parallel to ice flow. This passage was associated with a series of vertical shafts (Figure 4.6a), one of which was found to extend to the bed (68 m below the surface) via a complex series of shafts and steps where it continued as a low, wide and partially water-filled conduit with a till floor (Figure 4.6b). In corresponding to the location of existing fractures and displaying evidence of water flow, the vertical fractures are considered the product of hydrofracturing, with a series of early events creating blind, water-filled fractures that failed to establish a connection with the bed. The generation of these complex and interconnected drainage systems was initially associated with the formation of an ephemeral supraglacial lake at the confluence of the two glaciers that drained supraglacially and then englacially following the aforementioned process of cut and closure. On reaching the transverse fractures characteristic of this part of the glacier, this meltwater supply enabled the propagation of the hydrofractures through c. 60 m of cold ice to reach the glacier bed. The repeated formation of hydrofractures in the same location suggests a regular and predictable location for subglacial recharge associated with high summer meltwater input and longitudinal extension. In addition, the flooding of the system during the winter and the preservation of a series of shafts suggests that these networks can persist for a number of years and provide seasonal glaciological meltwater storage (Section 4.2.2). This hydraulic augmentation of crevasse formation provides a mechanism whereby fractures can penetrate through over 1 km of cold ice to reach the

(a)

(b)

Figure 4.6 Images of the drainage system of 'Crystal Cave' explored by Benn et al. (2009) illustrating an englacial vertical shaft (a) and the subglacial conduit (b) (Image Courtesy: Jason Gulley).

glacier bed, resulting in the interlinking of the supraglacial, englacial and supraglacial drainage networks (see Section 4.3.2). Alternatively, if production of meltwater volumes exceeds the capacity of englacial channels, hydraulic damming can occur leading to re-routing of drainage to higher englacial pathways or potentially even the glacier surface.

Geophysical surveys have provided additional insights into the nature, evolution and potential complexity of drainage systems in non-temperate glaciers. Relatively low frequency (50 and 100 MHz) Ground Penetrating Radar (GPR) surveys of Tellbreen, a small and predominantly cold valley glacier in Svalbard, were used by Bælum & Benn (2011) to determine both the thermal regime of the glacier and its substrate and also to delineate the position of englacial and subglacial channels. Survey results indicated the presence of discrete water-filled subglacial channels beneath the thin ice of the glacier margin that was otherwise frozen to the bed. These lacked any subglacial catchments and were instead fed by englacial channels located further up glacier. The process of cut and closure was considered crucial in enabling supraglacial channels created in the upper part of the glacier to transition to englacial channels down glacier and ultimately to reach the bed beneath the thin ice of the terminus. This study therefore corroborates the earlier findings of the speleological surveys at Hansbreen in revealing a surprisingly complex drainage system capable of storing, transporting and releasing meltwater throughout the year to produce a significant icing during the winter months (Section 5.3.5). Repeat GPR surveys of Stagnation glacier, a polythermal valley glacier on Bylot Island, revealed a series of temporal changes over the course of the ablation season that included the appearance and disappearance of englacial reflections in the terminus (transient englacial water bodies) and a reduction in the extent of an area of chaotic returns (interpreted as temperate ice) at a site further up-glacier (Irvine-Fynn *et al.*, 2006). In addition to mapping the glacier's thermal structure and the location of the cold-temperate ice transition surface, the results demonstrated the dynamic nature of the glacier's thermal and hydrological characteristics and the drainage of an estimated 17,000 m³ of water, again highlighting the transient storage and release of intraglacial water as a distinctive characteristic of polythermal glaciers.

Field observations by Boon & Sharp (2003) have helped to elucidate the temporal stages of the hydrofracturing process and to establish connections with the seasonal evolution of the supraglacial drainage networks. Initial observations showed supraglacial meltwater flowing into a crevasse located in the upper part of the ablation area of the John Evans Glacier, a non-temperate valley glacier located in the Canadian High Arctic. Within three days, the crevasse had filled to create a supraglacial pond, which continued to fill for another 11 days reaching a level 6.9 m above the channel floor. A widening of local crevasses accompanied by deep cracking noises provided early evidence of hydrofracturing. Four falls in lake level provided initial precursors to its complete drainage within 1 hour, which shortly thereafter resulted in the formation of a series of new crevasses at the glacier margin and the associated release of solute-rich waters. The abruptness of the lake drainage in conjunction with field evidence of new crevasses and fracturing noises led the authors to conclude that hydrofracturing at the base of a water-filled crevasse was the most likely cause. The precursory small drops in lake level were related to the initial formation of small fractures that quickly re-froze with the englacial temperatures being c. −5 to −10 °C. The resultant release of latent heat associated with these pre-emptory events in combination with a continued enlargement of the surface pond eventually resulted in more abrupt drainage capable of generating the turbulent heating required to establish a more permanent hydrological connection between the glacier surface and the bed. A subsequent increase in velocity down-glacier of the crevasse would have in turn increased the tensile stress in the vicinity resulting in a feedback that encouraged further crevasse widening and maintenance of the hydrological connection.

Even if it is accepted that supraglacial meltwater has the potential to reach the bed of non-temperate glaciers, traditional conceptual models have suggested that extensive marginal areas of cold basal ice will act as 'thermal dams' that limit or entirely preclude any active subglacial drainage whilst promoting the aforementioned water storage within the glacier. However, observations documenting the emergence of sediment and solute-rich meltwater from subglacial conduits at numerous non-temperate glaciers suggest that any such thermal obstructions can be breached, allowing subglacial drainage systems to remain active (e.g. Vatne *et al.* 1995, 1996; Copland *et al.*, 2003). Investigations undertaken as part of a paired study on Erikbreen & Hannabreen in Northern Spitsbergen have provided field evidence for the ability of meltwater to access the bed and create subglacial drainage systems beneath non-temperate glaciers. Hydrological investigations at Erikbreen involved the use of tracer experiments to investigate the potential nature and hydraulic efficiency of the drainage system feeding an ice-marginal portal (Vatne *et al.* 1995). In spite of its non-temperate basal thermal regime, the glacier was assumed to have a significant subglacial drainage component with the meltwater discharge at its margin exhibiting very high suspended sediment loads (see Section 4.2.3). The low measured through flow velocities of $0.08–0.32\,\mathrm{m\,s^{-1}}$ led the authors to suggest that subglacial drainage was active and associated with a distributed drainage system. These systems are characterised by numerous small channels with low discharges that are less efficient than the channelised systems typically found beneath temperate glaciers with higher through flow velocities of $0.4–1.0\,\mathrm{m\,s^{-1}}$. Significant variations in the residence time distribution curves taken only a few days apart also suggested that the drainage system was unstable. The combination of high levels of meltwater input during the initial stages of the ablation season coupled with the hydraulic inefficiency of the drainage system led to an increase in glacier flow that the authors related to the attainment of high basal water pressures and the activation of basal sliding (Section 4.3.1). This increased glacier flow was however short-lived with a subsequent reduction in velocity suggesting that the development of the subglacial drainage system and an enhancement in its capacity caused a subsequent fall in basal water pressure. This emphasises the importance of the connection between meltwater generation, glacier drainage evolution and ice dynamics that is explored further in Section 4.3.

Focusing more explicitly on the hydrological implications of glacier–permafrost interactions, the presence of the cold-temperate ice transition zone (CTZ) has been suggested to play a key role in determining both the hydrological pathways taken by meltwater (Figure 4.7) and as a consequence, the dynamic behaviour of the glacier (Irvine-Fynn *et al.*, 2011). If large and potentially stable conduits occur and are capable of draining the meltwater discharge through the cold terminal zone, then the basal water pressures up-glacier of the CTZ can remain low. If however the drainage capacity through the cold terminus is constricted and the supply of meltwater exceeds the drainage capacity, very high and potentially artesian pressures can be generated. This limits the development of hydraulically efficient subglacial drainage networks, which in turn necessitates the evacuation of the excess meltwater via alternative flow pathways. Annual variations in the extent and hydrological influence of cold marginal ice can therefore act to modulate the release of stored water and the development of drainage systems. This can in turn have dramatic impacts on the dynamic behaviour of the glacier at the start of the melt season (Section 4.3.1).

The generation of high basal water pressures during the start of the melt season or alternatively during summer storm events can in particular result in water being driven to the surface, creating dramatic supraglacial fountains that can persist for several days (Figure 4.8,

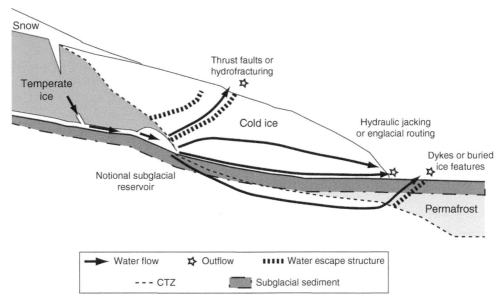

Figure 4.7 Potential hydrological pathways for meltwater to breach the cold ice margin of polythermal glaciers (Irvine-Fynn *et al.*, 2011, p17). Note the potential for water to exit via water escape structures through the ice-marginal permafrost.

Figure 4.8 Supraglacial upwelling on the surface of ice margin of Fountain Glacier that formed following a period of prolonged rainfall. The outlet was associated with the formation of a series of fresh ice fractures suggestive of hydraulic jacking (Image: Richard Waller).

Copland *et al.*, 2003). The development of these fountains can involve the exploitation of pre-existing discontinuities such as thrust faults or potentially the formation of new fractures through hydrofracturing. The development of high basal water pressures can also result in hydraulic jacking and an upward vertical displacement that can be recorded at the glacier surface. If this occurs over wide areas, it can trigger a dynamic response through the development of extensive ice-bed separation, a marked reduction in basal drag and potentially an increase in velocity leading to flow instabilities. Such impacts on glacier flow are however thought to be restricted to extreme drainage events, relating for example to the dramatic increases in discharge associated with jökulhlaup events. Brief periods of hydraulic jacking are instead more typically thought to promote the formation of new high-capacity drainage routes that allow the rapid evacuation of the pressurised subglacial water and cause an increase in basal drag as described at Erikbreen (Vatne *et al.*, 1995).

High basal water pressures can alternatively result in water being routed through the groundwater system. In this situation, the subglacial drainage conditions are controlled not only by glaciological parameters such as melt rates but also by the characteristics of the substrate. Within the specific context of permafrost environments, unfrozen taliks within the permafrost provide potential hydraulic pathways for glacially driven groundwater fluxes. Glacial drainage can consequently re-emerge in the glacier foreland to form upwellings or springs that can in turn promote the development of proglacial icings (Wainstein *et al.*, 2014) or in some cases open-system pingos (Section 5.3.5). The generation of artesian water pressures within confined subsurface aquifers can also result in the hydrofracturing of the overlying permafrost materials resulting in the fluidisation of sediments and the formation of clastic dykes or fracture fills (Section 5.4.2). This once again highlights the crucial importance of viewing the glacier and the sub-marginal permafrost as a coupled system, the hydrological and hydrogeological significance of which is explored in further detail in Section 4.2.5.

4.2.4 Suspended-Sediment and Solute Fluxes

With conceptions of the hydrology of glaciers occurring within permafrost environments traditionally assuming that meltwater has little access to the glacier bed, opportunities for the acquisition and flushing of subglacial material are typically thought to be limited. In combination with the assumptions of inactive basal processes and negligible subglacial erosion discussed in Section 3.4, subglacial sediment loads have in turn consequently been regarded as being very limited in comparison to temperate glaciers. However, as in the case of basal processes, recent research has provided mounting evidence that this need not necessarily be the case. Work at Erikbreen, a small but relatively fast moving polythermal valley glacier in northern Spitsbergen by Vatne *et al.* (1995) provides one case study illustration that suggests these assumptions are not universally applicable and that non-temperate glaciers in permafrost regions can discharge meltwater with unexpectedly high concentrations of suspended sediment. In this study, the authors recorded suspended sediment loads of between 0.1 and $4.2\,\mathrm{g\,l^{-1}}$ directly from a stream draining a subglacial portal. Significant seasonal and diurnal variations were also observed, with the suspended sediment loads broadly rising in line with increasing discharge during the course of the ablation season and daily peak concentrations being recorded in late afternoon or early evening. What is most noteworthy however is the recording of suspended sediment concentrations that were much higher than those recorded from temperate glaciers at equivalent meltwater discharges.

In considering the potential reasons for these unexpectedly high suspended sediment concentrations within the subglacial drainage, the authors suggested that the relatively high velocity of this glacier ($5–12\,cm\,day^{-1}$) and the development of a heavily crevassed surface precluded the development of an extensive supraglacial drainage system whilst also enabling surface drainage to access the glacier bed and the subglacial sediment stores. The authors also considered the potential implications of both historical climate change and the soft-bed paradigm, hypothesising that higher erosion rates associated with the glacier's greater extent and more active flow during the Little Ice Age may have resulted in the prior accumulation of subglacial sediments that have provided a longer-term sediment source that is being actively reworked by current subglacial drainage.

The authors subsequently concluded that the high-suspended sediment loads represent the combined product of extensive subglacial sediment availability coupled with the presence of a distributed drainage network capable of accessing and evacuating sediment from large areas of the bed. The limited drainage during the melt season combined with the high sediment availability resulted in a suspended sediment flux that did not decline during the melt season and instead simply varied in relation to fluctuations in discharge. This is distinctly different from the temporal patterns observed in temperate glaciers where the progressive rationalisation and channelisation of the subglacial drainage system during the melt season typically result in a rapid exhaustion of available sediment supplies and a concomitant decline in suspended sediment loads in spite of an increase in discharge (e.g. Gurnell *et al.*, 1994; Hodgkins, 1996). The authors hypothesised that the high sediment fluxes may also provide evidence of active subglacial sediment deformation capable of contributing significant amounts of sediments into the drainage system, providing a tantalising if somewhat speculative connection between basal processes and meltwater sediment fluxes.

Additional studies comparing the sediment yields from glaciers with contrasting basal thermal regimes have similarly challenged the assertion that non-temperate glaciers are invariably associated with reduced meltwater sediment fluxes. Bogen (1996) for example monitored the sediment yields of four contrasting Norwegian glaciers during the period 1989–1993: two temperate outlet glaciers (Engabreen and Nigardsbreen), a small and slow moving cold-based glacier (Øvre Beirbre) and a polythermal glacier on Svalbard (Brøggerbreen). Of these four, it was the polythermal glacier (Brøggerbreen) that displayed the highest average sediment yield of $613\,t\,km^{-2}\,yr^{-1}$ (equating to an erosion rate of $0.226\,mm\,yr^{-1}$). The results for this site also demonstrated significant interannual variations with the sediment yield in 1990 reaching in excess of $1200\,t\,km^{-2}\,yr^{-1}$. With the sediment monitoring station being situated 1 km downstream of the glacier margin, the author acknowledged however that the sediment yields in this instance are likely to have been influenced by proglacial sediment inputs.

As part of a broader consideration of the sediment fluxes associated with glaciers in both permafrost and non-permafrost environments (Section 4.4), Etzelmüller & Hagen (2005) highlight extensive recent work undertaken on polythermal glaciers on Svalbard occurring within permafrost environments that provide further insights into the potential roles played by proglacial processes. Whilst glacier velocities are lower, meltwater seasons shorter and the meltwater production more limited than in temperate environments, work in Svalbard suggests that these limitations can be more than compensated for by the greater availability of sediment stores within proglacial permafrost environments. The presence of extensive areas of dead ice within recently deglaciated forelands for example can contribute significant sediment fluxes to the fluvial systems that result in the development of strong positive correlations between discharge and sediment concentration that persist throughout the melt season.

In describing the seasonal changes in suspended sediment loads from Scott Turnerbreen, a cold-based glacier in Svalbard, Hodgkins (1996) further emphasises the key differences between non-temperate and temperate glaciers. In addition to the persistence of high suspended loads throughout the melt season, measured suspended sediment concentrations displayed less variability than the measured discharges, suggesting that discharge is only one of a range of controls. The potential existence of a distributed subglacial drainage system with access to extensive subglacial sediment as inferred at Erikbreen (Vatne *et al.*, 1995) was considered unlikely with Scott Turnerbeen being entirely cold-based. Instead, the sustenance of high-suspended sediment loads was considered more likely to relate to the transfer of sediment from ice-cored lateral moraines that provided a continuous supply of sediment into the subaerial ice-marginal channels that in turn constituted the glacier's principal hydrological pathways.

Interestingly, whilst the mechanisms promoting the more extensive availability of sediment stores in non-temperate glaciers described by Vatne *et al.* (1995) and Hodgkins (1996) are very different, both authors identify and emphasise the important role played by past conditions and process environments as well as the impacts of ongoing climate change. In both cases, the colder climatic conditions during the Little Ice Age were related to thicker, warmer and more active glaciers and the associated development of extensive sediment stores that are currently being reworked by subglacial and/or subaerial processes. In the case of Scott Turnerbreen where glacier recession and downwasting have exposed and subjected the sediment stores to subaerial reworking, the process is related to the concept of paraglacial denudation whereby deglaciation is associated with the destabilisation of glacial sediment stores and elevated rates of geomorphic activity (Church & Ryder, 1972; section 5.6.3).

Focusing more specifically on the dissolved loads associated with non-temperate glaciers, solutes are typically derived from one of two sources: atmospheric sources and crustal sources that are released via a varied suite of chemical weathering processes. The solutes present within snow cover in maritime locations such as Svalbard tend to be dominated by sea-salt species (e.g. Cl). The start of the ablation season commonly results in a concentrated early 'first flush' of solutes from these snowpack sources as they are leached by meltwater. As mentioned previously, this meltwater is initially stored supraglacially until saturation of the snowpack results in its runoff into ice-marginal channels. After this initial phase, the hydrochemistry of glacial runoff is typically dominated by sources related to chemical weathering (e.g. Ca, Na, K, HCO_3, SO_4 and Si). The presence of large amounts of freshly comminuted fine-grained material within glacial environments means that chemical weathering rates are comparable to those in temperate environments in spite of their colder temperatures and the limited time period available for rock–water interactions to occur.

Work at Hannabreen by Vatne *et al.* (1996) considered the hydrochemistry of meltwater as an alternative means of studying the nature and behaviour of the subglacial drainage system. The lack of crevassing and limited number of moulins at Hannabreen characteristic of non-temperate glaciers initially suggested a limited opportunity for supraglacial meltwater to access the bed. Utilising a mass balance approach developed by Collins (1977), the authors used measurements of electrical conductivity (EC) and SO_4 levels to differentiate the meltwater exiting the glacier into a rapid englacial quick flow component (characterised by low ECs and low SO_4 levels) and a delayed subglacial flow component (characterised by high ECs and high SO_4 levels). Whilst the results indicated that the quick flow englacial component dominated the glacial runoff, they inferred the occurrence of a relatively constant base flow input of delayed (subglacial) flow that was particularly prominent during low flow conditions when the englacial quick flow input was limited (Figure 4.9). This was interpreted as providing evidence for solute acquisition within a distributed subglacial drainage system. The lack of any seasonal changes in the delayed flow component also suggested that the distributed subglacial drainage system was stable and did not show the development into a channelised system commonly observed in temperate glaciers.

The seasonal storage of water in non-temperate glaciers over the winter months within subglacial, englacial or in some cases subaerial stores can increase residence times and promote the further enrichment of stored waters, the release of which can contribute to high solute concentrations during the early part of the meltwater season (Hodgkins, 2001). The first water to be released from a subglacial conduit at the John Evans Glacier, a polythermal glacier in the Canadian High Arctic for example displayed high solute concentrations with measured ECs of $>300\,\mu S\,cm^{-1}$. This was thought to represent the drainage of water stored over the winter season (Copland *et al.*, 2003), with the measured EC rapidly dropping over a few days to $<100\,\mu S\,cm^{-1}$ as the influence of supraglacially derived meltwaters became increasingly dominant.

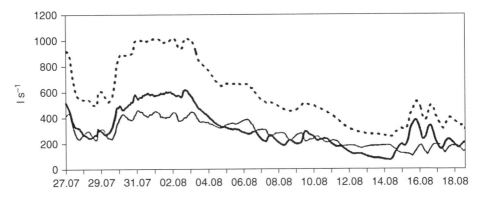

Figure 4.9 Flow separation of the runoff measured from a channel discharging from the terminus of Hannabreen, Spitsbergen (Vatne *et al.*, 1996, p71). The bulk flow (dashed line) features both quick flow (thick solid line) and delayed flow (thin solid line) components with the latter being related to active subglacial discharge.

Freezing and solute rejection can also result in an increase in solute concentrations during the winter months when the level of dilution associated with ablation is at a minimum. Studies from ice-marginal spring systems indicate that whilst suspended sediment fluxes typically drop during the winter, the solute concentrations commonly increase. Krawczyk (1992) used this distinction to differentiate the hydrochemical characteristics of meltwaters from Werenskioldbreen, Svalbard into a 'summer system' and a 'winter system'. The summer drainage contained a combination of Na^+ and Cl^- ions derived from precipitation and marine aerosols that were heavily diluted by surface melt. The spring systems studied also contained significant levels of Ca^{2+} and HCO_3 ions related to the chemical weathering of englacial debris as well as high-suspended sediment loads. In contrast, the winter drainage was associated with significantly reduced discharges and suspended sediment discharges, but higher solute concentrations.

In a relatively rare hydrochemical study of a cold-based glacier in the High Arctic that represents a companion work to that mentioned earlier on suspended-sediment fluxes, Hodgkins *et al.* (1998) interpreted the hydrochemistry of runoff from Scott Turnerbreen, Svalbard, in relation to the glacier's key drainage pathways. In contrast to Erikbreen, this small and relatively thin glacier is entirely cold-based and as such is thought to lack any subglacial drainage. Instead, the meltwater is discharged initially through subaerial lateral meltwater channels before flowing through a proglacial icing that contains interstitial waters with high concentrations of solutes (including SO_4, HCO_3 and Si). As a consequence, the hydrochemistry is controlled by seasonal variations in the delivery of solutes (e.g. SO_4) associated with the melt of snow and ice, subaerial weathering within ice-cored lateral moraines that feed solutes (primarily Ca^{2+}) into the lateral meltwater channels and finally, solute acquisition from the proglacial icing (that resulted in an elevation of Na^+ levels in particular). The authors again highlight the potential significance of recent environmental change and the abundant sediment supply relating to a thicker, warmer and more active glacier during the Little Ice Age. In this case, rather than contributing to an elevated subglacial suspended sediment flux, this glaciological legacy has resulted in the formation of widespread areas of ice-cored moraine. The high rock/water ratio in these settings combined with the presence of higher concentrations of reactive minerals than found within the meltwater channel resulted in this deglaciated area providing an important source of additional weathering-derived solutes.

Proglacial weathering processes in areas of permafrost surrounding non-temperate glaciers therefore have the potential to significantly augment the solute fluxes discharged from glaciated catchments in the same way that mass wasting in these environments can elevate suspended sediment loads (Section 4.4). The combined monitoring of bulk meltwater chemistry and discharge from gauging stations located at the eastern and western margins of Finsterwalderbreen, Svalbard and 2.5 km further downstream allowed Wadham *et al.* (2001) to quantify the role played by proglacial weathering. The fluxes of HCO_3, SO_4, Ca and Mg were observed to increase by 30–47% as a result of the chemical weathering of sulphates, carbonates and sulphides within the active layer of the proglacial permafrost. Moreover, the total chemical weathering rate calculated for 1999 was estimated to exceed the chemical weathering rate associated with the glaciated part of the catchment by a factor of 3.3. With the glacier foreland being associated with push moraines and ice-cored terrain associated with the glacier's recession from its LIA limit, this again illustrates the role of recent environmental change and glacier recession in providing large sediment stores that can be weathered and reworked.

More broadly, it emphasises the significant role played by permafrost-related processes operating in environments where glaciers and permafrost interact. In a related paper, Cooper *et al.* (2002) demonstrated that proglacial solute acquisition at Finsterwalderbreen is associated primarily with the weathering of highly reactive sediments within shallow active layer groundwater systems supplied by meltwater from proglacial snowpacks, buried ice and the glacier itself. Solute acquisition occurred principally through sulphide oxidation and carbonate dissolution to generate groundwaters that were sufficiently enriched to produce surface precipitates in response to both summer evaporation and winter freezing. The re-dissolution of these surface salts during the early part of the next melt season was then able to provide an additional early source of solutes.

As part of a long-term study on Austre Brøggerbreen, a glacier that has experienced a change in thermal regime from polythermal to cold-based as a consequence of its ongoing recession, Nowak & Hodson (2014) provide an insight into the emerging impacts of ongoing climate change on the meltwater chemistry of High Arctic catchments containing glaciers and permafrost. In comparing catchment runoff obtained from a proglacial sampling station between 1991 and 2010, the authors identified a range of significant hydrochemical changes resulting primarily from the increased chemical weathering taking place in ice-marginal and proglacial terrains recently exposed by glacier recession. Between 2000 and 2010 for example, pronounced increases were observed for HCO_3, Ca, K and SO_4 with mean concentrations doubling in some cases. The observed increase in these solutes derived from crustal weathering combined with a decline in pCO_2 values linked to the enhanced weathering of fine-grained, chemically reactive tills that were previously overlain by cold-based ice but which have now been exposed to subaerial conditions and processes as a consequence of glacier recession. The most significant hydrochemical changes observed related to a reorganisation of the glacier's principal drainage pathways between 2009 and 2010. The shutting down of a previously well-established drainage portal in the glacier's eastern margin resulted in meltwater bursting onto the glacier surface further to the west. This in turn resulted in the abandonment of a rock-walled proglacial channel with a new flowpath being eroded through an area of ice-cored moraines, providing access to highly reactive sediments and an abrupt increase in most of the crustal ion yields that occurred in the absence of any change in discharge.

4.2.5 Glacial Groundwater Fluxes in Permafrost Environments

Recent work within glaciology and glacial geology relating to the development of the soft-bed paradigm (Section 3.3.3) has emphasised the significance of groundwater flow beneath ice sheets. Subglacial groundwater flow can control the drainage of meltwater beneath the ice sheet, draining all the meltwater generated in areas of high hydraulic conductivity whilst requiring discharge via other means such as subglacial meltwater channels where hydraulic conductivities are low (Piotrowski, 1997). As a consequence, groundwater systems can control the porewater pressures within subglacial sediments, thereby exerting a strong influence on both basal processes and glacier dynamics (Boulton *et al.*, 1995; Piotrowski, 1997).

In the case of temperate glaciers, active subglacial drainage combined with the presence of permeable subglacial sediments provides the opportunity for the glacial hydrological system to couple with the subjacent groundwater system. Similarly, the presence of zones of warm-based ice beneath polythermal glaciers experiencing active basal melting combined with the occurrence of unfrozen taliks within the subglacial permafrost provides the potential for glaciers and permafrost to couple hydrologically and create an interconnected system. However, with frozen sediments typically displaying hydraulic conductivities orders of magnitude lower than the equivalent unfrozen sediments, particularly at low temperatures and when ice saturated (Williams & Smith, 1989), the presence of ice-marginal or subglacial permafrost can play an important role in regulating glacial groundwater fluxes. In acting as an aquitard, ice-rich permafrost can reduce the transmissivity of the groundwater system as a whole, resulting in an increased groundwater flux towards the ice-bed interface, increased water storage and the potential development of tunnels through catastrophic outburst events (Piotrowski, 1997).

As discussed previously in Section 3.6.3, the presence of ice-marginal permafrost in particular has been argued to prevent groundwater from upwelling at the ice margin. If this is the case, then in order to maintain the proglacial groundwater discharge, the piezometric surface grades to the outer limit of the permafrost generating artesian porewater pressures throughout the terminal zone (Boulton *et al.*, 1995) (Figure 4.10). This provides an influential hydrological glacier–permafrost interaction that in controlling subsurface fluid pressures, exerts an important control on the rheological behaviour of the glacier foreland and the style and location of failure in ice-marginal materials (Boulton & Caban, 1995) (Section 5.3.2).

Studies in permafrost hydrology indicate that rather than constituting a homogenous aquitard with universally low permeabilities, groundwater flow within permafrost can be maintained through unfrozen taliks that can occur within or below the permafrost layer (e.g. Woo *et al.*, 2008). These hydrological pathways coupled with the high hydraulic head gradients resulting from the presence of a glacier in the catchment can allow subglacially recharged groundwater fluxes to be driven into the proglacial area where they can upwell at the surface via high permeability zones to form discrete springs. Groundwater may also upwell and discharge into large lakes associated with through taliks although in these situations the evidence for their existence will be much less obvious. In studying the thermal changes associated with the propagation of a surge bulge through the polythermal Trapridge Glacier (Section 4.3.2), Clarke *et al.* (1984) provided indirect evidence for active groundwater flow in a talik located beneath a subglacial permafrost layer estimated to be 5–15 m thick. Borehole thermistor measurements at a site located down-glacier of the propagating surge bulge revealed a very large basal temperature gradient that when converted to a heat flux was approximately 10 times higher than the average geothermal heat flux. With ice deformation being limited and there being little evidence for channelised flow, the authors concluded that the only viable source of the elevated heat flux was active groundwater flow through the unfrozen sediment underlying the subglacial permafrost.

A study focusing specifically on glacier–permafrost hydrological interactions on Bylot Island by Moorman (2003) identified a series of features indicative of this interconnectivity. First, the presence of a large, perennial proglacial icing located in the foreland of Fountain Glacier was related to the presence of a persistent spring (Section 5.3.5). Up until 1999, this

(a) Proglacial permafrost

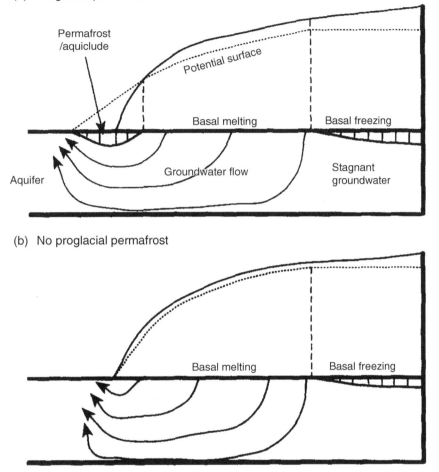

(b) No proglacial permafrost

Figure 4.10 Schematic diagram illustrating the movement of groundwater through an aquifer where proglacial permafrost is present (a) and absent (b) (Boulton & Caban, 1995, p564). Note how the presence of permafrost results in development of artesian porewater pressures in the ice-marginal zone.

spring occurred approximately 50 m beyond the glacier margin indicating groundwater flow through an unfrozen talik with the flow path being constrained by the glacier margin and surrounding proglacial permafrost. The water source was attributed to one or more ice-dammed lakes with the spring discharge varying according to their level. Scheidegger *et al.* (2012) describe a similar situation on Ellesmere Island where a series of relict spring mounds are thought to demonstrate a migration in the location of a glacier-fed spring due to a combination of glacier recession and permafrost aggradation. Second, a series of large relict channels up to several metres in diameter were observed within the lateral ice-cored moraines surrounding nearby Stagnation Glacier. These channels that are thought to have formed englacially have been preserved within the stagnant ice of the lateral moraine that has now become part of the ice-rich permafrost, with the slow creep rates associated with the cold permafrost temperatures suggesting that these inherited subsurface channels could be preserved for a considerable period of time.

Some of the earliest work on permafrost hydrology undertaken by Olav Liestøl (e.g. 1977) on Svalbard clearly demonstrates the landscape-scale hydrological interconnectivity of glaciers and permafrost within polar regions and some of the distinctive features that these connections can create (Section 5.3.5). His pioneering work emphasises how glaciers and permafrost should be conceived of as coupled components of a wider groundwater system rather than as separate entities. In regions such as Svalbard that are characterised by continuous permafrost, polythermal glaciers play a key role in these systems with the areas associated with warm-based conditions providing an important source of groundwater recharge (Figure 4.11). This groundwater can then re-emerge at the surface via springs, particularly within deep valleys where the permafrost is thin and the valley floor is located at an altitude well below the local potential groundwater level. The resultant artesian groundwater conditions can drive the groundwater to the surface through taliks to form springs and in some cases open-system pingos (Section 5.3.5) that primarily occur either in broad valleys or at the base of slopes facing coastal plains.

For water flow to be maintained through an unfrozen talik, sufficient heat must be generated to prevent the freezing of the porewater and the closure of the hydrological connection. In contrast to the 'conductive taliks' encountered in permafrost regions that relate simply to the thermal isolation commonly associated with large water bodies, these 'conductive-advective' taliks are maintained by heat advection related to active groundwater flow. This thermally erodes the permafrost, prevents freeze-back and is promoted by focused flow along discrete pathways such as fault zones (Scheidegger & Bense, 2014; Figure 4.12). The maintenance of flow can also be promoted by the development of high solute concentrations and associated freezing point depression. This can have a range of causes including dissolution of salts from marine sediments, chemical weathering

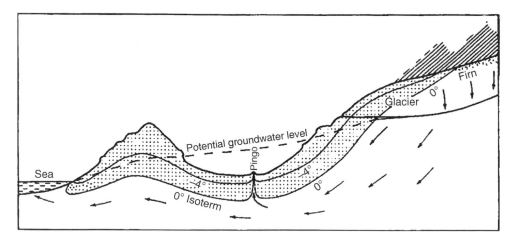

Figure 4.11 Schematic figure of the groundwater flows (shown by the arrows) within a glaciated permafrost environment (Liestøl, 1977, p9). The stipled area illustrates the extent of the permafrost whilst the blank area illustrates the unfrozen zone. Note how the glacier provides the principal source of groundwater recharge whilst pingos can form in the valleys where artesian groundwater pressures occur.

(a) Conductive taliks

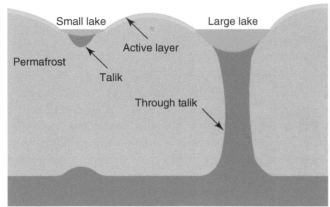

(b) Conductive-advective taliks

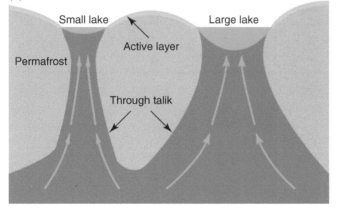

Figure 4.12 Schematic diagram illustrating the key characteristics of: (a) 'conductive taliks' formed through surface insulation and (b) 'conductive-advective taliks' formed via a combination of surface insulation and heat advection from groundwater flow, the creation of which is facilitated by glacier–permafrost interactions (Scheidegger & Bense, 2014, p2).

processes and the partial refreezing of meltwater and the accompanying rejection of solutes (Krawczyk, 1992). This is dramatically illustrated by Blood Falls, an iron-rich spring that discharges subglacial groundwater from the margin of the Taylor Glacier in the Dry Valleys region of Antarctica that is maintained by the partial refreezing (cryoconcentration) of saline porewater (e.g. Mikucki *et al.*, 2004).

The most active terrestrial springs observed in Svalbard that display discharge from subpermafrost, glacially recharged aquifer systems are universally found below the highest postglacial shoreline. This association has led to the suggestion that the springs were established at the end of the last glaciation when isostatic depression resulted in a higher sea level that allowed the establishment of a drainage pathway that may as a consequence be thousands of years old. As deglaciation progressed, the maintenance of flow through the associated groundwater systems allowed the most active springs to persist during a period of rapid permafrost aggradation that resulted in the formation of an extensive aquitard throughout large parts of Svalbard (Haldorsen & Heim, 1999).

The dependency of these sub-permafrost groundwater systems on polythermal glaciers as a key source of recharge results in an intriguing sensitivity to long-term climate change. In a study of two glacially fed groundwater systems near Ny-Ålesund, Haldorsen & Heim (1999) noted that spring discharges have decreased since the Little Ice Age, with the flow of one spring decreasing by approximately 50% since the 1920s. This is thought to relate to the influence of climate change and glacier recession on the basal thermal regime of the glaciers within the groundwater catchment. As the glaciers recede and downwaste, the reduction in ice thickness results in a reduction in thermal insulation and ice-flow velocities. These linked changes in turn result in a reduction in the extent of warm-based ice and eventually a switch from a polythermal to an entirely cold-based thermal regime (e.g. Lovell *et al.*, 2015).

In view of the emerging links between climate change, glacier thermal regimes and the behaviour of groundwater systems in polar environments, a series of recent papers have explored the ways in which these systems might evolve over longer time scales. Following their study of groundwater spring systems in Svalbard, Haldorsen *et al.* (2010) present a conceptual model that describes the changes thought to occur in groundwater systems during a transition from full glacial to interglacial conditions (Figure 4.13). The expansion of glaciers to full glacial conditions results in the development of large areas of warm-based ice. Associated melting of subglacial permafrost and the widespread supply of glacier-related recharge results in the establishment of new groundwater systems with aquifers becoming more extensive during this period. Groundwater flow systems will tend to be localised in valleys where warm-based ice is most prevalent, with discharge from springs initially occurring in a submarine setting and then in a terrestrial setting once glacier recession and isostatic rebound commence. Glacier recession during deglaciation results in permafrost aggradation and a progressive reduction in the extent of warm-based ice that limits the areal extent of both groundwater recharge and discharge. The activity of groundwater systems diminishes, reaching a minimum during the warmest part of the interglacial so long as the permafrost remains intact. Only groundwater springs related to glaciers retaining areas of warm-based ice will persist into the next glacial cycle. The renewed expansion of ice during the subsequent glaciation then results in a rejuvenation and expansion of the groundwater systems to recommence the cycle.

In developing this model, the authors highlight the critical differences between the groundwater systems that occur within glaciated permafrost regions and those found in temperate environments. Of particular relevance is the seemingly counter-intuitive relationship between climate change and the activity of these systems, with climate warming resulting in a reduction in groundwater flow and climate cooling resulting in an increase in groundwater flow. With this in mind, Haldorsen *et al.* (2010) suggest that the existing sub-permafrost groundwater systems on Svalbard should remain active during a period of warming climate so long as conditions do not become warmer than they did during the Holocene climatic optimum. Linking back to some of the earliest work by Liestøl, they hypothesise that a further reduction in groundwater discharges might see springs starting to form new open-system pingos with active flow being maintained by high hydrostatic heads and potentially high dissolved loads (Section 5.3.5).

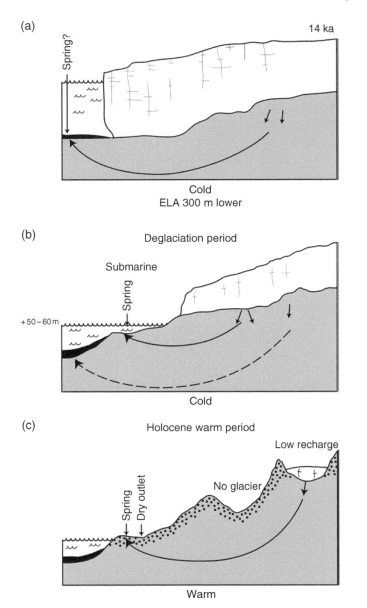

Figure 4.13 General model of the groundwater systems on Svalbard that illustrates their long-term changes in response to climate change (Haldorsen *et al.*, 2010, p400). Stipled area indicates the presence of permafrost. (a) Full glaciation with groundwater discharge into the fjord. (b) Deglaciation and groundwater discharge via a spring located below the highest postglacial shoreline. (c) Holocene interglacial associated with glacier recession, permafrost aggradation and more limited groundwater recharge and spring discharge.

Scheidegger & Bense (2014) use numerical modelling techniques to explore the influence of proglacial lakes on the pathways of groundwater movement during a period of ice-sheet recession and proglacial permafrost aggradation that reveal an additional source of complexity within these systems. Following deglaciation of a landscape, the formation of large proglacial lakes provides potential hydraulic pathways through which glacial groundwater can be discharged. The proglacial lakes initially result in the development of a through talik within the aggrading permafrost that provides a focus for groundwater upwelling, driven by the nearby glacier and related high hydraulic gradient. So long as the upward advective heat flow matches or exceeds the conductive heat flow from the talik to the bounding permafrost, the through talik and groundwater discharge is maintained. A reduction in the hydraulic head associated with continued glacier recession can however result in a reduction in the groundwater discharge and advective heat flow such that the talik shrinks and eventually disappears. This finally results in a rebound in the hydraulic head in the sub-permafrost aquifer as the discharge route is closed.

4.3   The Dynamic Behaviour of Non-Temperate Glaciers

4.3.1   Introduction

Non-temperate glaciers are widely considered to be slow moving and to lack the more dynamic behaviours of temperate glaciers as a consequence of the colder ice temperatures and more limited availability of the subglacial meltwater required to drive basal processes such as basal sliding and subglacial sediment deformation (Section 3.3). Recent field investigations into the basal boundary conditions and processes operating beneath cold-based glaciers resting on permafrost have however demonstrated that basal processes cannot only remain active at sub-freezing temperatures, but that they can continue to provide the dominant mechanism contributing to total surface velocities as in temperate glaciers (Section 3.5.2). This is unsurprising with the preceding section having demonstrated how in some situations, supraglacial meltwater can still access the bed of non-temperature glaciers where it can lead to the generation of transient high basal water pressures that can in turn induce basal motion and a wider dynamic response.

Having considered the nature and importance of basal boundary conditions and basal processes in Chapter 3, this section focuses on the dynamic behaviour of non-temperate glaciers underlain by permafrost. Section 4.3.2 considers the relatively small number of nonetheless informative studies that have been undertaken on modern-day, non-temperate glaciers that have indicated a greater complexity and diversity in their dynamic behaviour than was hitherto assumed to be the case. Section 4.3.3 focuses on the potential relationships between hydrological pathways, basal processes and ice-flow velocities emerging from research on modern-day ice sheets that are considered central to predicting their response to ongoing and future climate change and their contribution to sea level rise. Finally, Section 4.3.4 considers the outputs of ice-sheet models that have been used to elucidate the role and significance of basal thermal regime and glacier–permafrost interactions over larger spatial and longer temporal scales.

4.3.2 Modern-Day Non-Temperate Glaciers

Table 4.1 provides a summary of the velocity measurements of non-temperate glaciers reported within a selection of the published literature. Whilst the number of studies that have been undertaken remains limited in comparison to those for temperate glaciers, the table indicates that they cover a variety of glacier types with different thermal regimes and basal temperatures. This dataset therefore provides a useful insight into the range of velocities that have been recorded, which form the focus for discussion within this section. Where such data are available, velocity measurements have been subdivided according to position (surface or basal) and season (summer or winter).

Early field observations undertaken on the Meserve Glacier in Antarctica in the 1960s and 1970s by researchers from Ohio State University provide one of the earliest detailed investigations into the dynamic behaviour of non-temperate glaciers (e.g. Holdsworth & Bull, 1970; Holdsworth, 1974). As a small valley glacier 7 km long and <100 m thick located in the Wright Valley in the Transantarctic Mountains, this glacier can be considered a cold end member in terms of its thermal regime, with recorded mean basal ice temperatures of c. $-18.8\,°C$. With permafrost being ubiquitous within the McMurdo Dry Valleys (Bockheim *et al.*, 2007), the entire glacier almost certainly rests on permafrost.

Repeat surveys of surface markers yielded horizontal velocity estimates ranging from $0.30\,cm\,day^{-1}$ (c. $1.1\,m\,yr^{-1}$) near the glacier margin to $0.46\,cm\,day^{-1}$ (c. $1.7\,m\,yr^{-1}$) closer to the glacier centreline. The excavation of a 55-m long tunnel into the glacier margin allowed the contribution of deformation within the basal ice to be estimated using strain gauges and micrometers. Measurements within the ice immediately above the bed yielded velocities of $0.68 \times 10^{-3}\,cm\,day^{-1}$ (c. $2.5 \times 10^{-3}\,m\,yr^{-1}$) in an area of clear ice but a much higher value of $13.5 \times 10^{-3}\,cm\,day^{-1}$ (c. $5 \times 10^{-2}\,m\,yr^{-1}$) in an area of comparatively sediment-rich and solute-rich 'amber ice' (see Cuffey *et al.*, 2000). This 20-fold increase in the basal velocities recorded in the amber ice was related to the high solute concentrations that softened the ice and led to its more rapid deformation. Measurement of the deformation of boreholes and tunnels and the closure of drill holes also allowed the authors to constrain the flow law for the cold ice. Ice at c. $-18\,°C$ subject to low shear stresses (<0.5 bar) deformed according to a power law with an n value of 1.6–1.9 in contrast to the typical values of 3–4 that have been determined in relation to Glen's Flow Law from temperate glaciers. This demonstrates the much lower strain rates for a given shear stress and the higher viscosity of cold versus temperate ice (Section 2.5.1). Whilst this initial research was unable to detect any evidence of displacement across the ice-bed interface, subsequent work at the same site in 1995–1996 using more modern equipment recorded a small yet nonetheless non-zero sliding rate of 2–$8\,mm\,yr^{-1}$ (2–$8 \times 10^{-3}\,m\,yr^{-1}$) (Cuffey *et al.*, 1999) (Section 3.5.2).

With basal thermal regime widely being regarded as a key control on the dynamic behaviour of glaciers, a number of studies have suggested that areas of cold-based ice have the potential to influence the behaviour of the ice masses of which they form a part. Particular attention has focused on the velocity changes predicted to occur across basal thermal transitions (BTT) as a consequence of the traditional view that basal processes are by definition inactive beneath cold-based ice that can therefore be considered to be frozen to its bed (Section 3.4.1). The abrupt changes in basal processes, ice velocities and strain patterns

Table 4.1 Velocity measurements recorded at a selection of non-temperate glaciers. In the case of the polythermal glaciers, the reported velocity measurements are reported from areas of cold-based ice. See text for further details.

Name and location	Type	Thermal regime	Basal temperature (°C)	Velocity (m a^{-1})		Source(s)
				Surface[a]	Basal	
Meserve Glacier, Antarctica	Valley glacier	Cold	<−18.8	1.1–1.7	2.5 × 10^{-3b} 2–8 × 10^{-3}	Holdsworth & Bull (1970); Cuffey et al. (1999)
Urumqi Glacier No.1, Tien Shan	Valley glacier	Polythermal		3.7	>3.3^c	Echelmeyer & Zhongxiang (1987)
Storglaciären, Sweden	Valley glacier	Polythermal	−0.022	9.1 (mean Hz)	2.7–4.1	Moore et al. (2011)
John Evans Glacier, Canadian Arctic	Outlet glacier	Polythermal	<−1.75	12.8 mean Hz-winter 19.3 mean Hz-summer		Copland et al. (2003)
Trapridge Glacier, British Columbia, Canada	Valley glacier (surge type)	Polythermal	−1.5 to −3.4	<1–5.7 (pre-surge)		Clarke et al. (1984)
Bakaninbreen, Svalbard	Valley glacier (surge type)	Polythermal	−0.17	<1 (quiescent phase)		Murray et al. (1998); Murray et al. (2000).
Finsterwalder-breen, Svalbard	Valley glacier (surge type)	Polythermal		5–31		Nuttall & Hodgkins (2005)
Erikbreen, Svalbard	Valley glacier	Polythermal		1.8–44.9		Sollid et al. (1994)
Hannabreen, Svalbard	Valley glacier	Polythermal (mostly cold)		4.0–8.4		Sollid et al. (1994)

[a] Hz – Denotes measured horizontal component of total velocity.

[b] Rate of 0.25 m yr^{-1} was recorded in the ice immediately above the bed by Holdsworth & Bull (1970); whilst Cuffey et al. (1999) measured an actual sliding rate of 2–8 mm yr^{-1}.

[c] Cumulative deformation recorded within the subglacial ice-rich sediment.

believed to occur across these transitions have been suggested to have wider implications, with transitions from warm to cold-based conditions leading to deceleration of the glacier, horizontal compression, thrusting and debris transfer (e.g. Hambrey *et al.*, 1999). Focusing more explicitly on the characteristics of the subglacial materials and the potential role played by permafrost, Kleman & Hättestrand (1999) suggest that former thermal transitions beneath Pleistocene Ice Sheets have been recorded within the geomorphological record (Section 5.3.3), arguing for example that,

> . . .*ribbed moraines were formed by brittle fracture of subglacial sediments, induced by the excessive stress at the boundary between frozen- and thawed-bed conditions resulting from the across-boundary difference in basal ice velocity* (p63).

The potential connections between basal thermal regimes and glacier dynamics are dramatically illustrated through their association with states of fast ice-flow. Thermal measurements at the Trapridge Glacier in British Columbia in 1980 and 1981 for example demonstrated that a large wave-like bulge was actively forming at the boundary between warm-based ice (up-glacier) and cold-based ice (down-glacier) (Clarke *et al.*, 1984; Clarke & Blake, 1991). In considering the connections between thermal structure and surges, the authors hypothesised that the cold-based marginal ice would initially increase the resistance to basal sliding thereby promoting thickening of ice in the reservoir region up-glacier. Progressive thickening and an increase in shear stress would subsequently result in increased basal melting and rates of subglacial sediment deformation. This would in turn limit the development of an efficient subglacial drainage system, causing basal water pressures to rise further causing more rapid deformation and establishing a positive feedback loop that could initiate a surge event.

This potential influence of the basal thermal regime on basal boundary conditions, processes and ice dynamics is made most explicitly by Clarke & Blake (1991). In considering the role of the thermal boundary, they initially state that,

> *The collision zone is located at the boundary between unfrozen bed, over which sliding is possible, and frozen bed, over which sliding is inhibited. The cold-based frontal apron acts as a mechanical dam that causes the glacier to thicken up-flow from it* (p162).

It is important to note however that whilst Clarke & Blake (1991) initially hypothesised that a cold-based 'ice dam' may have helped to form the wave bulge, they ultimately concluded that their research findings provided no support for surge mechanisms based around the presence of thermal or mechanical dams.

Using a combination of geophysical surveys and down borehole measurements, subsequent research at Bakaninbreen, a soft-bedded, surge-type glacier location in southern Svalbard that surged in the 1980s, led Murray *et al.* (2000) to conclude that the location of the surge front was indeed thermally controlled and that the propagation of the surge front was associated with significant thermal changes and the degradation of subglacial permafrost. As at Trapridge Glacier, this surge event was associated with the formation of a steep surge front that reached up to 60 m in amplitude and was observed to propagate down glacier at a rate of 1.0–1.8 km yr^{-1} between 1985 and 1989. Radio echo soundings undertaken during the active surge phase demonstrated the existence of an internal reflecting horizon beneath the surge front that was interpreted to indicate the boundary between warm-based ice (up glacier) and cold-based ice (down glacier). Employing a combination of more targeted ground penetrating radar surveys and borehole investigations, the authors provided a valuable insight into the basal boundary conditions occurring both up and down-glacier of the surge front (Figure 4.14). Down glacier of the surge front where the ice was cold-based and comparatively thin (c. 55 m), the authors inferred the existence of a 5–22 m thick layer of debris-rich basal ice that was underlain by a layer of subglacial permafrost 10–15 m thick. Up glacier of the surge front, the glacier was warm based with the lower 31–57 m of the glacier comprising warm ice. The basal zone retained a layer of debris-rich basal ice 7–23 m thick, although the geophysical survey results suggested that this rested directly on unfrozen sediment (i.e. there was no subglacial permafrost present). The surge front was therefore thought to occur at the boundary between warm-based ice with unfrozen subglacial sediment (up glacier) and cold-based ice with subglacial permafrost (down glacier).

Down borehole measurements of the strain rates and inferred viscosities made using tilt cells and plough meters respectively suggested that the thermally defined surge front was also associated with marked changes in the rheology of the subglacial sediment (Porter & Murray, 2001). The strain rates recorded in the sediments immediately below the ice–bed interface were observed to be an order of magnitude higher up-glacier of the surge front in the area of warm-based ice ($0.44 \, a^{-1}$) than they were down glacier in the area of cold-based ice ($0.04 \, a^{-1}$). In addition, the effective viscosity of the subglacial sediment was also observed to be much higher down glacier (3.3×10^{12}–6.5×10^{13} Pa s^{-1}) than they were up glacier (1.1–4.3×10^{10} Pa s^{-1}), corroborating the geophysical interpretation that the sediment was frozen down glacier and unfrozen up glacier of the surge front. In relating these changes in boundary conditions to the surge event itself, the authors concluded that the propagation of the surge front was related to an expansion in the extent of warm-based ice. With the heat generated only being capable of melting a few centimetres or decimetres of ice-rich permafrozen sediment, the authors argued that the active surge was more likely to

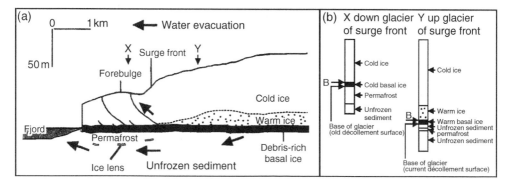

Figure 4.14 (a) Conceptual model of the basal boundary conditions associated with the surge of Bakaninbreen in 1985–1986; (b) Schematic logs through the glaciers at points X and Y (Murray *et al.*, 2000, p13502). Note how the surge front is located at the boundary between warm basal ice and unfrozen sediments (log Y) and cold basal ice and permafrost (log X) in the vicinity of the basal reflector (labelled B). See text for details.

have been associated with sliding or the deformation of a very thin layer of subglacial sediment, with the underlying permafrost acting as an aquitard that helped to maintain high basal water pressures. The subsequent leakage of this high-pressure water through basal freezing, injection into englacial faults or subglacial drainage led to the gradual termination of the surge event by 1994–1995 (Murray *et al.*, 2000).

A recent study specifically examining the influence of cold-based marginal ice on the ice flow of Storglaciären, a polythermal glacier in Sweden, has however challenged the assertion that cold-based ice is capable of adhering to an immobile bed such that it can generate large longitudinal stress and velocity gradients, thereby contributing to the wider debate concerning the connection between a glacier's basal thermal regime and its flow. Moore *et al.* (2011) argue that the basis for such a link is largely conjectural with the detailed empirical evidence being limited to the aforementioned studies at Bakaninbreen and Trapridge Glacier. The dynamic influence of cold ice thought to comprise a marginal zone 50–200 m wide at Storglaciaren has previously been inferred from the observation of an arcuate debris band that dips up-glacier within the glacier's terminus. With GPR surveys suggesting the feature emanates from the BTT, it has been considered the product of thrust faulting caused by the rapid deceleration of the glacier and associated longitudinal compression (Glasser *et al.*, 2003). It is worth noting here that Hambrey *et al.* (1996) similarly reported the occurrence of high-angle thrust faults within and below the surge front at Bakaninbreen containing a combination of fine-grained ice and basal debris. They suggested that the displacement along these thrusts, some of which may have been reactivated from previous surges, allowed the surge front to propagate down ice and to entrain and transfer basal debris to the glacier surface.

In order to test the hypothesis that the BTT is a slip/no-slip transition responsible for the development of substantial longitudinal compression, Moore *et al.* (2011) employed a combination of field measurements and numerical modelling. Thermistor measurements and GPR surveys confirmed the existence of cold-based marginal ice but rather than showing any abrupt changes, measured horizontal ice velocities declined smoothly along the two flow-parallel transects and were observed to remain significant right up to the ice margin, indicating that there was no stagnant ice. Subsequent modelling confirmed that the presence of a slip/no-slip transition should generate a steep decline in horizontal velocities around the position of the BTT as well as a peak in vertical velocities as the active warm-based ice is deflected upwards by the stagnant marginal cold-based ice. Neither of these features were however evident in the field survey data, which was instead most accurately modelled by a smooth reduction in the rates of basal sliding towards the margin. Force-balance analysis based around the surface velocity measurements did provide evidence that the glacier margin was more strongly coupled with the bed than the ice up-glacier. There was however no evidence to suggest that this increased ice-bed coupling was related to a thermal transition from warm to cold ice. Instead, this was related to the presence of an ice-marginal adverse bed slope. The authors therefore concluded that in situations where polythermal glaciers are underlain by unlithified sediments, changes in the stress field, ice-flow trajectories and sediment pathways are more likely to be affected by subglacial topography and hydrology than directly by basal thermal regime.

With this in mind, a series of field studies undertaken on the John Evans Glacier located on Ellesmere Island in the Canadian High Arctic were undertaken in the 1990s and 2000s with the express aim of developing a clearer understanding of the connections between the hydrology and ice dynamics of a non-temperate glacier. This body of work that has arguably not received the recognition it deserves has provided a comprehensive insight into the varied mechanisms that drive distinctive seasonal variations in the hydrological and dynamic behaviour of a non-temperate glacier. The John Evans Glacier is a predominantly cold polythermal glacier characterised by entirely cold ice within the accumulation zone, an extensive basal layer of warm ice averaging 20 m in thickness throughout much of the ablation zone, and cold lateral and frontal margins (Copland & Sharp, 2001). The early part of the melt season is associated with the generation of supraglacial meltwater that either ponds on the surface or flows to the glacier margin. As the meltwater season progresses, the drainage of meltwater into a series of moulins that occur in a crevasse field close to the ice margin results in the activation of englacial drainage pathways. Later in the melt season, the formation of moulins in a second crevasse field located further up-glacier results in up to 40% of the glacier's surface feeding meltwater to the glacier bed with dye tracer results confirming the link between the moulins and a major drainage portal at the glacier margin. This subglacial meltwater influx is initially trapped behind the frozen terminus at the start of the summer before it is released, initially via artesian fountains occurring on the glacier surface or in the foreland, and subsequently as an ice-walled subglacial channel.

Copland *et al.* (2003) coupled an investigation of the discharge variations with a survey of the surface velocities during the course of the 1998 and 1999 ablation seasons to examine the potential connections between the evolutionary changes in the hydrological system with the glacier's dynamic behaviour. These revealed the occurrence of short-term, high-velocity events lasting 2–4 days that were associated with periods of rapid meltwater input to the glacier bed, with horizontal velocities reaching four times their winter values. Early-season events associated with the re-establishment of drainage connections between the supraglacial and subglacial drainage systems resulted in an abrupt increase in meltwater inputs caused by a combination of increased ablation and the drainage of supraglacial and ice-marginal lakes that acted to pressurise the existing subglacial reservoir. The formation of artesian fountains indicated the occurrence of basal water pressures in excess of the ice overburden pressure that caused a reduction in basal friction and an increase in velocity. In addition, a clear spatial association between the areas of the highest velocity anomalies and the locations of artesian flow suggested that the hydraulic forcing was directly responsible for the enhanced basal velocities. A second high velocity event was observed to occur later in the year in 1999 during the middle of the melt season following a period of cold weather and snowfall. With the cold weather resulting in a widespread reduction in subglacial water pressures and a potential reduction in the subglacial drainage capacity, the recommencement of surface melt led to a significant meltwater pulse and a renewed pressurisation of the subglacial drainage system that triggered another increase in the horizontal and vertical velocities in the area associated with the principal subglacial drainage channel.

Follow-up research by Bingham *et al.* (2003, 2006) using similar methodologies examined the spatial and temporal propagation of these short-term, high-velocity events in order to clarify the relationship between the surface dynamics and the changes in the structure of the subglacial drainage system during the course of a melt season. Focussing on the lower ablation zone (LAZ) that is associated with warm basal ice, field investigations in 2000 and 2001 similarly indicated the occurrence of both spring events and mid-ablation season high-velocity events, both of which were associated with periods of vertical uplift. The spring event was associated with an increase in horizontal velocities across parts of the LAZ before a peak in velocities (reaching up to 240% of the mean annual value) propagated across the entire LAZ. This velocity anomaly subsequently dissipated in the lower parts of the LAZ, but was observed to persist in the upper part. The velocity anomalies were also observed to focus increasingly along the glacier's centreline over the course of the melt season. Mid-season high-velocity events were associated with a zone of fast ice flow that propagated down-glacier before causing an increase in surface velocities across the entire LAZ that was particularly pronounced along the glacier's centreline. As with the spring events, these later events resulted in velocity anomalies that persisted for longer in the upper parts of the glacier.

In interpreting these complex spatial and temporal changes in ice dynamics, Bingham *et al.* (2003) provide a conceptual model that illustrates the evolving connections between glacier hydrology, basal boundary conditions and ice dynamics that occur throughout the course of a melt season (Figure 4.15). At the start of the melt season in late May and early June, increased ponding of supraglacial meltwater causes the rapid propagation of crevasses through hydrofracturing that establishes hydraulic connections between the surface and the bed of the glacier (Boon & Sharp, 2003; section 4.2.3). These are subsequently exploited in late June by supraglacial drainage that drains into a cluster of moulins located c. 4 km up-glacier of the terminus. Dye tracer studies during this phase have demonstrated low subglacial water flow velocities of $<0.15\,\mathrm{m\,s^{-1}}$ indicative of a distributed subglacial drainage system with high basal water pressures. Over a period of 2–4 days, this zone of high basal water pressure propagates down-glacier causing the high-velocity anomalies characteristic of the spring events. Associated high vertical velocity anomalies suggest this relates either to the opening of subglacial cavities or to the dilation of subglacial sediment. The subsequent breaching of the thermal dam of cold ice and the release of pressurised subglacial meltwaters from the glacier margin that occurs in combination with a reduction in supraglacial meltwater inputs allows the subglacial drainage system to drain, basal water pressures to fall and the surface velocities to decline over most of the LAZ.

The subsequent occurrence of a summer flow pattern with velocities that are both higher than the annual mean and that feature relatively minor variations is similar to the hydrological forcings observed at many temperate glaciers, indicative of the development of a channelised drainage system. The occurrence of mid-season high-velocity events associated with rapid increases in the inputs of surface meltwater into the moulins however represents another distinctive component of the behaviour of non-temperate glaciers. The influence of exceptionally warm and strong winds in July 2000 for example resulted in a doubling in surface melt rates that exceeded the capacity of the nascent channelised drainage system, forcing subglacial meltwaters into adjacent distributed systems and causing a widespread increase in velocity (Bingham *et al.*, 2006). The localised focus of this high-velocity event around the glacier's centreline however indicates the presence of a large subglacial meltwater channel that is absent during the spring events.

The collation of a rare multi-year dataset of surface velocities at Finsterwalderbreen, a polythermal surge-type glacier in southwest Spitsbergen, provides evidence of variations in glacier flow over a range of different time scales (Nuttall & Hodgkins, 2005). As in the case of the John Evans Glacier, significant seasonal velocity variations were observed during the

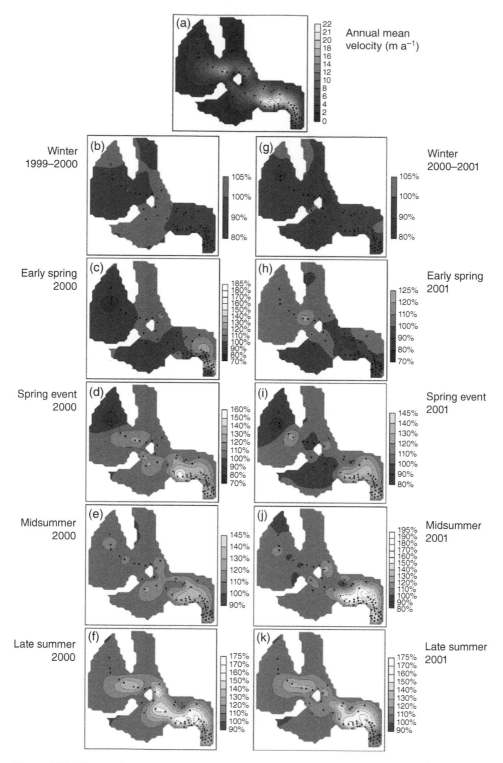

Figure 4.15 Seasonal variations in the distribution of mean annual velocities (m yr^{-1}) measured at the John Evans Glacier from 1999 to 2001 (Bingham *et al.*, 2003, p186). The black dots represent velocity stakes where velocities were measured. (a) indicates the annual mean velocity whilst (b) to (k) illustrate the seasonal velocity variations during 2000 and 2001 expressed as percentages of the mean annual velocity.

measurement period (1994–1997) with the winter velocities typically being only 10–60% of the spring-summer velocities and the velocities peaking in August. These marked variations in seasonal velocities are thought to provide evidence for active basal processes (sliding or subglacial sediment deformation) that are responsible for the enhanced summer velocities. This longer-term study also demonstrates intriguing evidence for interannual variations in velocity with the average velocities in 1996–1997 being up to 75% higher than the preceding years for almost all of the measured survey points. With the years displaying similar weather conditions, these variations are hypothesised to relate to short-term changes in the subglacial drainage system. Unfortunately, the absence of seasonal data for 1996–1997 meant that the authors were unable to determine whether these changes related to enhanced winter or summer velocities.

Research to date therefore indicates an association between glacier thermal regime and ice dynamics that is more complex than previously thought. Rather than directly influencing basal processes, recent work has tended to suggest that a glacier's thermal regime is more likely to play an important indirect role though its influence on the configuration of subglacial drainage systems, which in turn modulates the discharge of water and therefore controls subglacial water pressures. Paired investigations of Erikbreen and Hannabreen, two non-temperate glaciers located within adjacent valleys in northern Spitsbergen, demonstrate both the importance and the varied nature and influence of these connections (Sollid *et al.*, 1994). Whilst they are of similar aerial extent, the greater thickness of Erikbreen has resulted in the presence of more extensive temperate ice than Hannabreen that is largely cold-based. The contrasting thermal regimes have in turn resulted in the establishment of contrasting meltwater drainage systems and dynamic behaviours although in both cases they remain closely interconnected. Erikbreen is associated with more significant fluxes of meltwater to the bed that in conjunction with a hydraulically inefficient distributed drainage system results in the development of high basal water pressures, active sliding and velocities that are both relatively high and seasonally variable. These high velocities promote extensive crevassing that in turn facilitates the drainage of surface meltwater to the bed, establishing a positive feedback loop. With its colder thermal regime, Hannabreen is associated with very limited crevassing and a drainage system that is therefore dominated by supraglacial and ice-marginal flow pathways. Whilst there is evidence that meltwater is nonetheless able to access the glacier bed, the quantities remain insufficient to promote active basal movement. Hannabreen consequently displays slower velocities and lacks the seasonal flow variations of its warmer neighbour.

Recently published companion papers have proposed and tested what is described as the first general theory of glacier surging, capable of explaining ice flow oscillations in both temperate and polythermal settings (Benn *et al.*, 2019a,b). Rather than focusing explicitly on the basal thermal regime or the subglacial drainage system, the novelty of these contributions is their combination as part of a broader consideration of the mass flux and enthalpy budgets. In an incompressible material such as a glacier, enthalpy (internal energy content) is related to the temperature and liquid water content of the ice. Enthalpy can therefore be gained at the glacier bed through a combination of the geothermal heat flux and ice-flow related frictional heating, resulting in an increase in the temperature of the basal ice, thawing and meltwater generation, or a combination of the two. Enthalpy can subsequently be lost via conductive cooling or a loss of meltwater from the glacier system.

A broad balance between the associated enthalpy gains and losses will result in a stable steady state. However, certain drivers or perturbations relating to climate, geometry or the substrate can promote mass and enthalpy cycles that occur in the form of surges. Using their simplest model (excluding surface water inputs) as an illustration, intermediate rates of accumulation are identified as being unstable, with small perturbations and increases in ice thickness resulting in increased heating that cannot be balanced by enthalpy losses, thereby triggering a positive feedback and oscillating behaviour. During the quiescent phase, enthalpy becomes negative with conductive cooling resulting in frozen-bed conditions (and in theory the potential formation of subglacial permafrost). A gradual increase in ice thickness progressively reduces conductive heat losses eventually allowing the bed to thaw and basal meltwater to accumulate. The onset of basal motion then establishes a positive feedback between velocity and frictional heating that triggers a much more rapid increase in basal water storage and sliding velocity leading to a surge event. If surface water inputs are included, an additional crevassing-velocity feedback can allow supraglacial meltwater to reach the bed, augmenting both the rapid increase in enthalpy and the surge velocity. Finally, the termination of the surge is associated with enthalpy losses as the stored water is discharged more rapidly than it can be generated, after which the reduced ice thickness results in increased conductive cooling, a further loss of enthalpy and a return to a quiescent state.

In general terms, glaciers occurring in cold, dry (i.e. permafrost) climatic regimes are considered to display low balance fluxes and rates of enthalpy production that are easily balanced by the conductive heat losses promoted by low mean annual air temperatures (Benn *et al.*, 2019a). The model however correctly predicts that long, low-gradient glaciers (such as the John Evans Glacier) can display surge activity. The connection between surges and climate generates the associated prediction that the spatial distribution of surge type glaciers is likely to shift in response to climate change. Former surge type glaciers in Svalbard for example are highlighted as specific examples whereby an increasingly negative mass balance may preclude the accumulation of mass required to perpetuate the surge cycle such that 'these glaciers are no longer in quiescence, but a climatically induced senescence' (Benn *et al.*, 2019a, p12). This therefore serves as a specific illustration of the broader glaciological changes that are likely to result from ongoing and future climate change and a combination of recession, downwasting and deceleration (Section 6.5).

Rigorous testing of the theory in relation to detailed observations recorded prior to and during a recent surge of Morsnevbreen in Svalbard confirmed the ability of the model to predict the key spatiotemporal changes in elevation, velocity and enthalpy that occurred in five distinct phases during the surge cycle (Benn *et al.*, 2019b) (Figure 4.16). During the quiescent phase (1936–1990), velocities were low, ice thickness changes were driven solely by surface mass balance and enthalpy production increased slowly due to geothermal heating. A transition phase (1990–2010) saw the instigation of basal sliding and frictional heating in the upper reservoir and the onset of mass transfer. Surge initiation (2010–2014) was associated with an acceleration in the sliding-friction feedback with frictional heating rapidly becoming the dominant enthalpy source. In combination with crevasse formation and the transfer of surface water to the bed, this resulted in a rapid increase in sliding velocities in the glacier's central trunk (from <50 to c. 250 m a^{-1}). Peak velocities in 2016 coincided with a period of warm and wet weather increasing surface-to-bed inputs with a barrier of cold-ice limiting discharge. Finally, termination (2016–2019) was associated with the establishment of an efficient subglacial hydraulic connection and the loss of stored water that was initially rapid but which persisted over a multi-year period.

A feature of specific interest that was not predicted by the model was the formation of a 'barrier' of cold-based ice that is believed to have formed across the glacier's lower tongue during the transition phase. With an ice thickness of 175 m corresponding to a balance between geothermal heat gains and conductive heat losses, a reduction in thickness below this value resulted in negative enthalpy and basal freezing (i.e. subglacial permafrost

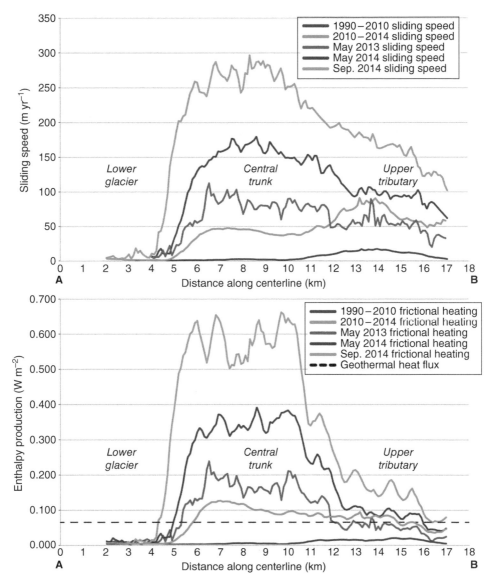

Figure 4.16 Sliding speed and enthalpy production rates at Morsnevbreen, 1990–2014 (Benn *et al.*, 2019b, p12).

formation). Whilst this did not cause the surge (which was triggered in the upper glacier), this thermal barrier and an associated steepening enthalpy gradient (positive up-glacier and negative down-glacier) influenced the evolution of the surge. In acting as an effective hydraulic barrier, the cold-based ice allowed the accumulation of large volumes of pressurised water that corresponded with the surge peak. Subsequent breaching of this dam (which was initially only transient) instigated the surge termination as enthalpy was lost through the discharge of this stored water.

In conclusion, this recent research demonstrates the complex and interconnected nature of the controls on glacier flow. In providing an integrated conceptual framework through which the combined influence of both the thermal and hydrological characteristics of a glacier can be considered, it allows the reconciliation of the seemingly contradictory findings of previous research. In emphasising the central role of enthalpy gradients, the work of Benn *et al.* (2019a, b) helps explain why boundaries in thermal regime are not universally associated with a glaciodynamic response. It is only in situations where such a thermal change corresponds with positive enthalpy and meltwater production up glacier, that its direct influence on subglacial drainage can exert an important yet indirect influence on glacier flow. This demonstrates the potential dangers inherent in overly simplistic assumptions regarding the roles played by basal thermal regime and of the pressure melting point in providing a clearly defined threshold between binary states as previously considered in Section 3.5.

4.3.3 Ice-Sheet Flow and Ice-Stream Dynamics

Ongoing glaciological research considered within the preceding section has clearly indicated that contrary to the traditional assumptions, hydrological connections can be established between supraglacial and subglacial drainage systems in non-temperate glaciers, even when they are largely composed of cold ice. Variations in supraglacial meltwater inputs combined with variations in the efficiency of the subglacial drainage system can consequently drive significant spatial and temporal changes in basal water pressure, resulting in the dynamics of these glaciers being surprisingly responsive to short-term changes in weather as well as longer-term changes in climate.

The publication of a highly influential paper by Zwally *et al.* (2002) postulated the existence of similar connections within much larger ice sheets that if proven, provides a mechanism whereby climate change and an increase in supraglacial melt rates can directly influence ice-flow velocities and subsequent rates of glacioeustatic sea-level rise. Using GPS measurements to derive surface velocities at a site located close to the equilibrium line of the west-central Greenland Ice Sheet, the authors recorded a summer acceleration in surface velocity that corresponded with the onset of surface melting and a subsequent deceleration that followed its cessation. They also demonstrated that the summer accelerations varied in magnitude according to the intensity of surface melting, with the highest rates of surface melting resulting in the most rapid accelerations. The demonstration of correlations between surface melt rates and ice-flow velocities combined with the observation of moulins in the vicinity of the study site provides two lines of evidence for the development of hydrological connections between the supraglacial and subglacial zones, with increased summer melt causing an increase in basal water pressures and

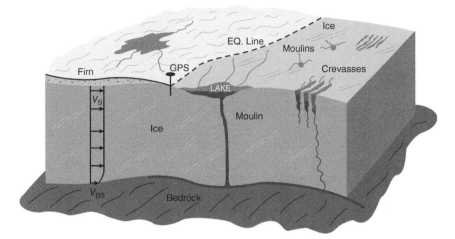

Figure 4.17 Schematic figure of the glaciological features occurring in the equilibrium and ablation zones and the nature of the hydrological connections between the ice sheet surface and the subglacial zone (Zwally *et al.*, 2002, p219).

sliding rates (Figure 4.17). Whilst the evidence provided for the linkages between surface melt and subglacial processes is indirect, this research clearly demonstrates the potential for the mechanisms that have been well studied on smaller ice masses to be applied at the scale of continental ice sheets.

This publication has helped to establish a new research agenda and a series of follow-up investigations that have sought to clarify the nature of the hypothesised connections between surface hydrology, subglacial drainage networks and ice dynamics. A series of research investigations focusing on the catchment areas of the Russell and Leverett Glaciers in Western Greenland have for example revealed hydromechanical responses for the Greenland Ice Sheet that are remarkably similar to those documented for much smaller glaciers in the previous section. In presenting ice velocity data for the 2008 melt season along a 35-km long GPS transect extending from the margin to the upper parts of the ablation area, Bartholomew *et al.* (2010) demonstrated that each of the four recording stations exhibited significant seasonal velocity changes, with a rapid acceleration in horizontal flow velocities to over double those of their winter values being followed by a more gradual decline to velocities that dropped below their winter values by the end of the melt season. The highest horizontal flow velocities coincided with the periods of most rapid surface uplift with subsequent periods of surface stability or lowering resulting in a reduction in horizontal velocities. In addition, this acceleration and the subsequent slowdown were observed to migrate up-glacier through the catchment over time.

These data led the authors to argue that the mechanisms involved in driving seasonal variations in motion are in essence identical to those previously described at alpine and High Arctic polythermal glaciers. Peak velocities relate to meltwater inputs to a glacier bed that is initially characterised by hydraulically inefficient drainage pathways, resulting in pressurisation, uplift and basal sliding. The progressive evolution to more efficient channelised systems subsequently results in a reduction in basal water pressures and a fall in velocity. Increases in sliding rates are then restricted to events associated with very high meltwater inputs that overwhelm the higher capacity drainage system, relating for example to the drainage of large supraglacial or ice-marginal lakes. Increases in velocity at one site were however not connected with the acceleration of the upstream sites, with an absence of longitudinal coupling at scales of >10 km suggesting that the enhanced velocities are primarily the result of localised hydrological changes.

Subsequent research has employed new tracers in efforts to provide more direct evidence for the changes in the subglacial drainage network considered responsible for the observed seasonal changes in velocity. Chandler *et al.* (2013) injected rhodamine and sulphur hexafluoride tracers into a series of moulins located between 7 and 57 km up-glacier of the margin of the same West Greenland catchment over three melt seasons. As in the preceding studies, the results demonstrated increasing drainage velocities during the course of the melt season consistent with an evolution from a slow and inefficient distributed drainage system to a fast and efficient channelised drainage system, with the recorded velocities being equivalent to those observed at temperate valley glaciers. The progressively later onset of increased tracer velocities with increasing distance up-glacier of the margin also suggested that the channelised drainage system propagates up-glacier during the course of the melt season. The absence of any increase in tracer velocities from the highest moulin located 57 km up-glacier of the margin suggests that the efficient drainage system may not have been able to extend this far up-glacier. In summary, the authors concluded that these direct measurements provide strong support for the use of surface velocities along flow lines to infer changes in basal hydrological conditions.

In considering this recent work, it is interesting to note that the discussions surrounding the causes of the seasonal ice-flow variations are limited to a consideration of the role played by changes in the subglacial drainage system. No measurement or explicit mention is made of any associated changes in the basal thermal regime. This appears surprising with the smaller-scale studies reported in the previous section suggesting that the basal thermal regime influences both the nature and efficiency of the subglacial drainage system, thereby providing an important contributory cause for the hydraulic forcing. It may be that thermal controls are less consequential in an ice-sheet setting, but in the absence of any such information, the potential role played by basal thermal conditions and glacier–permafrost conditions is as yet unclear. Nonetheless, this work does illustrate that the hydromechanical connections responsible for seasonal fluctuations in the surface velocity can operate in a broad range of glaciological settings.

Research focusing on the dynamic behaviour of Antarctic Ice Streams and in particular on the mechanisms responsible for their switching on and off suggests that basal thermal regime remains an important control on basal conditions and processes. As discussed previously in Section 3.5.4, investigations into the basal freezing and ice accretion occurring beneath stagnant ice streams led Christoffersen & Tulaczyk (2003a,b,c) to develop a model relating subglacial temperatures to ice-stream dynamics through its influence on basal ice formation and till rheology. In comparing borehole observations from two ice streams of the West Antarctic Ice Sheet, the active Whillans Ice Stream (formerly Ice Steam B) and the slow-moving Kamb Ice Stream (formerly Ice Stream C), Christoffersen & Tulaczyk (2003a) observed that the latter was characterised by (i) comparatively steep basal temperature gradients, (ii) a significant debris-rich basal ice layer c. 12–25 m thick, (iii) unfrozen till displaying supercooled temperatures (as low as −0.35 °C) and (iv) reduced till porosities close to the ice–sediment interface. They hypothesised that whilst the rapid motion of the Whillans Ice Stream was promoted by active basal melting and reduced effective pressures and shear strengths within the subjacent till, the stagnation of the Kamb Ice Stream was conversely associated with basal freeze-on and an associated consolidation and strengthening of the subglacial till.

In considering the liquid water content of the subglacial tills, Christoffersen & Tulaczyk (2003a,b) utilised theoretical treatments of frost heave commonly employed by permafrost engineers, notably the Clapeyron equation that provides a means of coupling pressure and temperature in freezing porous media (see Rempel 2007 and Christoffersen et al., 2006 for detailed discussions of the frost-heave modelling). In addition, whilst ice–water phase changes are often considered to be related solely to temperature and pressure, Christoffersen & Tulaczyk (2003b) also considered the roles played both by solute concentrations and interfacial effects. The latter are particularly important in fine-grained porous media such as the clay-rich till sampled beneath the nearby Whillans Ice Stream (Tulaczyk et al., 1998), in which the curvature of ice–water interfaces can lead to high interfacial pressures, a substantial reduction in the freezing point and the presence of supercooled water (Section 2.2.2).

Christoffersen & Tulaczyk (2003b) provide modelling results for both high interfacial pressures (i.e. fine-grained till) and low interfacial pressures (i.e. coarse-grained till) under two hydrogeological settings (closed and open-water systems). In the case of closed-water systems, freeze-on extracts pore water, resulting in till consolidation, increased shear strength and slower ice velocities. For both high and low interfacial pressures, till strength matches the driving stress after 65 years and the ice stream shuts down. Subsequently, the freezing front migrates into the subglacial till causing further till consolidation and the development of segregated ice lenses, whose thickness and spacing are determined by depth and by surface tension (i.e. particle size). In the case of open-water systems, the timing of ice-stream shutdown and the spacing of ice lenses are similar, although the individual ice lenses are thicker and the degree of till consolidation is more limited. Christoffersen & Tulaczyk (2003c) subsequently applied this model to the investigation of Pleistocene Ice Streams, suggesting that textural changes observed within a till sequence relating to the Baltic Sea Ice Stream in Denmark reflected ice-stream stoppage, basal freeze-on and till dewatering.

These model predictions are consistent with subsequent empirical observations with Vogel et al. (2005) observing that a basal ice layer 10–15 m thick beneath the Kamb Ice Stream featured alternating clear ice and debris-rich ice layers, with the latter increasing in thickness and sediment content towards the ice–sediment interface. The transitions from relatively clean ice to sediment-rich ice are proposed to reflect a progressive change from abundant water availability to limited water availability that culminated in a shutdown of the ice stream approximately 300 years ago. The presence of decimetre thick layers of clear ice close to the ice–sediment interface and a basal water layer are conversely thought to indicate imminent ice stream reactivation, with the current water supply exceeding that consumed by basal freeze-on, leading to a re-lubrication of the bed. In responding to and recording changes in the basal boundary conditions, it is possible that variations in the debris distribution within the basal ice layer therefore provide a potential archive record of the complex flow history of the ice stream (Engelhardt & Kamb, 2013).

Looking to the future and acknowledging the importance of subglacial hydrology, Bougamont *et al.* (2003) incorporated a basal drainage system into a model of the Whillans Ice Stream in order to determine whether its slowdown may result in its subsequent shut down, a situation that previously occurred in the neighbouring Kamb Ice Stream. Whilst the modelling results invariably saw a slowdown being followed by a complete shut down within 100 years, the authors suggest that this can be prevented by the delivery of significant amounts of meltwater (equivalent to 2–3 mm a^{-1}) to the base of the ice stream, the freezing of which can provide the heat flux required to prevent freezing of the subglacial till. This could be delivered either from subglacial sources up-glacier or from supraglacial sources as proposed by Zwally *et al.* (2002) and others, highlighting the need for further empirical research to constrain the nature of these drainage systems.

4.3.4 Pleistocene Ice Sheets and Cold-Based Ice

Having looked at the dynamic behaviour of modern-day glaciers and ice sheets, this section briefly considers a selection of studies that have considered the potential influence of basal thermal regime and cold-based ice on the dynamic behaviour of Pleistocene Ice Sheets. These studies are based primarily if not exclusively around the application of numerical ice-sheet models and as such are almost invariably associated with a series of assumptions and simplifications, the validity of which have been questioned in Section 3.5. Nevertheless, this work provides the opportunity to consider the significance of glacier–permafrost interactions on the behaviour of entire ice sheets over longer timescales and potentially entire glacial cycles.

As discussed previously in Section 2.4.2, the application of numerical ice-sheet models to both modern and ancient ice sheets has indicated that cold-based ice and subglacial permafrost are likely to have been extensive, particularly during the growth phase of large ice sheets. In concluding that 60–80% of the Laurentide Ice Sheet was cold based and 'frozen to the bed' at the LGM, Marshall & Clark (2002) suggested that these thermal conditions were in fact central to its preceding long-term growth. The crossing of an intrinsic thermal threshold associated with an increase in geothermal and strain heating subsequently led to a rapid expansion in the extent of warm-based conditions and the ice sheet's rapid collapse.

The development and utilisation of thermomechanical ice sheet models are based around the recognition of a fundamental coupling between the thermal and rheological properties of ice masses, such that small perturbations in ice flow can be magnified by creep instability feedbacks to create areas of fast ice flow (Section 2.5.1). In modelling the Scandinavian Ice Sheet using a thermomechanical model over a 150 ka time period to steady state, Payne & Baldwin (1999) observed the formation of cold-based ice under a central dome and a series of ridges radiating out from this area. Outside of this region, the ice sheet was characterised by warm wedges of ice extending 600 km inland of the margin that represented areas of fast ice flow. These resulted in the drawdown of ice and a further concentration of the ice flow. In robbing the adjacent ridges of ice, these areas become progressively colder and slower flowing, providing areas in which cold-based ice and by inference subglacial permafrost has the potential to persist throughout the majority of a glacial cycle. The authors were consequently able to demonstrate the formation and differentiation of regions of fast and slow moving ice with similar spacings and flow patterns to those observed in the geomorphic record through the influence of internal feedbacks alone.

In using a thermomechanically coupled numerical ice sheet model to simulate the evolution of the British Isles Ice Sheet during the last glacial cycle, Boulton & Hagdorn (2006) simulate the combined influence of basal thermal regime and substrate type on the dynamic behaviour and morphology of the ice sheet. Model runs prescribing uniform soft-bed conditions demonstrate the importance of thermo-mechanical coupling, with fast flow advecting cold ice from the summit area that in turn reduces creep rates and steepens the ice sheet so as to maintain high summit elevations (Section 2.3.2). The inclusion of a décollement parameter and the prescription of assumed 'slippery' patches in areas featuring significant accumulations of soft sediment allowed the authors to simulate the effect of spatial variations in boundary conditions (Section 3.3.1). With areas of basal freezing being associated with strong ice-bed coupling and high yield stresses, the formation of fast-flowing ice streams that control the flow and form of the ice sheet are limited to areas of warm-based ice with the velocities being proportional to the predicted melt rate. In contrast, cold-based conditions are by implication associated with slow flow and limited ice fluxes more likely to facilitate ice sheet accumulation, playing a key role for example in the initiation of the ice sheet over the high ground of Scotland. Cold-based ice is also considered to play an important role during the LGM, both in influencing ice stream location and defining their slow moving boundaries, as well as through the formation of an extensive peripheral zone associated with debris entrainment (Section 3.5.4) and the formation of extensive stagnation terrain (Section 5.3.2).

It is important to note that the general picture of cold-based ice playing a key role in promoting (i) ice accumulation in upland areas during the initial phases of ice-sheet development and (ii) debris entrainment and moraine formation in the marginal regions during deglaciation is broadly consistent with the glaciological inferences of geomorphological inverse models (e.g. Sollid & Sørbel, 1994). In using landform evidence from both modern and Pleistocene ice sheets to infer their subglacial thermal organisation, Kleman & Glasser (2007) identify evidence for frozen-bed patches at a range of spatial scales with the largest examples occurring in ice-dispersal centres and areas of high ground. In areas associated with active ice streams ('ice-stream webs'), these patches are considered an integral component with the fast flowing ice 'shifting between alternative routes through the maze of frozen-bed patches over time' (p588). In constraining the lateral margins of the ice streams and isolating them from the peripheral drainage basins, frozen-bed patches are therefore argued to have played a key role in promoting the stability of the parent ice sheets.

4.4 Sediment Fluxes Within Glaciated Permafrost Catchments

Examining the sediment pathways and fluxes through glaciated drainage basins containing permafrost provides the opportunity to integrate the preceding considerations of the thermal, dynamic, hydrological and physical characteristics of non-temperate glaciers and to consider the influence of glacier–permafrost interactions at the catchment scale. The geomorphological significance of these complex interactions is clearly emphasised by

Etzelmüller & Hagen (2005) who state that 'the distribution of sub-zero temperatures within glacier bodies and in surrounding landscapes is crucial for understanding geomorphological processes with respect to sediment production, transport and deposition in modern Arctic and high-alpine glacierized catchments' (p23). Their conceptual modelling of these catchments as 'sediment cascades' provides a useful systems-based approach to their investigation via the systematic identification of distinctive inputs (weathering and erosion), throughputs (entrainment and sediment transfer) and outputs (deposition and storage) and of the associated sediment stores and transport processes (Figure 4.18). This in turn accentuates the key differences between polythermal and temperate glacier systems and the elements of the system directly influenced by permafrost. It also once again highlights the need to consider glaciers and permafrost from a more holistic viewpoint as parts of an integrated system.

Focusing initially on the processes of sediment production, the sediment cascades of both polythermal and temperate glaciers located in mountainous environments are fed by a combination of subglacial sediment inputs relating to the processes of erosion and external or 'extraglacial' inputs associated primarily with weathering and mass movement on the slopes above the glacier surface (limited to sporadic nunataks in ice-sheet settings). Whilst the generalities are the same, cold climatic regimes are commonly thought to promote higher rates of extraglacial sediment production through mechanical weathering processes (e.g. Etzelmüller, 2000). Hall *et al.* (2002) caution however against the uncritical acceptance of what they consider to be questionable assumptions regarding cold-climate weathering that have persisted within the literature for over a century, most notably the presumption that freeze-thaw weathering is of prime importance. In specific relation to permafrost environments however, they acknowledge the effectiveness of ice segregation operating within a 'frozen fringe' at the boundary between the active layer and the subjacent permafrost in brecciating bedrock (Walder & Hallet, 1985) (Figure 4.19). Physical modelling involving the bidirectional freezing of wet chalk blocks by Murton *et al.* (2006) has provided clear empirical evidence for this mechanism in the progressive development of micro and macro cracking within the upper part of the permafrost layer during simulated seasonal freeze-thaw cycles. This developed a layer of ice-rich fractured bedrock similar to that observed within frost-susceptible lithologies in Arctic permafrost regions (limestones, shales and sandstones) with fracture patterns resembling those commonly found in the upper 0.5 m to several metres of bedrock affected by Pleistocene permafrost (e.g. Murton, 1996). The horizon of segregated ice formation was also shown to be highly thaw sensitive with an increase in simulated summer air temperatures leading to its rapid melting. In addition to promoting subaerial weathering, the cold climatic conditions associated with polythermal glaciers and subglacial permafrost are commonly associated with erosion through plucking or block removal that are facilitated by cold-based ice (Section 3.4.2) and capable of providing an input of coarse material. In contrast, the production of fine-grained material associated with abrasion and subglacial shearing beneath temperate glaciers (e.g. Boulton, 1978) is considered limited if not entirely absent beneath cold-based ice (Section 3.5.3).

Returning to the sediment cascade model of Etzelmüller & Hagen (2005) (Figure 4.18), the aforementioned processes can lead to the development of a subglacial sediment stores that can be entrained into the basal zone of the glacier. In addition, extraglacial sediment can be transferred to the glacier margin via lateral (and potentially medial) moraines. As

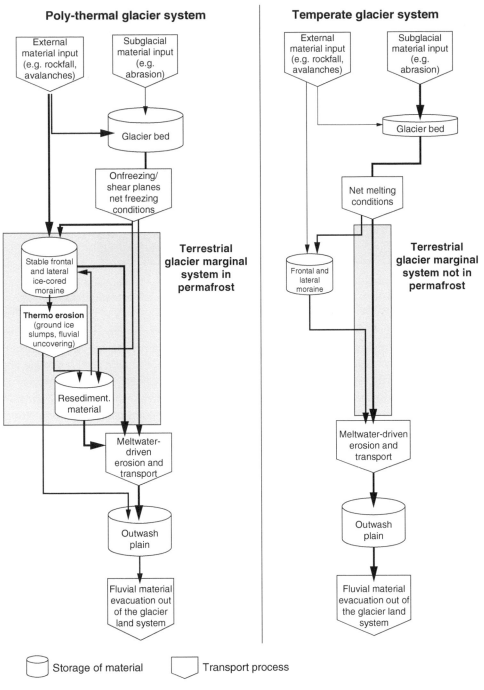

Figure 4.18 Schematic figure illustrating storage and transfer of sediment through both polythermal and temperate glacier systems (Etzelmüller & Hagen, 2005, p22). The grey shaded areas represent the glacier body whilst the thickness of the lines symbolises the relative amounts of sediment transport. Note how ice-marginal sediment transfer in the polythermal marginal setting (grey area) involves significant transport through ice-cored moraine systems with permafrost. In contrast, sediment evacuation from temperate glaciers is dominated by glaciofluvial systems. See text for details.

Figure 4.19 Segregated ice lens located in gneiss bedrock beneath the margin of Fountain Glacier, Bylot Island. Note the network of fractures above and to the left of the ice lens. Coin for scale (Image: Richard Waller).

discussed in the previous chapter, basal thermal regime is widely regarded as an important control on the processes, pathways and quantities of material entrained and transported by glaciers (e.g. Bogen, 1996; Hallet *et al.*, 1996; Hambrey & Glasser, 2012). In considering some of the key distinctions, temperate glaciers are generally considered to exhibit thin basal ice layers with limited basal debris loads due to the predominance of basal melting, with surging glaciers associated with the marginal tectonic thickening of basal ice providing one notable exception (e.g. Sharp *et al.*, 1994). In contrast, polythermal glaciers terminating in areas of permafrost that are associated with thermal transitions from warm-based interiors to cold-based margins commonly feature basal ice layers several metres in thickness associated with the mass transfer and net basal adfreezing of subglacial meltwater (Section 3.5.4). In calculating the discharge of debris through basal ice in Western Greenland for example, Knight *et al.* (2002) describe a basal ice sequence of up to 30 m in thickness at the margin of the Russell Glacier featuring volumetric debris concentrations of up to 50%. This provides a debris flux estimated at between 12 and 45 m^3 m^{-1} a^{-1} that is an order of magnitude higher than values reported from temperate valley and outlet glaciers in Norway, Iceland and Switzerland.

Even entirely cold-based glaciers are capable of entraining basal material to generate surprisingly significant basal debris loads, although the slow ice velocities mean that the debris fluxes are thought to remain very low. An additional distinction relating back to the absence of abrasion and comminution is a tendency for the material to inherit and retain its original characteristics and to experience little if any modification during transport (e.g. Hambrey & Fitzsimons, 2010). In reviewing the key characteristics of Antarctic glacigenic sediments, this led Licht & Hemming (2017) to observe that 'glacigenic sediments generated in arid polar environments are largely products of mechanical breakdown of bedrock or remobilisation of unlithified deposits in the subglacial environment' (p5) (see Section 5.4.1).

It is the formation of extensive tracts of ice-cored moraine at the frontal and lateral margins of polythermal glaciers terminating in permafrost that are widely considered to provide one of the most influential and distinctive elements of their sediment cascades (Figure 4.18) (see Sections 5.3.2 and 5.6.3 for more detailed coverage of their formation and degradation, respectively). Whereas temperate glaciers are conceptualised as involving the simple transfer of material from ice-marginal moraine and meltwater systems, the ice-cored moraines characteristic of the termini of polythermal glaciers constitute what Etzelmüller (2000) refer to as 'sediment magazines'. In other words, they constitute significant transient stores of sediment the release of which is governed by the climatic, hydrological and wider environmental conditions.

The stagnant debris-rich ice that provides the key sediment store within these terrestrial glacier marginal systems can itself be regarded as a legacy of past environmental conditions and of larger Little Ice Age glaciers with more extensive warm-based ice and higher rates of erosion and subglacial sediment delivery to the marginal zone where it was entrained via basal adfreezing (e.g. Lovell *et al.*, 2015). Subsequent permafrost aggradation accompanying glacier recession has the potential to promote the longer-term preservation of these sediment stores over centennial or possibly millennial timescales (Section 5.3.4). There is increasing evidence however that a complex suite of processes related to ongoing climate change are destabilising these stores that have previously been relatively stable (Section 5.6.3). The exposure and thermal erosion of these marginal ice-cored moraines by fluvial drainage systems in particular can dramatically augment the sediment supply into the forefield, thereby compensating for the reduced subglacial sediment supply discussed previously in Section 4.2.4. Following the study of four proglacial areas in Svalbard, Etzelmüller (2000) for example concluded that the mobilisation of sediment by thermal erosion from these moraine tracts alone was of the same order of magnitude as the total annual suspended sediment transfer from the entire upstream catchments. The final stages of the cascade proposed by Etzelmüller & Hagen (2005) involve the temporary storage of material in the proglacial outwash plain prior to its subsequent evacuation from the glacial landsystem by fluvial processes. Hodgkins *et al.* (2003) have shown that these zones can act as both sources and sinks of sediments during individual melt seasons and between different years.

The ability of proglacial fluvial systems to export substantial amounts of material and to provide an aggregated estimation of sediment yields from glaciated drainage basins has resulted in the measurement of associated sediment fluxes becoming a key focus of studies seeking to determine the sediment yields and erosion rates of glaciated regions more broadly. An influential paper by Hallet *et al.* (1996) compiled the empirically determined sediment yields from 62 study sites in glaciated catchments encompassing Alaska, Norway, Svalbard, the Swiss Alps, Central Asia, Canada, Greenland, New Zealand and Iceland. In addition to confirming that glaciated drainage basins are generally associated with significantly higher sediment yields than their non-glaciated equivalents, the review estimated effective rates of glacial erosion that varied by several orders of magnitude from 0.01 mm yr^{-1} for polar glaciers and thin temperate plateau glaciers on crystalline bedrocks, to 0.1 mm yr^{-1} for temperate valley glaciers on crystalline bedrocks, to 1.0 mm yr^{-1} for temperate valley glaciers on more erodible substrates, peaking at $10–100 \text{ mm yr}^{-1}$ for large and fast-moving glaciers in the tectonically active mountains of south-east Alaska. Whilst the relationships between sediment yields and erosion rates and basal thermal regimes are not entirely clear within the data employed, non-temperate glaciers are considered to provide some of the lowest values with the authors repeating the traditional assumption that, 'erosion of any kind vanishes as basal temperatures drop well below the pressure melting point' (p231). In reconsidering the empirical basis for the modelling of glacial erosion rates, a recent paper by Cook *et al.* (2020) has in contrast emphasised the role of precipitation, with this parameter providing a stronger (positive) relationship with erosion rates than temperature. A combination of precipitation and temperature provided the strongest relationship with erosion rates at the Meserve Glacier providing a notable outlier with very low values resulting from its location within a polar desert environment.

A detailed examination of the specific influence of glaciers on the sediment yields of High Arctic catchments is provided by Bogen & Bønsnes (2003) who measured and compared the suspended sediment fluxes of one unglaciated and two glaciated catchments in Svalbard over several years. The unglaciated catchment was characterised by the lowest mean sediment yield of $82.5 \, t \, km^{-2} \, yr^{-1}$, which was nonetheless significantly higher than those recorded from unglaciated catchments in mainland Norway, reflecting the influence of sparse vegetation and elevated rates of bedrock weathering. The mean values recorded from glaciated catchments (Bayelva and Endalselva) where significantly higher ($359 \, t \, km^{-2} \, yr^{-1}$ and $281 \, t \, km^{-2} \, yr^{-1}$ respectively) with particularly high sediment yields being derived from the glaciers and ice-cored moraines ($586 \, t \, km^{-2} \, yr^{-1}$ and $1077 \, t \, km^{-2} \, yr^{-1}$ respectively). Significant interannual variations in the sediment yields reflected differences in the magnitude and frequency of summer storms that resulted in abrupt increases in the suspended sediment concentrations, with an event in September 1990 generating a value of $4000 \, mg \, l^{-1}$ in the Bayelva catchment (containing the glaciers Austre Brøggerbreen and Vestre Brøggerbreen), an order of magnitude higher than the usual range of $100–300 \, mg \, l^{-1}$. These sediment pulses were caused by the flushing of sediments from the glaciated parts of the catchments with the observed release of turbid meltwater from the glacier margins coupled with a strong correlation between discharge and suspended sediment concentrations indicating a plentiful sediment supply and the inference that the subglacial drainage systems in the largely non-temperate glaciers were still active. In considering the influence of basal thermal regime, whilst the sediment yields remained at the lower end of the range, the glaciated Svalbard catchments displayed a similar range of values to those derived from catchments in mainland Norway associated with temperate glaciers (Figure 4.20). This highlights the importance of bedrock characteristics and the weathering rates, which can generate a plentiful sediment supply capable of compensating for the limited basal sliding rates and ice velocities.

An earlier study focusing on Scott Turnerbreen, a small glacier with an entirely cold thermal regime, similarly identified an increase in sediment availability throughout the meltwater season. Hodgkins (1996) argued however that its evacuation via well-developed subglacial meltwater channels was unlikely and that supraglacial and ice-marginal channels provided more plausible sediment transfer pathways. As with Etzelmüller (2000), marginal ice-cored moraines were viewed as the dominant sediment sources with persistent mass wasting and an increase in the fluvial reworking of material explaining an observed increase in sediment availability during the melt season. An important additional control worth emphasising is the significance of landscape history and the influence that past

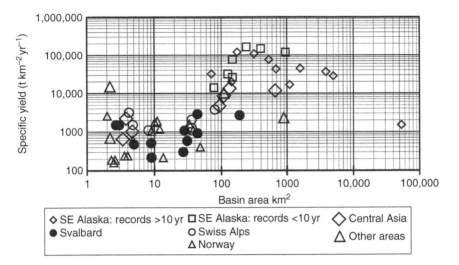

Figure 4.20 Specific sediment yields from a range of glaciated catchments plotted against basin area (Bogen & Bønsnes, 2003, p186). Includes data originally provided by Hallet (1996).

processes can have on modern-day sediment availability. Hodgkins (1996) concluded that the suspended sediment load of the proglacial stream of Scott Turnerbreen 'is not indicative of glacial erosion since the glacier is frozen to its bed, but is derived from the sediment stored following glacial erosion in an earlier time period' (p151). This highlights the need for caution when using proglacial sediment fluxes as a means for estimating modern-day rates of glacial erosion. It also emphasises the influential role played by paraglacial processes and the de-icing of ice-cored moraines that represent the combined product of deglaciation and permafrost dynamics that are explored in more detail in Section 5.6.

4.5 Summary

- The hydrological systems of non-temperate glaciers are commonly considered to be dominated by supraglacial and lateral drainage pathways with the development of subglacial drainage being limited by the short melt season and a perceived inability of surface meltwater to access the bed (due for example to a lack of crevassing).
- There is growing evidence however that subglacial meltwater activity beneath non-temperate glaciers is more commonplace than previously thought, particularly in relation to polythermal glaciers with significant areas of warm-based ice. Surface meltwater has been observed to access the bed though cold ice via the processes of cut and closure and hydrofracturing whereby the formation of supraglacial lakes results in cracking and the formation of new drainage pathways along pre-existing structural discontinuities. Subglacial meltwater drainage has consequently been observed beneath glaciers that are almost entirely cold-based.
- Recent research has also acknowledged the important role played by subglacial groundwater systems that transfer water from areas of recharge (warm-based ice beneath

polythermal glaciers) through subsurface aquifers and taliks into glacier forelands to produce perennial springs that can create large icings and open-system pingos. The activity of these systems varies in response to the extent and thermal regime of the glacier that provides the water source, with emerging evidence that current glacier recession is causing these systems to shut down as the areas of warm-based ice become more restricted in extent.

- Where subglacial drainage does occur, suspended and dissolved loads can be surprisingly high. High-suspended sediment loads are promoted by the development of distributed drainage systems that in combination with available subglacial sediment stores, mean that these systems do not necessarily display the progressive exhaustion of supply through the melt season characteristic of temperate glaciers. High solute levels are promoted by the combination of slow transit times through the subglacial zone coupled again with the widespread availability of subglacial sediments. Solute concentrations associated with the onset of the melt season and the drainage of subglacial waters stored over the winter have been observed to be particularly high.

- The temporal evolution of non-temperate glacier drainage systems is initially associated with the supraglacial ponding of water before it is able to establish hydraulic connections with the bed. Water can then be temporarily stored at the glacier bed, generating high water pressures illustrated by the formation of supraglacial fountains prior to the re-activation of subglacial drainage routes that typically occurs as a result of pronounced spring snowmelt flooding occurring a few weeks after the start of the melt season. Summer storm events can also result in the over-pressurisation of subglacial drainage systems.

- The development of high basal water pressures during both the spring snowmelt season as well as other periods of elevated meltwater inputs (e.g. warm and wet weather) coupled with hydraulically inefficient subglacial drainage systems can have a marked influence on dynamic behaviour. Variations in subglacial meltwater fluxes result in non-temperate glaciers frequently displaying significant seasonal velocity variations with summer velocities being higher than the winter velocities and peak velocities occurring at the start of the snowmelt season. The observation of close connections between meltwater inputs and surface velocities beneath glaciers and ice sheets indicates that basal sliding and subglacial sediment deformation can remain active beneath cold-based ice.

- Basal thermal conditions are still considered an important control on glacier flow. However, the evidence for a clear and direct connection between basal thermal regime and glacier velocity relating to the activation and cessation of basal processes around a clearly defined temperature threshold remains subject to debate. Recent research indicates that the influence of basal thermal regime and therefore of subglacial permafrost is more likely to be indirect in nature via its influence on the nature and evolution of the subglacial drainage systems that dictate basal water pressures and in turn glacier velocity.

- The velocity of non-temperate glaciers is therefore now understood to be far more responsive to variations in weather, climate and meltwater inputs than previously thought. The investigation of connections between surface melt and ice dynamics in major ice sheets as a critical link between climate change, glacier flow and sea-level change now constitutes a major current research focus in glaciology. Whilst the specific role of basal thermal regime remains unclear, its influence on till consolidation provides a potential explanation for ice-stream shutdowns. The recent use of enthalpy gradients has provided an exciting new conceptual framework through which these complex and seemingly contradictory associations between climate, thermal regime, glacier hydrology and dynamic behaviour can be better understood.

- In the context of Pleistocene Ice Sheets, cold-based ice and glacier–permafrost interactions are considered to have been widespread during their growth phases with thermo-mechanical models suggesting that these boundary conditions facilitated their initial accumulation. In the later stages of a glacial cycle, areas of cold-based ice are likely to have constrained the locations of fast-flowing ice streams and to have encouraged debris entrainment around their peripheries.
- Sediment fluxes within glaciated permafrost catchments are surprisingly high and in some cases equivalent to those recorded in more temperate glaciated settings. In mountain environments, the limited glacial erosion rates are compensated for by the enhanced extraglacial sediment inputs associated with weathering and mass movement. In addition, the formation of significant volumes of debris-rich ice and associated ice-cored moraines provide potent 'sediment magazines' that can augment fluvial sediment effluxes from these catchments. Finally, landscape history can play an influential role with previously greater glacier extents resulting in the formation of glacial sediment stores that can be reworked in the modern day.

References

Bælum, K. & Benn, D.I., 2011. Thermal structure and drainage system of a small valley glacier (Tellbreen, Svalbard), investigated by ground penetrating radar. *The Cryosphere*, 5, 139–149.

Bartholomew, I., Nienow, P., Mair, D., *et al.*, 2010. Seasonal evolution of subglacial drainage and acceleration in a Greenland outlet glacier. *Nature Geoscience*, 3, 408–411.

Benn, D.I., Gulley, J., Luckman, A., Adamek, A. & Glowacki, P.S., 2009. Englacial drainage systems formed by hydrologically driven crevasse propagation. *Journal of Glaciology*, 55, 513–523.

Benn, D.I., Fowler, A.C., Hewitt, I. & Sevestre, H., 2019a. A general theory of glacier surges. *Journal of Glaciology*, 65, 701–716.

Benn, D.I., Jones, R.L., Luckman, A., *et al.*, 2019b. Mass and enthalpy budget evolution during the surge of a polythermal glacier: a test of theory. *Journal of Glaciology*, 65, 717–731.

Bingham, R.G., Nienow, P.W. & Sharp, M.J., 2003. Intra-annual and intra-seasonal flow dynamics of a High Arctic polythermal valley glacier. *Annals of Glaciology*, 37, 181–188.

Bingham, R.G., Nienow, P.W., Sharp, M.J. & Copland, L., 2006. Hydrology and dynamics of a polythermal (mostly cold) High Arctic glacier. *Earth Surface Processes and Landforms*, 31, 1463–1479.

Bockheim, J.G., Campbell, I.B. & McLeod, M., 2007. Permafrost distribution and active-layer depths in the McMurdo Dry Valleys, Antarctica. *Permafrost and Periglacial Processes*, 18, 217–227.

Bogen, J., 1996. Erosion rates and sediment yields of glaciers. *Annals of Glaciology*, 22, 48–52.

Bogen, J. & Bønsnes, T.E., 2003. Erosion and sediment transport in High Arctic rivers, Svalbard. *Polar Research*, 22, 175–189.

Boon, S. & Sharp, M., 2003. The role of hydrologically-driven ice fracture in drainage system evolution on an Arctic glacier. *Geophysical Research Letters*, 30, doi:10.1029/2003GL018034.

Bougamont, M., Tulaczyk, S. & Joughin, I., 2003. Numerical investigations of the slow-down of Whillans Ice Stream, West Antarctica: is it shutting down like Ice Stream C? *Annals of Glaciology*, 37, 239–246.

Boulton, G.S., 1978. Boulder shapes and grain-size distributions of debris as indicators of transport paths through a glacier and till genesis. *Sedimentology*, 25, 773–799.

Boulton, G.S. & Caban, P., 1995. Groundwater flow beneath ice sheets: part II—its impact on glacier tectonic structures and moraine formation. *Quaternary Science Reviews*, 14, 563–587.

Boulton, G.S. & Hagdorn, M., 2006. Glaciology of the British Isles Ice Sheet during the last glacial cycle: form, flow, streams and lobes. *Quaternary Science Reviews*, 25, 3359–3390.

Boulton, G.S., Caban, P.E. & Van Gijssel, K., 1995. Groundwater flow beneath ice sheets: part I—large scale patterns. *Quaternary Science Reviews*, 14, 545–562.

Bukowska-Jania, E. & Szafraniec, J., 2005. Distribution and morphometric characteristics of icing fields in Svalbard. *Polar Research*, 24, 41–53.

Chandler, D.M., Wadham, J.L., Lis, G.P., *et al.*, 2013. Evolution of the subglacial drainage system beneath the Greenland Ice Sheet revealed by tracers. *Nature Geoscience*, 6, 195–198.

Christoffersen, P. & Tulaczyk, S., 2003a. Thermodynamics of basal freeze-on: predicting basal and subglacial signatures of stopped ice streams and interstream ridges. *Annals of Glaciology*, 36, 233–243.

Christoffersen, P. & Tulaczyk, S., 2003b. Response of subglacial sediments to basal freeze-on 1. Theory and comparison to observations from beneath the West Antarctic Ice Sheet. *Journal of Geophysical Research: Solid Earth*, 108(B4), doi:10.1029/2002JB001935.

Christoffersen, P. & Tulaczyk, S., 2003c. Signature of palaeo-ice-stream stagnation: till consolidation induced by basal freeze-on. *Boreas*, 32, 114–129.

Christoffersen, P., Tulaczyk, S., Carsey, F.D. & Behar, A.E., 2006. A quantitative framework for interpretation of basal ice facies formed by ice accretion over subglacial sediment. *Journal of Geophysical Research: Earth Surface*, 111(F1), doi:10.1029/2005JF000363.

Church, M., 1974. Hydrology and permafrost with reference to northern North America. In: Demers, J. (Ed.), *Permafrost Hydrology*. Proceedings of Workshop Seminar. Canadian National Committee, International Hydrological Decade, Environment Canada, Calgary, Alberta, 7–20.

Church, M. & Ryder, J.M., 1972. Paraglacial sedimentation: a consideration of fluvial processes conditioned by glaciation. *Geological Society of America Bulletin*, 83, 3059–3072.

Clarke, G.K. & Blake, E.W., 1991. Geometric and thermal evolution of a surge-type glacier in its quiescent state: Trapridge Glacier, Yukon Territory, Canada, 1969–89. *Journal of Glaciology*, 37, 158–169.

Clarke, G.K., Collins, S.G. & Thompson, D.E., 1984. Flow, thermal structure, and subglacial conditions of a surge-type glacier. *Canadian Journal of Earth Sciences*, 21, 232–240.

Collins, D.N., 1977. Hydrology of an Alpine glacier as indicated by the chemical composition of meltwater. *Zeitschrift für Gletscherkunde und Glazialgeologie*, 13, 219–238.

Cook, S.J., Swift, D.A., Kirkbride, M.P., Knight, P.G. & Waller, R.I., 2020. The empirical basis for modelling glacial erosion rates. *Nature Communications*, 11, 1–7.

Cooper, R.J., Wadham, J.L., Tranter, M., Hodgkins, R. & Peters, N.E., 2002. Groundwater hydrochemistry in the active layer of the proglacial zone, Finsterwalderbreen, Svalbard. *Journal of Hydrology*, 269, 208–223.

Copland, L. & Sharp, M., 2001. Mapping thermal and hydrological conditions beneath a polythermal glacier with radio-echo sounding. *Journal of Glaciology*, 47, 232–242.

Copland, L., Sharp, M.J. & Nienow, P.W., 2003. Links between short-term velocity variations and the subglacial hydrology of a predominantly cold polythermal glacier. *Journal of Glaciology*, 49, 337–348.

Cuffey, K.M., Conway, H., Hallet, B., Gades, A.M. & Raymond, C.F., 1999. Interfacial water in polar glaciers and glacier sliding at −17°C. *Geophysical Research Letters*, 26, 751–754.

Cuffey, K.M., Conway, H., Gades, A.M., *et al.*, 2000. Entrainment at cold glacier beds. *Geology*, 28, 351–354.

Echelmeyer, K. & Zhongxiang, W., 1987. Direct observation of basal sliding and deformation of basal drift at sub-freezing temperatures. *Journal of Glaciology*, 33, 83–98.

Engelhardt, H. & Kamb, B., 2013. Kamb Ice Stream flow history and surge potential. *Annals of Glaciology*, 54, 287–298.

Etzelmüller, B., 2000. Quantification of thermo-erosion in pro-glacial areas - examples from Svalbard. *Zeitschrift für Geomorphologie*, 44(3), 343–361.

Etzelmüller, B. & Hagen, J.O., 2005. Glacier-permafrost interaction in Arctic and alpine mountain environments with examples from southern Norway and Svalbard. In: Harris, C. & Murton, J.B. (Eds.), *Cryospheric Systems: Glaciers & Permafrost*. Geological Society, London, Special Publications, 242, 11–27.

Glasser, N.F., Hambrey, M.J., Etienne, J.L., Jansson, P. & Pettersson, R., 2003. The origin and significance of debris-charged ridges at the surface of Storglaciären, northern Sweden. *Geografiska Annaler*, 85A, 127–147.

Gulley, J.D., Benn, D.I., Müller, D. & Luckman, A., 2009. A cut-and-closure origin for englacial conduits in uncrevassed regions of polythermal glaciers. *Journal of Glaciology*, 55, 66–80.

Gurnell, A.M., Hodson, A., Clark, M.J., *et al.*, 1994. Water and sediment discharge from glacier basins: an arctic and alpine comparison. *Variability in Stream Erosion & Sediment Transport*. IAHS Publication number 224, 325–334.

Haldorsen, S. & Heim, M., 1999. An Arctic groundwater system and its dependence upon climatic change: an example from Svalbard. *Permafrost and Periglacial Processes*, 10, 137–149.

Haldorsen, S., Heim, M., Dale, B., *et al.*, 2010. Sensitivity to long-term climate change of subpermafrost groundwater systems in Svalbard. *Quaternary Research*, 73, 393–402.

Hall, K., Thorn, C.E., Matsuoka, N. & Prick, A., 2002. Weathering in cold regions: some thoughts and perspectives. *Progress in Physical Geography*, 26, 577–603.

Hallet, B., Hunter, L. & Bogen, J., 1996. Rates of erosion and sediment evacuation by glaciers: a review of field data and their implications. *Global and Planetary Change*, 12, 213–235.

Hambrey, M.J. & Fitzsimons, S.J., 2010. Development of sediment–landform associations at cold glacier margins, Dry Valleys, Antarctica. *Sedimentology*, 57, 857–882.

Hambrey, M.J. & Glasser, N.F., 2012. Discriminating glacier thermal and dynamic regimes in the sedimentary record. *Sedimentary Geology*, 251, 1–33.

Hambrey, M.J., Dowdeswell, J.A., Murray, T. & Porter, P.R., 1996. Thrusting and debris entrainment in a surging glacier: Bakaninbreen, Svalbard. *Annals of Glaciology*, 22, 241–248.

Hambrey, M.J., Bennett, M.R., Dowdeswell, J.A., Glasser, N.F. & Huddart, D., 1999. Debris entrainment and transfer in polythermal valley glaciers. *Journal of Glaciology*, 45, 69–86.

Hodgkins, R., 1996. Seasonal trend in suspended-sediment transport from an Arctic glacier, and implications for drainage-system structure. *Annals of Glaciology*, 22, 147–151.

Hodgkins, R., 1997. Glacier hydrology in Svalbard, Norwegian High Arctic. *Quaternary Science Reviews*, 16, 957–973.

Hodgkins, R., 2001. Seasonal evolution of meltwater generation, storage and discharge at a non-temperate glacier in Svalbard. *Hydrological Processes*, 15, 441–460.

Hodgkins, R., Tranter, M. & Dowdeswell, J.A., 1998. The hydrochemistry of runoff from a 'cold-based' glacier in the High Arctic (Scott Turnerbreen, Svalbard). *Hydrological Processes*, 12, 87–103.

Hodgkins, R., Cooper, R., Wadham, J. & Tranter, M., 2003. Suspended sediment fluxes in a high-Arctic glacierised catchment: implications for fluvial sediment storage. *Sedimentary Geology*, 162(1-2), 105–117.

Hodgkins, R., Tranter, M. & Dowdeswell, J.A., 2004. The characteristics and formation of a high-arctic proglacial+++ icing. *Geografiska Annaler*, 86A, 265–275.

Holdsworth, G., 1974. *Meserve Glacier Wright Valley, Antarctica: Part I. Basal Processes.* Research Foundation and the Institute of Polar Studies, Report No. 37, The Ohio State University Research Foundation.

Holdsworth, G. & Bull, C., 1970. The flow law of cold ice: investigations on Meserve Glacier, Antarctica. *IASH Publication*, 86, 204–216.

Irvine-Fynn, T.D.L., Moorman, B.J., Williams, J.L.M. & Walter, F.S.A., 2006. Seasonal changes in ground-penetrating radar signature observed at a polythermal glacier, Bylot Island, Canada. *Earth Surface Processes & Landforms*, 31, 892–909.

Irvine-Fynn, T.D., Hodson, A.J., Moorman, B.J., Vatne, G. & Hubbard, A.L., 2011. Polythermal glacier hydrology: a review. *Reviews of Geophysics*, 49, RG4002.

Kleman, J. & Glasser, N.F., 2007. The subglacial thermal organisation (STO) of ice sheets. *Quaternary Science Reviews*, 26, 585–597.

Kleman, J. & Hättestrand, C., 1999. Frozen-bed Fennoscandian and Laurentide ice sheets during the Last Glacial Maximum. *Nature*, 402, 63–66.

Knight, P.G., Waller, R.I., Patterson, C.J., Jones, A.P. & Robinson, Z.P., 2002. Discharge of debris from ice at the margin of the Greenland ice sheet. *Journal of Glaciology*, 48, 192–198.

Krawczyk, W.E., 1992. Chemical characteristics of water circulating in the Werenskiold Glacier (SW Spitsbergen). In: Eraso, A. & Pulina, M. (Eds.), *Proceedings of the 2nd International Symposium of Glacier Caves and Karst in Polar Regions*. Silesian University, Sosnowiec, 65–80.

Licht, K.J. & Hemming, S.R., 2017. Analysis of Antarctic glacigenic sediment provenance through geochemical and petrologic applications. *Quaternary Science Reviews*, 164, 1–24.

Liestøl, O., 1977. Pingos, springs and permafrost in Spitsbergen. *Norsk Polarinstitutt Årbok*, *1975*, 7–29.

Liestøl, O., Repp, K. & Wold, B., 1980. Supra-glacial lakes in Spitsbergen. *Norsk Geografisk Tidsskrift*, 34, 89–92.

Lovell, H., Fleming, E.J., Benn, D.I., *et al.*, 2015. Former dynamic behaviour of a cold-based valley glacier on Svalbard revealed by basal ice and structural glaciology investigations. *Journal of Glaciology*, 61, 309–328.

Marsh, P. & Woo, M.K., 1981. Snowmelt, glacier melt, and high arctic streamflow regimes. *Canadian Journal of Earth Sciences*, 18, 1380–1384.

Marshall, S.J. & Clark, P.U., 2002. Basal temperature evolution of North American ice sheets and implications for the 100-kyr cycle. *Geophysical Research Letters*, 29, 67–61, doi:10.1029/2002GL015192.

Mikucki, J.A., Foreman, C.M., Sattler, B., Lyons, W.B. & Priscu, J.C., 2004. Geomicrobiology of Blood Falls: an iron-rich saline discharge at the terminus of the Taylor Glacier, Antarctica. *Aquatic Geochemistry*, 10, 199–220.

Moore, P.L., Iverson, N.R., Brugger, K.A., *et al.*, 2011. Effect of a cold margin on ice flow at the terminus of Storglaciären, Sweden: implications for sediment transport. *Journal of Glaciology*, 57, 77–87.

Moorman, B.J., 2003. Glacier-permafrost hydrology interactions, Bylot Island, Canada. In: Phillips, M., Springman, S.M. & Arenson, L.U. (Eds.), *Proceedings of the 8th International Conference on Permafrost*. International Permafrost Association, Zurich, Switzerland, 783–788.

Müller, D. (2007), Incision and closure processes of meltwater channels on the glacier Longyearbreen, Spitsbergen. M.S. thesis, Tech. Univ. Braunschweig/Univ. Cent., Svalbard, Norway, 118 pp.

Murray, T., Stuart, G.W., Miller, P.J., *et al.*, 2000. Glacier surge propagation by thermal evolution at the bed. *Journal of Geophysical Research: Solid Earth*, 105(B6), 13491–13507.

Murton, J.B., 1996. Near-surface brecciation of chalk, isle of thanet, south-east England: a comparison with ice-rich brecciated bedrocks in Canada and Spitsbergen. *Permafrost and Periglacial Processes*, 7, 153–164.

Murton, J.B., Peterson, R. & Ozouf, J.C., 2006. Bedrock fracture by ice segregation in cold regions. *Science*, 314(5802), 1127–1129.

Nowak, A. & Hodson, A., 2014. Changes in meltwater chemistry over a 20-year period following a thermal regime switch from polythermal to cold-based glaciation at Austre Brøggerbreen, Svalbard. *Polar Research*, 33, 22779.

Nuttall, A.M. & Hodgkins, R., 2005. Temporal variations in flow velocity at Finsterwalderbreen, a Svalbard surge-type glacier. *Annals of Glaciology*, 42, 71–76.

Payne, A.J. & Baldwin, D.J., 1999. Thermomechanical modelling of the Scandinavian ice sheet: implications for ice-stream formation. *Annals of Glaciology*, 28, 83–89.

Piotrowski, J.A., 1997. Subglacial hydrology in north-western Germany during the last glaciation: groundwater flow, tunnel valleys and hydrological cycles. *Quaternary Science Reviews*, 16, 169–185.

Porter, P.R. & Murray, T., 2001. Mechanical and hydraulic properties of till beneath Bakaninbreen, Svalbard. *Journal of Glaciology*, 47, 167–175.

Rempel, A.W., 2007. Formation of ice lenses and frost heave. *Journal of Geophysical Research*, 112(F2), doi:10.1029/2006JF000525.

Scheidegger, J.M. & Bense, V.F., 2014. Impacts of glacially recharged groundwater flow systems on talik evolution. *Journal of Geophysical Research: Earth Surface*, 119, 758–778.

Scheidegger, J.M., Bense, V.F. & Grasby, S.E., 2012. Transient nature of Arctic spring systems driven by subglacial meltwater. *Geophysical Research Letters*, 39, L12405, doi:10.1029/2012GL051445.

Sharp, M., Jouzel, J., Hubbard, B. & Lawson, W., 1994. The character, structure and origin of the basal ice layer of a surge-type glacier. *Journal of Glaciology*, 40, 327–340.

Sollid, J.L. & Sørbel, L., 1994. Distribution of glacial landforms in southern Norway in relation to the thermal regime of the last continental ice sheet. *Geografiska Annaler: Series A, Physical Geography*, 76, 25–35.

Sollid, J.L., Etzelmüller, B., Vatne, G. & Ødegard, R.S., 1994. Glacial dynamics, material transfer and sedimentation of Erikbreen and Hannabreen, Liefdefjorden, northern Spitsbergen. *Zeitschrift für Geomorphologie, Supplementband*, 97, 123–144.

Tulaczyk, S., Kamb, B., Scherer, R.P. & Engelhardt, H.F., 1998. Sedimentary processes at the base of a West Antarctic ice stream; constraints from textural and compositional properties of subglacial debris. *Journal of Sedimentary Research*, 68, 487–496.

Vatne, G., Etzelmüller, B., Sollid, J.L. & Ødegård, R.S., 1995. Hydrology of a polythermal glacier, Erikbreen, northern Spitsbergen. *Hydrology Research*, 26, 169–190.

Vatne, G., Etzelmüller, B., Ludvid Sollid, J. & Ødegård, R.S., 1996. Meltwater routing in a high arctic glacier, Hannabreen, northern Spitsbergen. *Norsk Geografisk Tidsskrift*, 50, 66–74.

Vogel, S.W., Tulaczyk, S., Kamb, B., *et al.*, 2005. Subglacial conditions during and after stoppage of an Antarctic Ice Stream: is reactivation imminent? *Geophysical Research Letters*, 32(14), doi:10.1029/2005GL022563.

Wadham, J.L., Cooper, R.J., Tranter, M. & Hodgkins, R., 2001. Enhancement of glacial solute fluxes in the proglacial zone of a polythermal glacier. *Journal of Glaciology*, 47, 378–386.

Wainstein, P., Moorman, B. & Whitehead, K., 2014. Glacial conditions that contribute to the regeneration of Fountain Glacier proglacial icing, Bylot Island, Canada. *Hydrological Processes*, 28, 2749–2760.

Walder, J. & Hallet, B., 1985. A theoretical model of the fracture of rock during freezing. *Geological Society of America Bulletin*, 96, 336–346.

Williams, P.J. & Smith, M.W., 1989. *The Frozen Earth: Fundamentals of Geocryology*. Cambridge University Press.

Woo, M.K., 1986. Permafrost hydrology in North America. *Atmosphere-Ocean*, 24, 201–234.

Woo, M.K., Kane, D.L., Carey, S.K. & Yang, D., 2008. Progress in permafrost hydrology in the new millennium. *Permafrost and Periglacial Processes*, 19, 237–254.

Zwally, H.J., Abdalati, W., Herring, T., *et al.*, 2002. Surface melt-induced acceleration of Greenland ice-sheet flow. *Science*, 297, 218–222.

5

The Landscape Expression of Glacier–Permafrost Interactions: Landforms, Sediments and Landsystems

5.1 Introduction

The detailed examination of landscape evidence to reconstruct the presence and key characteristics of former glaciers is a long-standing research activity with roots that can be traced back to the very origins of glaciology. Building on the pioneering work of Perraudin, Venetz and Charpentier, Louis Agassiz employed numerous lines of landscape evidence to deduce that glaciers had been much more extensive during the recent geological past as part of his 'Glacial Theory' (e.g. Agassiz, 1841). In making use of contemporary observations in the Swiss Alps to interpret enigmatic landforms in areas lacking glaciers, this work provides an early illustration of the use and importance of process-form connections and 'space-time analogues' (e.g. Schumm, 1991). Upon being shown scratches and groves in bedrock surfaces near Edinburgh during his first visit to Scotland for example, Agassiz is reported to have noted their similarity to glacial striations near contemporary glaciers and to have exclaimed 'that is the work of ice!' (Gordon, 1995, p66).

   Research undertaken on formerly glaciated terrains since the 1960s has sought to develop, test and refine these lines of landscape evidence in order to enable the reconstruction of the basal boundary conditions of former glaciers and ice sheets. This more focused geomorphological work has followed the recognition that boundary conditions such as basal thermal regime are central to the prediction of a glacier's form, thickness and dynamic behaviour (e.g. Beget, 1987; section 3.3). The cornerstone of this 'inverse modelling' involves the establishment of potentially diagnostic connections between environment, process and form - an approach clearly illustrated by the widespread use of the aforementioned striations for example to infer the operation of basal sliding, the occurrence of basal melting and the presence of warm-based ice (Section 3.4.2). However, as previously discussed (Section 3.5.3), the subsequent realisation that such features can be created beneath non-temperate ice suggests that connections between process and form are likely to be more complex than commonly assumed. In addition, the landscape impact of cold glacier ice remains the subject of active discussion (e.g. Slaymaker & Embleton-Hamann, 2018) as part of a broader debate concerning the geomorphic impact of glaciers stimulated by the publication of Agassiz's seminal work.

Glacier–Permafrost Interactions, First Edition. Richard I. Waller.
© 2024 John Wiley & Sons Ltd. Published 2024 by John Wiley & Sons Ltd.

This chapter starts by exploring the origins and implications of the long established and still widely held view that cold-based glaciers are associated with little or no landscape impacts, expressed within the concept of 'glacial protectionism' (Section 5.2). Section 5.3 considers a range of distinct landforms that have been explicitly related to cold-based ice and glacier–permafrost interactions before the same potential connections are explored with specific reference to the sedimentology and structural characteristics of glacigenic sediments in Section 5.4. These key components are integrated in Section 5.5 as part of a broader consideration of the landform-sediment assemblages and 'glacial landsystems' in which glacier–permafrost interactions are thought to play an influential role. Finally, Section 5.6 considers the paraglacial processes and products associated with the landscape adjustments that accompany deglaciation, in which glacier–permafrost interactions are thought to play a prominent role.

5.2 Cold-Based Glaciers and Glacial Protectionism

With basal processes playing a key role in driving glacial erosion, entrainment, transport and deposition, the assumption that they are largely inactive beneath cold-based ice has led to the supposition that the geomorphological impact of these types of ice masses is limited by definition (Section 3.4). Cold-based ice is therefore commonly regarded as a landscape preservative. Rather than causing the significant landscape change which has become the classic 'textbook' view of glaciers more generally, cold-based glaciers are widely believed to protect pre-glacial landscapes, landforms and other delicate features from the subaerial processes that would otherwise lead to their modification or complete destruction.

This concept of 'glacial protection' is by no means new and some of the earliest debates in glacial geomorphology revolved around the fundamental question of whether glaciers are more or less effective at eroding landscapes than the 'normal' erosive agents typical of subaerial conditions (described as weathering, wind, frost, rain and running water) (e.g. Bonney, 1871; Garwood, 1910). Following on from the pioneering glaciological work of figures including Tyndall and Ramsay, well-known geomorphologists such as William Morris Davis & Walther Penck suggested that ice was more effective and therefore that glacial erosion was chiefly responsible for the creation of classic Alpine features such as glacial troughs and cirques. Bonney (1871) and Garwood (1910) however argued the opposite, establishing an alternative school of thought that suggested that ice in general erodes less rapidly than other denuding agents and therefore tends to play a protective role. As part of a systematic consideration of the classic Alpine landforms attributed to glacial activity, Garwood (1910) consistently argues that their formation relates primarily to the protective influence of glaciers. In explaining the formation of hanging valleys for example, he argues that the erosion of the main trough valley was caused principally by fluvial erosion that was particularly effective during interglacial periods when glaciers had retreated into the tributary valleys. In retaining a protective ice cover throughout the interglacial, these tributary valleys consequently experienced more limited denudation than the trough valley resulting in their establishment as hanging valleys. In concluding, Garwood acknowledges that 'under certain other conditions ice may erode more vigorously that water' (p336), although he does not elaborate on what these other conditions might involve.

Subsequent research identified the influential role played by basal thermal regime in determining the rate of glacial erosion and confirmed the ability of warm-based glaciers in particular to create a broad range of distinctive glacial landforms. However, a protective role was still often ascribed to ice masses that were cold and slow moving. Observations made in and around tunnels excavated into the north-western margin of the Greenland Ice Sheet near the Thule air base provided convincing evidence for a more protective role. Swinzow (1962) for example reported that patterned ground visible in the foreland extended under the glacier margin. In addition, Goldthwait (1960) documented the presence of dead moss on subglacial boulders observed at the bottom of a vertical shaft that were radiocarbon dated to be approximately 200 years old (Figure 5.1). Repeat observations were in turn made around the receding margin of the Tiger Ice Cap in northern Baffin Island by Falconer (1966) who described the exposure of vegetated and undisturbed patterned ground with mosses that were radiocarbon dated to 335 ± 75 years B.P.

More recent work on Ellesmere Island has provided some of the most remarkable evidence that attests both to the potential for delicate features to be preserved beneath cold-based ice and to the resilience of biological life within permafrost environments. Field investigations at Twin Glacier in central Ellesmere Island by Bergsma *et al.* (1984) for example revealed the presence of a well-preserved High Arctic plant community within a recently deglaciated 'light-coloured release zone' (p50) dominated by dead and weathered lichens. Plants released by the ablating glacier included arctic blueberry with intact foliage and in places the vegetation released was sufficiently abundant to allow the authors to quantitatively estimate the cover and species frequency. Radiocarbon dating indicated that the vegetation had been buried by ice for approximately 400 years suggesting that it pre-dated the Little Ice Age and had been inundated by glacier ice during this subsequent cold

Figure 5.1 Boulder located at the margin of the Fountain Glacier on Bylot Island recently exposed by recession. The presence of both the lichens on the boulder surface and layer of turf suggest that the surface has experienced little if any glacial modification beneath the marginal cold-based ice (Image: Richard Waller).

period. Follow-up research by La Farge *et al.* (2013) similarly identified intact tundra surfaces around the margins of the cold-based Teardrop Glacier, also located in central Ellesmere Island. Radiocarbon dating confirmed the earlier findings, demonstrating that the bryophyte assemblages had remained in pristine condition in spite of being entombed beneath glacier ice for 400 years. Most remarkably, as a consequence of their remarkable structural preservation, the bryophytes were found to remain biologically viable and were observed to regenerate when collected for laboratory growth experiments (Figure 5.2).

The ability for new plants to regenerate from a small number of cells ('totipotency') preserved within subglacial permafrost clearly has significant implications for our understanding of the biological recolonisation of landscapes that have been inundated by ice during previous advances. The remarkable preservation of vegetation beneath cold-based ice masses has also opened up new opportunities for detailed palaeoclimatic reconstructions within Polar regions more generally. Margreth *et al.* (2014) for example has radiocarbon dated mosses recently exposed around a series of cold and polythermal ice masses on the Cumberland Peninsula on Baffin Island in order to provide a high-resolution record of their Neoglacial expansion and new insights into late Holocene ice dynamics. Subglacial environments in these settings containing dead vegetation comprising mosses, liverworts, leaves and lichens consequently represent a new type of palaeoenvironmental archive the potential value of which has only recently been discovered. The time window available to study these archives is however limited with the exhumation required for access resulting in their exposure to subaerial processes that typically lead to their rapid destruction.

The discovery of a diverse array of well-preserved archaeological artefacts exposed by the ongoing recession of glaciers and perennial snow patches in Scandinavia, North America and the European Alps provides a new and novel illustration of the long-term preservation potential of cold-based ice masses. This has led to the development of 'glacial archaeology'

Figure 5.2 Vegetation regrowth on recently exposed ground located a few metres from the receding margin of Fountain Glacier, Bylot Island (Image: Richard Waller).

as a new archaeological sub-discipline devoted to the study of frozen sites, objects and artefacts with its own dedicated journal (Dixon *et al.*, 2014). The frozen conditions and lack of significant basal motion beneath what are typically referred to as 'ice patches', but which in many instances can be regarded as the degraded remnants of former glaciers, are considered ideal for the preservation of artefacts (Nesje *et al.*, 2012). Their systematic study in central Norway by researchers such as Lars Pilø and Espen Finstad has resulted in the recovery of ancient hunting equipment and 'scaring sticks' as well as occasional human remains dating back as far as 4000–3700 years BC (Finstad & Pilø, 2019). Between 1914 and 2011, no less than 234 artefacts had been recovered from 28 snow patches providing a detailed insight into Neolithic and Bronze Age culture and hunting practices in this mountain region (Callanan, 2014).

The potential long-term preservation of delicate features is not limited to ice-marginal locations however with a study of the sediment contained within the base of the GISP2 ice core at Summit, Greenland, suggesting that the ice beneath the central part of the Greenland Ice Sheet has been non-erosive for the entire Pleistocene (Bierman *et al.*, 2014). In measuring the meteoric ^{10}Be, organic carbon and total nitrogen levels within a basal silty ice zone, the authors concluded that the entrained material comprised pre-glacial soil that formed prior to the establishment of the present Greenland Ice Sheet with the ^{10}Be data suggesting exposure of at least 200 ka and potentially over 1 Ma, similar to ancient preglacial regoliths identified in Arctic Canada. The presence of the soil in the basal ice suggests that the central part of the ice sheet has experienced extremely low erosion rates (albeit with its entrainment into the cold-based ice) leading the authors to suggest that 'an ancient landscape underlies 3000m of ice at Summit' (p405).

A characteristic feature of permafrost landscapes glaciated by cold-based ice masses can therefore include the preservation of pre-glacial landforms such as tors, felsenmeer, patterned ground and subaerial drainage systems that largely reflect the long-term operation of deep weathering, mass wasting and fluvial erosion under ice-free conditions. Indeed, the absence of primary glacial features has led to the conclusion that,

> *An entire cycle of glaciation need result in no landscape modification save that resulting from isostatic effects* (Dyke, 1993, p224).

This view (which incidentally is not held by the quotation's author) poses significant challenges to those undertaking glaciological reconstructions and has led to a long-standing debate about the correct interpretation of areas within formerly glaciated regions dominated by pre-glacial features. The absence of classic glacial landforms led early workers to interpret such areas as representing ice-free enclaves or nunataks (e.g. Linton, 1950) that also served as biological refugia (Dahl, 1955). However, an alternative and increasingly accepted hypothesis is that these areas were glaciated by cold-based and therefore minimally erosive ice (e.g. Sugden, 1968; Kleman, 1994). The nature of the evidence and key lines of argument employed in this debate are well illustrated in specific relation to the interpretation of tors found extensively in parts of the Cairngorm Mountains of Scotland for example (Figure 5.3). The presence of these deeply weathered tors was originally believed by Linton (1950) to indicate that the summit plateau had escaped glaciation such that they represented a much older series of landforms inherited from warmer and wetter

Figure 5.3 Glacially modified granite tors on the Cairngorm plateau located close to the summit of Ben Avon, Scotland (Image: Richard Waller).

times existing prior to the Quaternary Period. However, whilst Sugden (1968) agreed that the general morphology of the summit plateau was largely inherited, closer scrutiny of the tors led to the discovery of erratics and a subtle smoothing of their west-facing aspects. This demonstrated that the plateau had been glaciated by ice with a discernible albeit limited geomorphic impact. This is consistent with our understanding of glacial erosion (Section 3.4.2) with Boulton's (1972) model predicting that some plucking and abrasion can occur beneath cold-based ice and therefore some landscape modification is possible. Boulton (1974) and Drewry (1986) similarly suggested that bedrock protuberances (such as tors) might be affected by basal debris being sheared within the basal part of a cold-based ice mass even if little or no sliding is taking place.

In studying the geomorphic impacts of small plateau icefields in northern Norway in order to assess their ability to erode pre-existing landscape features, Gellatly *et al.* (1988) identified an altitudinal threshold relating to differences in thermal regime that determined the degree of glacial modification. Three small ice caps located at elevations >1500 m in the southern Lyngen Alps featured recently deglaciated forelands dominated by intact patterned ground, blockfields or bedrock outcrops devoid of abrasional features, a geomorphic signature considered indicative of dry-based conditions and glacial protection. In contrast, the forelands of lower elevation plateau icefields (900–1200 m) displayed abundant scoured and striated bedrock and boulder moraines suggestive of active basal sliding, wet-based conditions and significant glacial modification that had removed pre-existing landforms relating to bedrock weathering.

This demonstrates both the significant spatial variations in glacial modification that can occur over small areas as well as the profound difficulties associated with the recognition and accurate interpretation of landscapes that lack clear evidence of glacial modification. Distinguishing areas that have been ice free from those that have been glaciated by cold-based ice is however essential for accurate ice-sheet reconstructions as the two hypotheses are likely to predict markedly different ice-sheet elevations, profiles and volumes. They are also associated with differing reconstructions of the basal boundary conditions required to validate numerical ice-sheet models. The recent development of cosmogenic exposure (CE) dating that can be applied to land surfaces and to any erratics present has helped to resolve this particular debate and to confirm the ability of ancient, pre-glacial features to be preserved for long time periods beneath a cover of cold-based ice. Davis *et al.* (2006) for example, described the preservation of a large raised marine delta on the East Coast of Baffin Island. Radiocarbon dating of *in situ* bivalves within the constituent marine sediments suggested an age of >54 ka B.P. However, CE dating of a series of large erratics resting on the delta surface using ^{10}Be and ^{26}Al yielded a much younger modal age of 14 ka B.P., suggesting that whilst the area had been glaciated during the last glacial period, the primary landforms had been left largely undisturbed.

Kleman and his co-workers have undertaken extensive research in Scandinavia that has demonstrated the widespread preservation of preglacial landforms, the mapping of which has been used to constrain the presence and spatial distribution of cold-based ice and subglacial permafrost. Kleman & Borgström (1990) for example described areas of patterned ground on the summit plateau of Mt. Fulufjället in west-central Sweden that were argued to have been created during a preceding interstadial and therefore to have survived Late Weichselian glaciation largely unaffected. It is worth noting here that this view of glacial inactivity accords with that of Garwood (1910) who stated that,

> . . .*an ice-covered surface which is on the whole horizontal will be relatively protected, and that the tablelands and plateaux in a glaciated district will have their pre-glacial features to a great extent preserved* (p311).

The use of the widely held assumption that basal processes are inactive beneath cold-based ice (Section 3.4) has led to these areas being viewed as 'frozen-bed patches' associated with subglacial permafrost (Kleman *et al.*, 1999). In addition, the presence of cross-cutting subglacial lineations and meltwater-related landforms is thought to belie the preservation of older glacial features relating to former glaciations associated with different ice-sheet configurations and flow directions (Kleman 1992). The resultant superimposition of non-glacial features and glacial landforms associated with successive glaciations results in the development of palimpsest landscapes in which the frozen-bed patches provide 'windows' back in time to earlier periods of landscape development (Kleman, 1992, 1994). Detailed mapping of these relict land surfaces has revealed a complex and patchy distribution that varies according to spatial scale and contrasts with the simpler concentric zonation described by Dyke (1993) and others (Section 3.4.2). These patches form part of a broader geomorphological inverse model comprising a 'dry-bed system', where basal processes are inactive and older landforms are preserved, and a 'wet-based system', where basal processes are active and glacial lineations are generated (Kleman & Borgström, 1996)

(Section 5.5.1). The mapping of these landform assemblages has consequently enabled the reconstruction of the basal thermal regime of the Fennoscandian Ice Sheet and the elucidation of its controlling factors (Kleman *et al.*, 1999) (Section 2.3.2).

Kleman & Borgström (1994) elaborate on the geomorphological evidence for 'frozen-bed patches', suggesting a distinctive spatial patterning of basal process around these zones (Figure 5.4). They described a 'frozen patch' landform assemblage featuring five distinct elements: (i) a relict surface (with well-preserved pre-glacial landforms), (ii) a lateral sliding boundary (marked by an abrupt contact between the relict surface and adjacent glacial lineations), (iii) a lateral shear moraine (associated with a ridge sub-parallel to ice flow), (iv) a stoss-side moraine (single- or multi-crested located at the proximal edge of the relict surface) and (v) a transverse lee-side till scarp. Targeted cosmogenic exposure dating by Fabel *et al.* (2002) on a selection of frozen-bed patches in northern Sweden provided

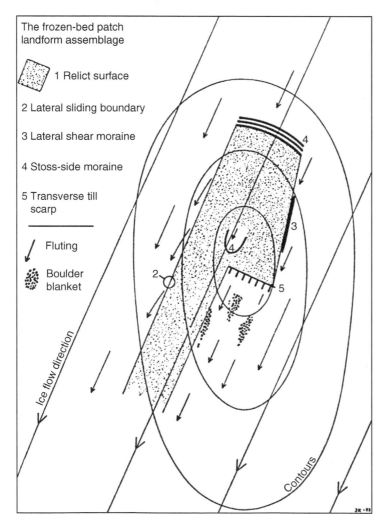

Figure 5.4 Schematic diagram illustrating the relative position of the landforms characteristic of frozen-bed patch environments in an idealised mountain environment (Kleman & Borgström, 1994, p261). See text for details.

contrasting exposure ages of c. 34–61 kyr for the relict surfaces and c. 8–13 kyr for the till scarps and erratic blocks resting on the surfaces, corroborating the long-term survival of these relict surfaces through at least one glacial cycle.

These features delineate the boundaries of the former frozen-bed patch and related 'phase-change surfaces' that separate thawed from frozen regions. Associated changes in subglacial processes across these surfaces or boundaries are considered central to the development of the landforms. On the stoss-side of the frozen patch, a switch from unfrozen to frozen conditions leads to compressive flow and the detachment and stacking or folding of available sediment to produce a moraine. Conversely, on the lee-side, a switch from frozen to unfrozen leads to extensive flow, detachment of the till sheet and the formation of a transverse till scarp. Kleman & Hättestrand (1999) have subsequently suggested that a switch from frozen- to thawed-bed conditions can result in the extension, fracture and boudinage of frozen till sheets to produce much larger ribbed or Rogen moraines that are in turn considered diagnostic of thermal boundaries at the ice-sheet scale (Section 5.3.3).

Returning to the historical debate concerning the geomorphic influence of glaciers considered at the start of this chapter, Slaymaker & Embleton-Hamann (2018) suggest that, 'the balance between glacial erosion and glacial protection remains an unanswered question' (p237). In considering the creation of Quaternary landforms within their recent review of advances in global mountain geomorphology, they cite the development of landscapes of selective linear erosion (Figure 5.5, Sugden & John, 1976) as an illustration

Figure 5.5 Landscape of selective linear erosion on the east coast of Baffin Island. Note the prominent, deeply incised fjords (containing sea ice) and the more limited relief associated with the intervening plateau surfaces that are still glaciated (Image: Richard Waller).

of the complex interactions that can occur between the processes of uplift and erosion, resulting in striking spatial variations in the evolution of glaciated mountain landscapes (Section 5.5.3). These landscapes display significant differences in the erosion between low- and high-altitude areas that result in a 'regional altitude-dependent continuum of glacial modification' (p237). Whilst the low-altitude areas are dominated by thick, warm-based ice that is more erosive, the high-altitude areas are dominated by thin, cold-based glaciers that are minimally erosive. Uplift can subsequently remove the low-altitude surfaces from the influence of warm-based glaciation, limiting the rates of erosion and generating relict land surfaces characterised by slow rates of geomorphic change.

The identification and mapping of distinct altitudinal boundaries between areas of glacially eroded bedrock and more heavily weathered bedrock has been regularly employed to reconstruct the characteristics of former glaciers within mountainous areas for over a century. In their review of these 'glacial trimlines', Rootes & Clark (2020) document their use as far back as 1878 by James Geikie, one of the pioneers of glacial geology. In exploring the slopes of Hecla, a mountain on South Uist in the Western Isles of Scotland, he commented that,

> *Nothing can be more distinct than the line of demarcation between its sharp, jagged, peaked summit and its rounded, softly outlined shoulders, with their moutonée surface* (Geikie, 1878, p851; quoted in Rootes & Clark, 2020, p3).

In locating a process boundary between glacial erosion and subaerial weathering, these trimlines have proven invaluable in enabling the reconstruction of both local ice thicknesses and consequently regional ice surface slopes. A particular focus on the evidence for periglacial weathering, of what in some instances would have represented 'palaeo-nunataks', has led to the use of the term 'periglacial trimline' in UK-based research, with Ballantyne (1997) for example using a combination of bedrock weathering contrasts, gibbsite deposits and cosmogenic isotope dates to identify an upper weathering limit thought to represent the maximum altitude of the last ice sheet in Scotland.

In recognising the ability of cold-based glaciers to preserve periglacial phenomena (Section 5.2) and employing the same process-form connections used more broadly within geomorphological inverse models (e.g. Kleman, 1994; Kleman & Borgström, 1996), work in Scandinavia has proposed an alternative palaeoglaciological interpretation - namely that these trimlines represent process boundaries between warm (wet) ice and cold (dry) ice. Consequently, rather than describing ice limits and former nunataks, it has been suggested that trimlines can instead delineate the former extent of frozen-bed zones at a range of spatial scales that experienced continuous cold-based conditions throughout the glacial cycle (Kleman *et al.*, 1999). Following an ensuing period of debate and the recognition that both types can co-exist, the use of the terms 'thermal trimline' and 'ice-marginal trimline' has been recommended (Rootes & Clark, 2020). The more routine use of cosmogenic isotope dating techniques has demonstrated that many of these features are likely to have been incorrectly interpreted in the past, calling into question the former mapping of ice limits in Scotland for example (McCarroll, 2016). It also demonstrates the ongoing wider challenges in identifying landscape evidence for cold-based ice and glacier-permafrost implications with the author commenting that,

It is now clear that even very experienced geomorphologists cannot necessarily tell the difference between terrain that has been recently glaciated and terrain that has not, because cold-based ice can leave virtually no trace (p130).

5.3 Specific Landforms Related to Cold-Based Glaciers and Glacier–Permafrost Interactions

5.3.1 Lateral Meltwater Channels

The lateral meltwater channels that form at the cold-based margins of cold-based or poly-thermal glaciers are considered by many to constitute one of the most distinctive types of landform associated with these types of ice mass that are therefore potentially diagnostic of cold-based ice and glacier–permafrost interactions. The association of this particular type of meltwater channel with cold-based glaciers relates to well-established assumptions regarding their hydrological behaviour and key drainage pathways (Section 4.2.1). The combination of limited surface crevassing with cold and impermeable surface ice is thought to prevent supraglacial meltwater from draining englacially, promoting its flow to the glacier margins to form ice-marginal meltwater channels. The presence of relatively thin and cold marginal ice that is frozen to its bed is then thought to prevent meltwater flowing back under the glacier margin leading to the progressive erosion of lateral meltwater channels (Figure 5.6), although the thermal erosion of the meltwater into the glacier margin can result in the formation of submarginal channels (Benn & Evans, 2009).

Figure 5.6 Active and relict lateral meltwater channels in the foreland of Fountain Glacier, Bylot Island. These mark the current and former positions of this polythermal glacier that features a cold-based margin (Image: Richard Waller).

One of earliest detailed descriptions of lateral meltwater channels was made by Schytt (1956) following a glaciological research expedition to the area around Thule in north-west Greenland. Observations made primarily using aerial photographs revealed the presence of a series of lateral meltwater channels approximately 30 m deep along the northern margin of the Moltke Glacier. The combination of their substantial depth and their lateral spacing of 100–400 m led the author to discount the suggestion made previously by Mannerfelt (1945) that the channels were formed on an annual basis. The glacier's thermal regime was however considered central to their formation with a 'sub polar' basal thermal regime promoting supraglacial drainage that could then flow for substantial distances along the glacier margin to create the channels observed. In contrast, temperate glaciers are thought to be dominated by englacial and subglacial drainage systems with lateral drainage channels being relatively minor as a consequence. Interestingly, Schytt (1956) suggested a possible connection between the formation of lateral meltwater channels and 'shear' moraines, with ice-marginal drainage being focused towards the troughs observed to form between the shear planes (or 'inner' moraines) (see Sections 3.5.4 and 5.3.2).

In describing the formation of lateral meltwater channels at the margins of cold-based glaciers in the Dry Valleys of Antarctica, Atkins & Dickinson (2007) demonstrated that these landforms can form in the most extreme glaciological environments. Even though the supply of meltwater is very limited, with ephemeral streams flowing for a maximum of two months a year, their active formation in ice-marginal environments indicates that even the coldest ice masses are capable of causing some landscape modification (Section 5.5.4). Short (<500 m long) and shallow (<10 m deep) channels were observed to incise into the till-covered slopes surrounding the margins of numerous glacier margins within the Dry Valleys. Detailed study of the Packard and Lower Wright Glaciers led to the production of a schematic model in which supraglacial meltwater cascades from the glacier surface to erode distinct channels into permafrozen till at the contact between the ice aprons characteristic of the ice margins (Section 3.6.1) and the adjacent hillside. Subsequent recession of the ice margin leads to the abandonment of older channels and the formation of new channels, thereby creating a distinctive series of nested channels that chart the progressive retreat of the ice margin over time (Figure 5.7).

Whilst lateral meltwater channels provide what Atkins & Dickinson (2007) consider to be the 'most obvious and persistent geomorphological signature of cold-based glacier activity' (p47), the authors noted that they were by no means ubiquitous within the landscape with their formation being dependent upon elevation, topography and substrate. An elevation of <1500 m was required to generate a sufficient supply of supraglacial meltwater whilst a hillside sloping towards the ice margin was also required to channel the meltwater. In addition, a sediment-covered surface was required with the limited meltwater discharges being incapable of eroding into exposed bedrock. In considering the limited subglacial erosion associated with these very cold glaciers (see previous section), the authors make the important point that any channels formed during ice retreat may be preserved subglacially during a re-advance. A subsequent retreat may therefore result in a complex and seemingly disordered record of advance and retreat with channels of different ages being juxtaposed upon each other.

In describing the landscapes created by the cold-based central regions of a series of ice caps that formed in the Canadian High Arctic during the Late Wisconsinan, Dyke (1993) similarly identified nested sequences of lateral meltwater channels visible within valleys

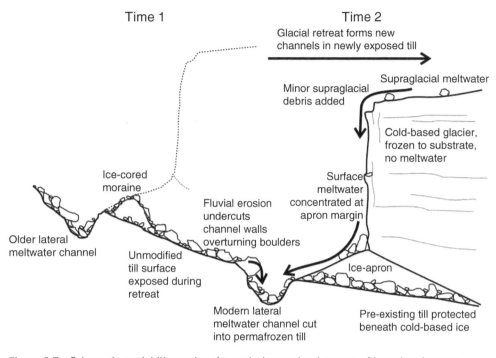

Figure 5.7 Schematic model illustrating the evolutionary development of lateral meltwater channels in response to glacier recession and the associated abandonment of the ice-cored moraine and ice apron (Atkins & Dickinson, 2007, p53).

occurring above the limit of postglacial marine submergence (Figure 5.8). With these areas lacking 'classic' landforms such as scoured bedrock or end moraines, the channels were again thought to provide the clearest evidence for past glacial activity. The lateral meltwater channels were extensive and were observed to persist to what was seen as the final remnants of the ice caps prior to their eventual disappearance. In the absence of moraine sequences, the extensive channel networks led Dyke (1993) to suggest that they can provide an alternative geomorphological record of ice retreat with a potentially very high temporal resolution that may even illustrate the impact of individual melt events.

The ability of lateral meltwater channels to provide important palaeoglaciological information has been widely recognised in Scandinavia in particular, where they are routinely used to reconstruct both the pattern of glacier recession and the thermal regime of the ice mass (e.g. Klemen *et al.*, 1997; Sollid & Sørbel, 1994). In spite of a long history of their mapping in Scandinavia, Greenwood *et al.* (2007) identified the patchy nature of their mapping and a lack of clarity and consistency in their interpretation as reasons for their comparative underutilisation within reconstructions of the British-Irish Ice Sheet. The authors have consequently developed diagnostic criteria that can be used within inversion models to distinguish between subglacial, lateral, proglacial and supraglacial/englacial channels (Table 5.1). In facilitating the identification of marginal and submarginal lateral meltwater channels in particular, the development of this formal methodology has paved the way for their more routine use in reconstructing patterns of deglaciation (e.g. Clark *et al.*, 2012).

Figure 5.8 Extensive 'staircase' of lateral meltwater channels delineating the recession of a former glacier margin in a valley located on the Borden Peninsula, northern Baffin Island (Map data: Google Earth, Landsat/Copernicus).

The observation of 'flights' of nested lateral meltwater channels on the flanks of nunataks in the Glacier Bay region of Southern Alaska has led Syverson & Mickelson (2009) to question whether such features are necessarily indicative of cold-based thermal regimes. The authors describe a series of lateral meltwater channels that have been observed to form along the receding margin of the stagnant Burroughs Glacier. The combination of a perched water table within the stagnant ice with high ablation rates and precipitation totals is thought to drive water to the ice margin where it flows along the margin before descending into the ice via subglacial chutes. This has led to the development of sequences of long, linear lateral meltwater channels typically up to 200 m long with between 2 and 8 channels being observed to form each year. Whilst the authors conclude that such features do not therefore provide definitive proof of a non-temperate thermal regime, they do provide ways in which the lateral meltwater channels formed around temperate ice masses might be differentiated from those associated with polythermal or cold-based ice. First, the ability of subglacial chutes to capture ice-marginal drainage means that the lateral meltwater channels associated with temperate glaciers are comparatively short. Second, they are associated with fluted till surfaces and other features indicative of active basal processes that are typically absent from non-temperate glacier forelands (see Section 5.3.3).

5.3.2 End Moraines

End moraines constitute one of the best known glacial landforms formed within ice-marginal environments. In view of the diverse usage of the term, it should be noted at the outset that the term 'moraine' is employed here to refer to the distinctive landforms created

Table 5.1 Diagnostic criteria for the classification of meltwater channels devised by Greenwood *et al.* (2007). See original paper for details.

Subglacial	Lateral		Proglacial	Supraglacial/englacial
	Marginal	Submarginal		
• Undulating long profile[1,2]	• Parallel with contemporary contours[4,7]		• Regular meander bends[10]	• Meander forms crescentic valley on face of hill[1]
• Descent downslope may be oblique[1,3]	• Forms 'series' of channels parallel to each other[1,8,9]		• Occasional bifurcation[10]	• Low gradient[7]
• Descent downslope may form steep chutes[1]	• Approximately straight[4]		• Flows direct downslope[7]	• Sinuous[7]
• Complex systems—bifurcating and anastomosing[3,4]	• Perched on valley sides[10]		• *Large dimensions – wide and deep*[10]	• *Approximately constant width*[7]
• High sinuosity[4]	• May terminate in downslope chutes[1,8]	• May form networks[1]	• *Crater chains*[10]	
• *Abandoned loops*[4]	• Absence of networks[4]	• Steeper gradient (oblique downslope)[1]		
• *Abrupt beginning and end*[1,2]	• Gentle gradient[1]	• Sudden change in direction[1]		
• *Absence of alluvial fans*[1]	• Parallel for long distance[1]			
• *Cavity systems and potholes*[3]	• *May terminate abruptly*[10]			
• *Ungraded confluences*[3]	• *May be found in isolation from all other glacial features*[6,9]			
• *Variety of size and form within the same connected system*[5]				
• *Association with eskers*[6]				

N.B. Criteria in italics of limited use/not suitable for use with DEM (see text).
Sources: [1]Sissons (1961); [2]Glasser and Sambrook Smith (1999); [3]Sugden *et al.* (1991); [4]Clapperton (1968); [5]Sissons (1960); [6]Kleman and Borgström (1996); [7]Price (1960); [8]Schytt (1956); [9]Dyke (1993); [10]Benn and Evans (1998).

at glacier margins rather than a type of sediment deposited directly by ice. Research on these landforms undertaken over several decades has revealed a remarkable diversity of end moraine types and an equally diverse range of mechanisms associated with their formation. Some of this research has suggested that ice-marginal permafrost can play an important role in their formation leading some authors to suggest that certain types of end moraine may represent the explicit product of interactions between glaciers and permafrost.

Etzelmüller & Hagen (2005) provide a useful conceptual model of the geomorphological processes associated with glacier–permafrost interactions and the ice-marginal and proglacial moraines that can be created as a consequence. Their model highlights the broader connections between prevailing climate, glacier thermal regime (considering the glacier and the permafrost as a coupled system) and the generation of different types of moraine. Ice-cored moraines (Section 5.3.2.1) are associated both with cold and polythermal glaciers terminating in permafrost. Composite moraines however (Section 5.3.2.2) are thought to be limited to polythermal glaciers with the high strength of permafrost limiting their formation around entirely cold glaciers located in the coldest, driest environments. Evans (2009) has similarly presented a process-form continuum for the production of end moraines in which their characteristics, ice content and genesis are constrained by the basal thermal regime (Figure 5.9).

Both Etzelmüller & Hagen (2005) and Evans (2009) suggest that even close to or beyond the latitudinal or altitudinal limit of permafrost, the presence of a marginal fringe of cold-based ice can promote the adfreezing and entrainment of subglacial sediment (Section 3.5.4), its short-distance transport, and its subsequent deposition to form relatively small and asymmetrical annual moraines (Figure 5.9c). A number of field studies have observed this process-form connection with Krüger (1993, 1996) for example demonstrating the important role played by the basal adfreezing of subglacial sediment during the winter at Myrdalsjökull. In combination with small winter readvances, this led to the formation of end moraines up to 4 m high composed of stacked slabs of frozen sediment. Matthews et al. (1995) proposed a similar model of terminal moraine formation at Styggedalsbreen in Norway in which the repetition of winter freeze-on of subglacial debris coupled with a small readvance, followed by ice wastage and the fluvial deposition of sands and gravels led to the incremental formation of moraine ridges up to 6 m high. Evans & Hiemstra (2005) have suggested that this process can lead to the incremental formation of thick till sequences in addition to morainic landforms. Whilst subglacial freezing in these situations can be likened to subglacial permafrost aggradation, the seasonal nature of the process coupled with the likely absence of subaerial permafrost in these environments means that such end moraines do not represent the product of glacier–permafrost interactions *sensu stricto*. In developing a process-form continuum of end moraines controlled by basal thermal regime, Evans (2009) consequently places these at the warm-based end of the continuum (Figure 5.9c).

5.3.2.1 Ice-Cored Moraines

Glacial environments featuring polythermal glaciers terminating in areas of thicker and more extensive permafrost such as Svalbard, Greenland and the Canadian Arctic provide some of the clearest insights into the moraine types and mechanisms of moraine genesis most likely to be associated with glacier–permafrost interactions (e.g. Kälin, 1971; Etzelmüller et al., 1996; Boulton et al., 1999; Waller & Tuckwell, 2005). The widespread

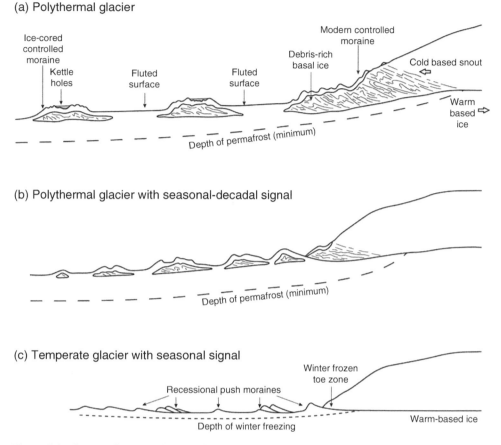

Figure 5.9 Process-form continuum of end moraine production associated with different basal thermal regimes (Evans, 2009, p202). Note how frozen ground is associated with all of the process environments with permafrost being explicitly associated with ice-cored moraine formation at polythermal glaciers (a and b). Temperate glaciers are associated with seasonal winter freezing (c) and the production of smaller recessional push moraines.

formation of ice-cored moraines within these glacial environments represents the first of two main types of end moraine that have been explicitly related to glacier–permafrost interactions. Numerous studies have argued that polythermal basal thermal regimes and the associated thermal transitions from warm-based interior ice to cold-based marginal ice are central to the formation of extensive tracts of ice-cored moraine as they promote the entrainment of significant volumes of sediment through the formation of thick basal ice sequences or debris rich folia (Section 3.5.4) (e.g. Sollid & Sørbel, 1988, Etzelmüller *et al.*, 1996; Lyså & Lønne, 2001). Horizontal compression of the decelerating marginal ice and the formation of tightly folded debris-rich foliations and discrete debris-rich thrusts and debris bands are then argued to facilitate the englacial transfer of this material to the glacier surface where it can accumulate to produce an extensive cover of supraglacial sediment (Goldthwait, 1951) (Figure 5.10). In considering end moraine formation at the termini of Arctic glaciers, Dyke & Savelle (2000) conclude that 'ice-cored moraines seem to represent the norm, rather than the exception' (p616) with their formation requiring the

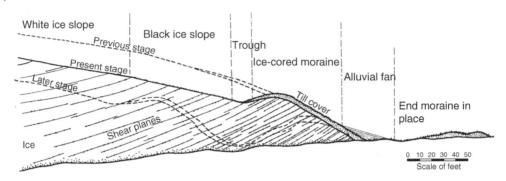

Figure 5.10 Schematic cross section of the margin of the Barnes Ice Cap that illustrates the progressive formation of ice-cored moraine at the ice margin where upwards ice flow leads to the concentration of supraglacial sediment (Goldthwait, 1951, p570). The end moraine on the right illustrates its final form when the ground ice has melted.

upward flow of ice at a glacier margin that is stable or retreating slowly enough for the margin to be buried in supraglacial melt out. The connection of these processes with the basal thermal regime and with glacier–permafrost interactions is made most explicitly by Evans (2009) who states that, 'polythermal conditions are crucial to the concentration of supraglacial debris and controlled moraines in glacier snouts via processes that are most effective at the glacier-permafrost interface' (p183).

Subsequent glacier recession can result in these areas of debris-covered ice becoming decoupled from the active part of the glacier (mechanically if not physically) resulting in widespread ice stagnation and the formation of extensive areas of ice-cored moraine (e.g. Sollid & Sørbel, 1988; Ham & Attig, 1996; Dyke & Savelle, 2000). A variety of processes including mass wasting, the seasonal ablation of exposed ice and thermal erosion and sediment reworking by rivers and lakes can result in the progressive 'de-icing' of the moraine complex and the reworking of the material released (Figure 5.11). The presence of debris-rich structures within the downwasting ice commonly results in the initial development of distinct linear ridges within the evolving moraine complex. Early observation of these ridges within the debris-covered marginal regions of polythermal ice sheets where they were associated with extensive debris bands resulted in them being referred to as 'inner moraines' or 'shear moraines' (Weertman, 1961). More recently, the use of the term 'controlled moraines' (originally introduced by Gravenor & Kupsch, 1959) has been recommended by Evans (2009) in view of their inheritance of debris-rich structures present in the ice margin. The melt-out of these structures imparts a characteristic linearity that remains prominent whilst the underlying buried ice persists. However, the progressive loss of the ice results in its deterioration over time, with its eventual removal resulting in the production of hummocky moraine characterised by a more chaotic assemblages of hummocks and hollows (e.g. Reinhardy *et al.*, 2019). The occurrence of distinctive end moraines comprising broad belts of disordered mounds and ridges in northern Norway that contrast morphologically with the end moraines found in southern Norway that feature more distinct ridges orientated transverse to ice flow led Sollid & Sørbel (1988) to consider the potential role played by thermal regime and permafrost. They suggested that the hummocky end moraines found in northern Norway represent the degraded remains of former ice-cored

Figure 5.11 Large area of degrading ice-cored moraine situated at the margin of Skeiðarárjökull, south-eastern Iceland. Note the prominent exposures of stagnant ice and the reworking of sediment via both fluvial processes and mass movement (Image: Richard Waller).

moraines that were in turn formed by the debris entrainment that occurred at the thermal transition of sub-polar glaciers terminating in regions of permafrost. Extensive studies focusing on the deglacial landform records of both the north-western and southern Laurentide Ice Sheets have similarly related areas of hummocky moraine to thermal transitions and the degradation of stagnant ice (Section 5.5.2).

Within temperate proglacial environments that lack permafrost, the buried ice is generally thought to melt out within the space of a few decades (e.g. Krüger & Kjær, 2000), although the preservation of unusually thick sequences of debris-rich ice associated with surge events may allow the buried ice within ice-cored moraines to persist for a century or more (e.g. Everest & Bradwell, 2003) (Figure 5.12). Within permafrost regions, however, the formation or thickening of permafrost during deglaciation means that 'debris in buried glacier ice is entombed, rather than released, as the base of the permafrost descends' (Dyke & Savelle, 2000, p616). So long as the thickness of the debris cover exceeds the depth of the active layer, then the buried ice can remain intact and stable over much longer time scales (Section 5.3.4). As such, Etzelmüller & Hagen (2005) consider large and stable ice-cored moraines to be indicative of glacier–permafrost interactions. If deglaciation is associated with a renewed phase of permafrost aggradation in ice-marginal environments, this can promote the preservation of buried ice over millennial timescales in mid-latitude settings (Section 5.6.3) (Figure 5.9a). In modern-day permafrost environments, this process has resulted in the continued preservation of significant masses of buried glacier ice within major belts of ice-cored terrain that have formed within parts of the Canadian Arctic

Figure 5.12 Stagnant debris-rich basal ice preserved within the ice-proximal part of a large end moraine complex on Skeiðarársandur, south-east Iceland, thought to have been created during a surge event in the late nineteeth century. Note the thick cover of supraglacial sediment key to its preservation and the debris flows formed after it was exposed during a jökulhlaup event (Image: Richard Waller).

(e.g. French & Harry, 1990; St. Onge & McMartin, 1999; Dyke & Savelle, 2000; Lakeman & England, 2012, Evans *et al.*, 2021) and Western Siberia (e.g. Astakhov & Isayeva, 1988; Kaplyanskaya & Tarnogradskiy, 1986) following the recession of continental scale ice sheets and the postglacial aggradation of permafrost. The nature and formation of these products of so-called 'incomplete' or 'retarded' deglaciation are explored further in Section 5.3.4 and their wider contributions to distinctive ice-sheet landsystems in Sections 5.5.2 and 5.6.3.

   Ice-walled lake plains comprise a distinctive landform type that has been extensively described within areas of hummocky moraine relating to terminal positions of ice lobes of the southern Laurentide Ice Sheet (Ham & Attig, 1996; Clayton *et al.*, 2008). These frequently overlooked landforms are of particular relevance and interest with their genesis having been attributed by some to the influence of permafrost during deglaciation. Both of the aforementioned studies describe extensive ice-walled lake plains up to 5 km in diameter that occur within areas of hummocky moraine related to the detachment and stagnation of large masses of debris-rich ice (Figure 5.13). The lake plains feature flat or gently convex profiles and sequences of lake sediment up to tens of metres in thickness that extend through to the local bedrock and which are surrounded by rim ridges of coarser sediment (Clayton *et al.*, 2008). Their most distinctive aspect however is their position in elevated positions up to 30 m above the surrounding landscape that reflects the significant thickness of the lacustrine infills. Ham & Attig (1996) relate the ice-walled lake basins to stable lakes that initially formed on the surface of the stagnant ice mass. Thermal erosion resulted in them

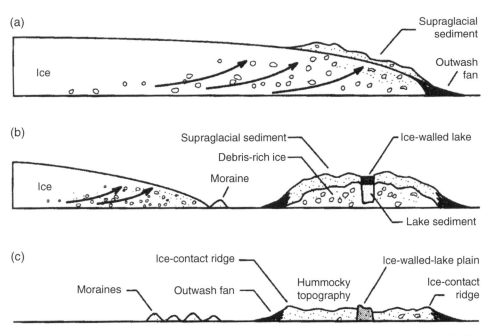

Figure 5.13 Schematic figure illustrating the formation of hummocky moraine and ice-walled lake plains in Wisconsin (Ham & Attig, 1996, p183). (a) Surge event into an area of permafrost leads to the formation of a supraglacial sediment cover. (b) Stable phase associated with glacier recession, the preservation of debris-rich ice beneath a thick cover of supraglacial sediment and the formation of ice-walled lakes. (c) Unstable phase associated with rapid melting of buried ice, hummocky moraine formation and topographic inversion.

progressively melting downwards through the stagnant ice to the local bedrock, thereby providing the accommodation space for their thick sediment infills derived from the reworking of supraglacial sediment. They argue that permafrost conditions during the first phase of their development promoted their long-term stability by preserving the buried ice within which they formed, whilst preventing the establishment of the hydrological pathways required for their drainage. A subsequent amelioration in climate resulted in a second phase of landscape development and instability during which the buried ice melted more rapidly, resulting in the formation of hummocky moraine and the drainage of the ice-walled lakes. Evidence of postglacial ice-wedge formation coupled with palaeoecological analyses suggest that permafrost aggradation during deglaciation may have delayed the melting of the buried ice for 7000–9000 years with the ice-walled lake plains providing landscape evidence for this early and prolonged stable phase. Even when the climate warmed, molluscan evidence from an ice-walled lake basin in North Dakota suggests the lakes were able to persist for several centuries in a climate similar to modern-day Minnesota with the insulation provided by a thick cover of supraglacial sediment preserving the constraining ice (Clayton *et al.*, 2008).

5.3.2.2 Push Moraines, Thrust Moraines and Composite Ridges

Push moraines, thrust moraines and composite ridges that are broadly associated with the compressive deformation of material by advancing ice margins represent the second type of end moraine the formation of which has been explicitly linked to the interactions of glaciers with frozen ground (e.g. Mackay & Mathews, 1964; Rutten, 1960; Kupsch, 1962; Haeberli, 1979; Bennett *et al.*, 2005; Etzelmüller & Hagen, 2005). It is worth noting at the outset that some authors have argued that the resultant glacitectonic features are *not* end moraines (e.g. Rutten, 1960, Etzelmüller *et al.*, 1996). Nonetheless they are covered in this section as ice-marginal features that are routinely used to identify the terminal limits of glaciers and ice sheets.

As mentioned previously (Section 4.3.1), Mackay (1956, 1959) undertook one of the earliest field studies that documented a number of exposures in the western Canadian Arctic featuring tilted, folded and thrust frozen sediments situated within areas characterised by arcuate ridges and intervening lakes. Herschel Island provides a particularly dramatic example of what was interpreted as an ice-thrust landform, comprising an elevated thrust block over 100 km² in extent and 150 m in elevation featuring prominent arcuate ridges and extensive folds and thrusts. The association of the island with a large marine depression located up ice that is similar in volume and situated within a seafloor that is otherwise flat, provides supporting evidence for an origin through proglacial thrusting to produce a 'hill-hole pair'.

Identifying a key role for permafrost in push-moraine formation is self-evident in the Western Canadian Arctic where permafrost is thought to have persisted for over 40,000 years until the present day (Mackay *et al.*, 1972). However, subsequent work by Rutten (1960) and Kupsch (1962) on ice-thrust ridges in western Europe and western Canada respectively, advocated the influence of permafrost in mid-latitude locations where permafrost is now absent, suggesting therefore that its influence may have been more widespread during Quaternary glacial periods. Kupsch (1962) notes features similar to those described by Mackay in various parts of the Missouri Coteau in Saskatchewan and Alberta, comprising sharply crested arcuate ridges up to 125 m high associated with unconsolidated sediments of Cretaceous age. Rutten (1960) likewise describes straight or arcuate ridges 50–200 m high with single ridges extending for up to 100 km. Both studies also describe similar structural characteristics with the ridges comprising imbricate thrust stacks of unconsolidated sediment. It is the rigidity and integrity of the thrust blocks coupled with a surprising level of preservation of the primary bedding that leads both authors to advocate the presence of ice-rich permafrost. This is thought to cement the otherwise unconsolidated sediments, promoting the preservation of small-scale sedimentological structures that would otherwise have been destroyed during deformation (Section 5.4.2). The role of permafrost is neatly summarised by Rutten (1960) who states that,

> The permafrost layer formed a hard, brittle sheet. When it broke, whole slices were pushed out of the way, with their original sedimentary structures intact. After postglacial thawing the pushed-up blocks remained largely intact structurally because of the internal friction of the layers of sand, gravel and clay (p296).

With the ice-pushed ridges in both cases involving a series of sharply defined imbricate thrusts originating from a single detachment or décollement, both authors also argue that this surface is likely to have corresponded with the lower boundary of the permafrost, which in the case of the European examples is estimated to have been approximately 150 m in thickness. Rutten (1960) goes further by arguing that the rigidity of the permafrost layer was dependent upon river systems draining towards the ice margin and forming an area of waterlogged ground and ice-rich permafrost, thereby explaining why such features were formed during the Riss but not the Würm glaciation when the rivers drained away from the margin. Whilst such a scenario might also have generated large lake systems inimical to the formation and preservation of permafrost, this hydrological inference nonetheless highlights the hydrogeological significance of ice-marginal permafrost that has been investigated in greater detail in recent literature and is covered later in this section.

Consideration of ice-thrust ridges within their broader geomorphological setting can help to elucidate their potential association with ice-marginal permafrost (Section 5.5.2). Moran *et al.* (1980) for example describes ice-thrust ridges within an ice-marginal landform assemblage commonly observed throughout the Prairies of North America and related to the southern margin of the Late Wisconsinan Laurentide Ice Sheet. These landscapes comprise two types of terrain: (i) an outer zone of glacial-thrust blocks and source depressions, and (ii) a more extensive zone of drumlinised terrain located further up-glacier. The glacially thrust terrain features arcuate hills and ridges up to >100 km^2 in area composed of pre-existing material that has been translocated from circular or elongate source depressions located up-ice. As with the previously work of Rutten (1960) and Kupsch (1962), available exposures confirmed that the materials (that include sedimentary bedrock) had been thrust into imbricate slabs or overfolded, with deformation depths reaching up to 200 m. These features occur within 2–3 km of the reconstructed ice margin and are related to a marginal frozen-bed zone within which active ice is thought to have frozen to the substrate with continued ice motion leading to evacuation and deformation of the displaced materials. Moran *et al.* argue that the initial stages of ice advance were associated with 'frozen-bed conditions', with the development of a firm bond between the ice sheet and the pre-existing ice-rich sediments allowing the transmission of shear stresses into the substrate. Failure could have been induced in areas of the bed characterised by either low shear strengths (e.g. lake clays) or favourably orientated planes of weakness.

Field investigations in modern-day glacial environments in which glacier margins have been observed to advance into and couple with permafrost has provided more detailed insights into the morphological and structural diversity of these ice-marginal landforms and the roles permafrost can play in their development. Kälin (1971) for example studied the formation of an active push moraine at the Thompson Glacier on Axel Heiberg Island that at the time of investigation was advancing at c. 6.3 cm day^{-1} into a proglacial outwash plain characterised by continuous permafrost (Figure 5.14). The 'bulldozing' influence of the advancing glacier created a half-moon shaped push moraine 2.1 km long, 0.7 km wide and 45 m high (in comparison to the surrounding outwash plain) with the constituent frozen outwash sediments fracturing into blocks c. 5–10 m thick separated by thrust faults that were often filled with ice. Research in the high mountains of the Swiss Alps by Haeberli (1979) has suggested that smaller but equivalent forms can be found at the margins of small glaciers with frozen margins, whereby the cold-based ice can propagate the basal shear stresses into the frozen proglacial sediments, the deformation of which is in turn facilitated by their ice-rich nature.

Similar thrust-block moraines have been described throughout the permafrozen terrains of the Canadian High Arctic with Kälin (1971) going on to identify 35 examples on Ellesmere and Axel Heiberg Island alone. The majority of the push moraines are associated

Figure 5.14 Push moraine complex located at the margin of Thompson Glacier, Ellesmere Island that is still clearly visible in July 2022 (Image Courtesy: Mike Hambrey).

with outlet glaciers (88%) with most examples being composed largely of outwash (94%). This particular type of push moraine associated with the thrusting and stacking of slabs of bedrock or sediment has consequently been related to glacier–permafrost interactions (Evans, 1991), with the frozen sediments being particularly prone to brittle failure and thrusting at the base of the permafrost providing an ideal décollement as noted previously by Kupsch (1962). With permafrost sequences in the High Arctic commonly reaching several hundreds of metres in thickness, most of the thrust-block moraines occur below the postglacial marine limit where permafrost thicknesses are more limited (Evans, 1991). However, thrust block moraine formation has also been described at the margins of the Suess and Wright Lower Glaciers in the Dry Valleys of Antarctica, indicating the potential for these features to form in the most extreme glaciological environments (Fitzsimons, 1996). In this particular situation, the presence of proglacial lakes was considered key to allowing both the accretion of ice and debris to the cold-based glacier and also in providing a zone of transient wet-based conditions that facilitated the entrainment and translocation of blocks of frozen sediment.

A series of studies in Svalbard have examined the formation of large multi-crested push moraines or composite ridges in response to the advance of polythermal glaciers into areas of pre-existing permafrost. Boulton *et al.* (1999) for example describe the formation of a push moraine at Holmstrømbreen comprising an arcuate belt of ridged terrain extending across the entire valley and measuring c. 1.5 km from its proximal to its distal extremity, created by a surge advance into an ice-contact scarp of proglacial outwash (Figure 5.15). Detailed mapping of the push moraine revealed three structural zones that illustrated a

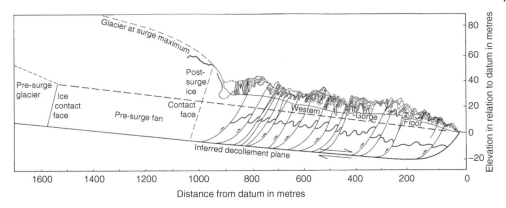

Figure 5.15 Schematic figure illustrating the formation of the push moraine complex at Holmstrømbreen that indicates the position of the pre-surge fan and the form and structure of the resultant push moraine (Boulton *et al.*, 1999, p362). Note the heavily deformed nature of the proglacial sediments and the inferred location of the décollement plane (which in reality may have contained small irregularities).

progressive reduction in cumulative strain with increasing distance from the ice margin. Moving from proximal to distal, they comprised (i) an internal zone dominated by thrusts and nappes, (ii) an intermediate zone characterised by asymmetric folds and thrusts and (iii) an external zone dominated by parallel folds and small-scale thrusts. Hambrey & Huddart (1995) also describe a push moraine complex at Uvêrsbreen with an elevation of 45 m above the surrounding proglacial area. Whilst the deformation style was again dominated by brittle failure and thrusting, which the authors relate to the predominance of outwash sediments, some ductile deformation was also observed where diamictons and debris-rich basal ice had been incorporated into the push moraine sequence.

Whilst there are clear differences in the morphology and structural signatures of these moraine complexes, there are noteworthy similarities that highlight the conditions and processes related to their formation and the role played by permafrost. In both cases, the push moraines are associated with extensive zones of stagnant, ice-cored moraine that separate them from the exposed glacier margin. As in previous investigations, both authors relate the formation of the push moraines to the presence of a single décollement horizon along which the majority of the deformation has taken place. At Holmstrømbreen, the detachment is located at a depth of c. 30–50 m that is coincident with the former ice-bed interface and which is also suggested to provide an indication of the depth of permafrost at the time of glacier advance (Figure 5.15). In considering the mechanics of push moraine formation, Boulton *et al.* (1999) identify two key requirements. First, they calculate that the large-scale displacement of the pushed sediments in front of the advancing ice margin would have required an artesian head of around 40 m and that this excess porewater pressure would need to have progressively risen as the push moraine shortened and thickened. Second, the glacier must have been able to transmit compressional stresses through a thin layer 1 km long and only a few tens of meters thick. Both of these requirements are argued to require the presence of ice-marginal permafrost that acted as an aquiclude. This promoted the development of artesian porewater pressures and allowed the décollement to extend into the foreland whilst simultaneously enhancing the stiffness of the surface sediments and their ability

(a)

(b)

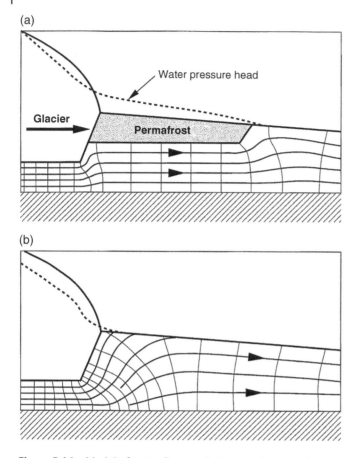

Figure 5.16 Model of water flow at glacier margins featuring proglacial permafrost (a) and no proglacial permafrost (b). Note how in acting as an aquitard and preventing the upwelling of groundwater at the margin, the permafrost generates a much larger area of artesian porewater pressures (Boulton *et al.*, 1999, p368).

to transmit stress (Figure 5.16). Hambrey & Huddart (1995) emphasise the significance of the basal thermal regime suggesting that in the case of surge of Uvêrsbreen, the glacier itself constituted an integral part of the deforming sequence, with a cold-based margin allowing the glacier to mechanically couple with the proglacial permafrost and thereby transmit the compressive stresses required to form the push moraine. They similarly recognise the important role played by permafrost in generating high porewater pressures although they suggest the décollement's position may have been lithologically rather than thermally controlled, occurring at the interface between surficial outwash gravels and underlying glaciomarine muds. In demonstrating significant differences in the structural signatures of two surge-related push moraines associated with small sub-polar glaciers in the same fjörd in Svalbard, Hart & Watts (1997) have also highlighted the influential role played by the nature and rheological properties of the proglacial sediments and the glacial (surge) history.

In connecting the formation of push moraines with the presence of permafrost, these modern-day investigations have led to the suggestion that their development could be indicative of glacier–permafrost interactions. Haeberli (1979) for example contended that, 'there is no doubt that push-moraines are morphological expressions of the deformation of permafrost' (p43). In reviewing the morphology, structural evolution and wider significance of push moraines, Bennett (2001) however cautions that the establishment of any such process-form connection is ambiguous and contentious. The formation of large push moraines at temperate surge type glaciers such as the Taku Glacier in Alaska (e.g. Motyka & Echelmeyer, 2003) for example indicates that the rapid application of stress to unconsolidated, unfrozen materials can lead to styles of deformation that some authors have argued require permafrost conditions as a prerequisite. Bennett (2001) suggests that an area of genuine consensus relates to a recognition of the importance of the presence, rheology and hydrogeological conditions of basal décollements. If correct, this suggests that whilst permafrost is not essential to the development of push moraines, its presence can at least facilitate push moraine formation through the generation of extensive artesian porewater pressures at the base of the permafrost layer, enabling it to act as a low friction décollement for the overlying permafrozen materials.

As part of a study of the ice-cored moraines and push moraines formed at two valley glaciers on Svalbard (Erikbreen and Usherbreen), Etzelmüller *et al.* (1996) explore the connections between glacier thermal regime and glacier dynamics and the role played by permafrost in the genesis of both moraine types covered in this section. In addition to arguing that both types of moraine formation require permafrost conditions, they identify the potential for a crossover or 'hybridisation' of ice-cored moraines and push moraines, with a surge of Usherbreen into an ice-cored moraine system resulting in its reactivation and the deformation of the ice masses they contain. The recognition of and differentiation between ice-cored moraines and push moraines can therefore prove challenging in some circumstances, with the moraine types displaying close spatial associations and being genetically related in so far that they are both subject to ice-marginal compressive deformation (Haeberli, 1979). With glacially induced movement being recorded over 300 m in front of the advancing glacier margin, Etzelmüller *et al.* (1996) once again emphasise the role of permafrost in controlling the mechanical properties of the proglacial sediment and facilitating the transmission of the compressive stresses into the glacier foreland. Somewhat controversially, they conclude that with push moraines forming in the absence of a direct physical connection with the glacier margin, they should be viewed principally as periglacial features or 'deformed and disturbed sediments in permafrost where the stress applied is of glacial origin' (p61).

5.3.3 Subglacial Landforms

Glacial geomorphologists have tended to assume that cold-based or dry-based glaciers are by definition associated with a cessation of basal processes (Section 3.4) with Kleman & Borgström (1996) for example asserting that a 'frozen-bed condition leads to a hiatus in landform development' (p898), thereby leading to the preservation of pre-glacial surfaces as discussed in Section 5.2. Whilst there is extensive evidence for glaciers resting on permafrost playing this protective role, the recognition that basal processes can remain active at sub-freezing temperatures has led to a growing recognition that cold-based glaciers resting on permafrost can also play a more active geomorphological role than has hitherto been assumed to be the case (Section 3.5.3). Whilst rates of subglacial erosion and landscape modification are likely to be extremely low in comparison with temperate glaciers, early observations suggested that they are not entirely inactive. In searching for evidence of basal motion beneath the cold-based Meserve Glacier in Antarctica, Holdsworth (1974) for example described 'rock smears' or 'streaks' of powdered quartz and feldspar on the surface of embedded subglacial boulders, which he related to interactions between rocks held within deforming basal ice with the tops of boulders embedded within the static substrate.

More recent field mapping of deglaciated terrain in the Allan Hills region in East Antarctica by Atkins *et al.* (2002) has revealed a surprisingly diverse range of subglacial landforms and sediments produced by cold-based glaciers. With a mean annual air temperature of $-30\,°C$ and an estimated basal temperature of c. $-24\,°C$ beneath the Manhaul Bay Glacier, this indicates the potential for geomorphic activity to remain active beneath glaciers located in some of the coldest environments on Earth (Section 5.5.4). The authors describe distinctive erosional features occurring within 50 m of the receding ice margins, primarily comprising abrasion marks with variable sizes, groupings and orientations. These are subdivided into: broad 'scrapes' (up to 500 mm in width, 40 mm in depth and 1200 mm in length) that commonly comprise assemblages of smaller striae or grooves, individual striae or grooves typically centimetres in width and depth and decimetres in length, scraped boulders whose desert varnish has been removed by abrasion and which have sometimes been overturned, and 'ridge and groove lineations' with an appearance similar to slickensides. Associated depositional features were observed to include: sandstone and siltstone breccia, isolated sandstone boulders (up to 3 m in diameter) that sometimes formed poorly defined trains, ice-cored debris cones (up to 3 m high and 8 m in diameter) and concentrations of sandstone debris located on the lee-sides of bedrock ridges and escarpments as well as in surface depressions. Taken collectively, these features are thought to have formed subglacially following the initial entrainment of blocks of bedrock from the stoss-sides of bedrock outcrops (Figure 5.17). These entrained blocks are then dragged across the bed to create the abrasional features, a process that results in their comminution and deposition as breccias on the lee-side of bedrock outcrops as isolated boulders or ice-cored cones. In addition, the ridge and groove lineations are believed to relate to the glacitectonic deformation of bedrock in response to the stress gradients imposed on bedrock promontories with displacement occurring along weak shale layers.

Follow-up research by Lloyd-Davies *et al.* (2009) has provided additional detail on both the features and mechanisms involved. The authors highlight the important role played by the ice-block aprons characteristic of cold-based glacier margins (Section 3.6.1), the overriding of which enables the ice margin to both override bedrock obstacles and to entrain boulders and other debris that can enable subsequent erosion. Variations in the contact pressure between the stoss-side of any obstacles and their lee-sides where cavitation is believed to occur results in significant glacitectonic deformation of the horizontally bedded bedrock leading to its brecciation. This material is subsequently rotated and dragged down-glacier as part of a 'rolling carpet' of ice blocks to produce boulder trains tens to hundreds of metres long. The authors also describe distinctive ploughed boulders close to the margin of the Odell Glacier associated with ridges of friable tillite on their lee sides.

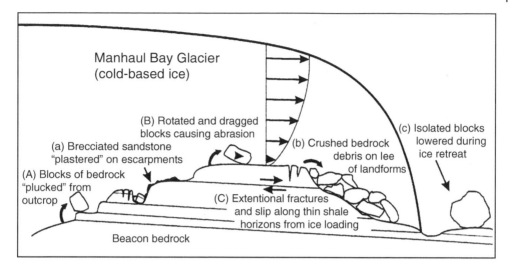

Figure 5.17 Schematic figure of the range of erosional and depositional features observed in the Allan Hills in Antarctica (Atkins *et al.*, 2002, p661). Erosional processes and features are signified by capital letters and depositional processes and features by the lower case letters. See text for details.

As well as demonstrating the potential for geomorphological activity to remain active beneath very cold glaciers, these studies highlight notable differences between the equivalent features generated beneath warm-based and cold-based ice. Beneath cold-based ice for example, many scrapes, striations and grooves were observed to terminate abruptly and display levées composed of the original striator, suggesting the eroding clast had disintegrated during simple shear that occurred in the absence of meltwater. In addition, Atkins *et al.* (2002) noted that 'the marks are variable in shape, size, and grouping, and are unlike the more consistent, uniform sets of parallel striae and grooves found commonly on bedrock abraded by warm-based sliding ice' (Atkins *et al.*, 2002, p66). These distinctive differences provide potentially diagnostic criteria that could be used in relict glacial environments in conjunction with other lines of evidence to reconstruct the thermal regime of former ice masses and the occurrence of glacier–permafrost interactions (see Sections 5.4.1 and 6.3). However, the subtle nature of many of these features coupled with the intense subaerial weathering associated with permafrost environments raises questions of their preservation potential with Lloyd-Davies *et al.* (2009) suggesting that only the displaced boulders and larger tectonic features are likely to survive over long timescales.

Work in past glacial landscapes has also provided insights into the processes and landscape signatures of erosion beneath cold or dry-based ice. Detailed mapping of different types of tor on the Cairngorm Plateau in Scotland by Hall & Phillips (2006) for example revealed that that their geomorphological character (Figure 5.3) and the related degree of glacial modification displayed consistent spatial variations across the plateau (Section 5.2). The authors consequently proposed a five stage model involving increasingly significant degrees of glacial modification, in turn related to differences in basal thermal regime (Figure 5.18). Stage 1 tors showed no sign of modification and are considered to have been covered by non-erosive, dry-based (i.e. cold-based) ice. Stage 2 tors showed subtle

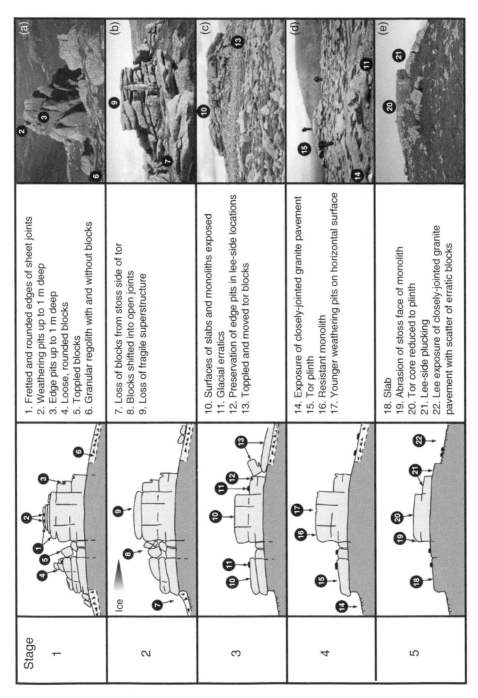

Figure 5.18 Model of the glacial modification of tors on the Cairngorm plateau in Scotland (Hall & Phillips, 2006, p823/John Wiley & Sons, Inc.). The figure provides details on the characteristic features of the stages that illustrate a progressive increase in the level of modification and a transition from cold to warm-based ice.

Stage		
1	(a)	1. Fretted and rounded edges of sheet joints 2. Weathering pits up to 1 m deep 3. Edge pits up to 1 m deep 4. Loose, rounded blocks 5. Toppled blocks 6. Granular regolith with and without blocks
2	(b)	7. Loss of blocks from stoss side of tor 8. Blocks shifted into open joints 9. Loss of fragile superstructure
3	(c)	10. Surfaces of slabs and monoliths exposed 11. Glacial erratics 12. Preservation of edge pits in lee-side locations 13. Toppled and moved tor blocks
4	(d)	14. Exposure of closely-jointed granite pavement 15. Tor plinth 16. Resistant monolith 17. Younger weathering pits on horizontal surface
5	(e)	18. Slab 19. Abrasion of stoss face of monolith 20. Tor core reduced to plinth 21. Lee-side plucking 22. Lee exposure of closely-jointed granite pavement with scatter of erratic blocks

modifications including the toppling of blocks and their localised transport thought to indicate more significant rates of ice motion via internal creep. Stages 3 and 4 involve the more extensive removal of the tor suprastructure and its eventual reduction to a stump, with lee-side plucking in stage 4 providing evidence for transient warm-based conditions. Finally, stage 5 sees the wholesale modification of the tors to produce whalebacks and roche mountonnées, indicative of more rapid and long-lived basal sliding and erosion beneath warm-based ice. In relating the degree of modification to basal processes and the basal thermal history of the locality, the authors create a geomorphological inverse model in which the presence or tors in stages 1 and 2 provides evidence for persistent frozen-bed conditions (i.e. glacier–permafrost interactions).

With drumlins and other forms of classic glacial lineations typically being regarded as the product of active basal processes and warm-based conditions, Rogen or ribbed moraines constitute a specific type of flow-transverse glacial bedform that has been related to cold-based ice and more specifically to subglacial thermal transitions (Figure 5.19). In a paper examining the extent of frozen-bed conditions beneath the Fennoscandian and Laurentide Ice Sheets during the Last Glacial Maximum (LGM), Kleman & Hättestrand (1999) suggest that in the absence of direct measurements, these landforms can be used to identify and map larger-scale spatial variations in basal thermal regime and subglacial rheology. They argue that ribbed moraines are the result of the brittle fracture of 'frozen drift sheets' (p63), with the fracturing occurring at the boundary between frozen and thawed-bed conditions. An increase in the basal ice flow velocity generates strong longitudinal tensile stresses that

Figure 5.19 Ribbed or Rogen moraine at Lake Rogen in Sweden. The transverse ridges and intervening lakes are orientated transverse to ice flow that occurred from right (north west) to left (south east) (Image: Richard Waller).

in combination with a reduction in the depth of the phase-change surface to the bedrock–sediment interface is thought to create a detachment zone along which the frozen subglacial sediment fractures to produce the transverse ribbed moraines (Figure 5.20). These detachment zones consequently migrate up-ice during deglaciation creating landform zones dominated by ribbed moraine that illustrate the time-transgressive location of the thermal boundary between thawed bed zones (marked by glacial lineations) and frozen bed zones (marked by the preservation of relict features). Mapping of these landform zones at the ice-sheet scale can in turn be used to infer the location of large areas of frozen-bed conditions beneath the ice dispersal centres of the Laurentide and Fennoscandian Ice Sheets during the LGM.

If this interpretation is correct, then ribbed moraines provide important insights into both the geomorphological expression of the glacier–permafrost interactions and their potential influence on both the dynamic behaviour and morphology of continental-scale ice sheets. However, the origin of ribbed moraines remains the subject of ongoing debate. Alternative hypotheses include their association with the development of flow-related

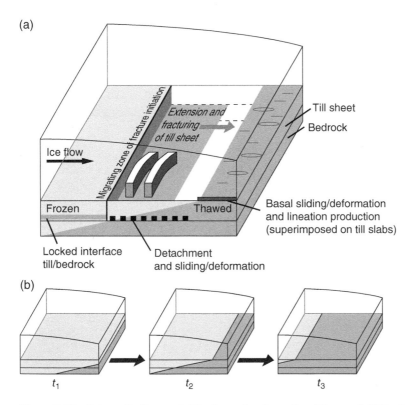

Figure 5.20 Schematic figure of ribbed moraine formation (Kleman & Hättestrand, 1999, p65). (a) illustrates the fracturing central to their formation occuring within a narrow, thermally controlled zone (marked by the dashed line) in which the phase-change surface rises to intersect with the till-bedrock interface. This enables the extension and fracturing of the till sheet and the formation of the transverse features. (b) illustrates the up-glacier recession of the boundary between the thawed and frozen bed thought to have accompanied the final decay of the ice sheet.

instabilities within actively deforming subglacial sediments (Dunlop *et al.*, 2008) that would traditionally be considered reflective of entirely warm or wet-based subglacial conditions. Möller & Dowling (2018) meanwhile have argued that they are the product of equifinality and as such, any search for a singular connection between process and form is misleading.

With the model of Kleman & Hättestrand (1999) being based around the assumption of an abrupt change in basal boundary conditions, subglacial processes and basal ice flow velocities across a thermal boundary, it is important to note that as discussed previously (Section 4.2), studies on modern-day glaciers undertaken with the express aim of examining the influence of thermal transitions have as yet failed to provide clear or consistent evidence of a significant influence on patterns of ice flow and strain fields (e.g. Moore *et al.*, 2011). A more recent focus on hydrological controls has culminated in the combination of thermal and hydrological changes within the concept of enthalpy gradients (Benn *et al.*, 2019a, b). This predicts that it is only where a thermal change corresponds with enhanced meltwater production up-glacier, that it is likely to cause an important yet still indirect influence on glacier flow. Further consideration of the potential geomorphic expression of significant inputs of meltwater at thermal boundaries may therefore provide additional opportunities to test this novel hypothesis.

5.3.4 Massive Ground Ice

The recession of glaciers commonly results in the deposition and burial of stagnant ice masses within the proglacial area (e.g. Attig & Clayton, 1993; Attig & Rawling, 2017). Such buried ice masses are relatively transient features within more temperate deglacial landscapes where they typically melt out within a few decades or centuries (Section 5.3.2). During the course of his first exploration of Glacier Bay in Alaska in 1879, the renowned naturalist John Muir provided an early and remarkably astute account of the processes and landforms associated with the burial of ice. Whilst examining an un-named glacier terminating on land he observed that,

> *Many large masses, detached from the wasting front by irregular melting, were partly buried beneath mud, sand, gravel, and boulders of the terminal moraine. Thus protected, these fossil icebergs remain unmelted for many years, some of them for a century or more, as shown by the age of the trees growing above them, though there are no trees here as yet. At length melting, a pit with sloping sides is formed by the falling in of the overlying moraine material into the space at first occupied by the buried ice. In this way are formed the curious depressions in drift-covered regions called kettles and sinks*
> (Muir, 1988, p128).

Within colder climatic regimes, glacier recession is associated with the aggradation of permafrost which can promote the longer-term preservation of buried ice located below the depth of seasonal thaw, potentially over geological timescales (Sugden *et al.*, 1985; Clayton *et al.*, 2001). The long-term preservation of buried ice within deglaciated permafrost landscapes can therefore be considered a distinctive landscape feature associated with glacier–permafrost interactions with the ice itself becoming a distinctive depositional product (Section 5.4.1). When viewed from the perspective of permafrost environments, these buried ice masses represent a particular type of ground ice, a term used to refer more broadly to any type of ice formed in freezing or frozen ground (Section 2.2.3) (French, 2017). Large bodies of so-called massive ice constitute one of the most spectacular forms of ground ice that have been widely documented in the western Canadian Arctic (e.g. French & Harry 1990; Mackay & Dallimore 1992; Lacelle *et al.*, 2018) and western Siberia (e.g. Vtyurin & Glazovskiy 1986; Astakhov & Isayeva 1988; Astakhov *et al.*, 1996). Massive ice is defined by a gravimetric ice content exceeding 250% by weight (Harris, 1986) and ordinarily occurs as large, tabular bodies that can reach over 10 m in thickness. Whilst the distribution, stratigraphic setting and physicochemical characteristics of massive ice have been increasingly well constrained, its mode of origin and palaeoenvironmental significance remain subject to debate. Although the principal mechanisms proposed to account for the generation of massive ice differ, both relate to the interactions between glaciers and permafrost and are therefore worth exploring in detail.

The first hypothesis is that massive ice represents exceptionally large bodies of segregated-intrusive ice created by a combination of subaerial permafrost aggradation and high sub-permafrost porewater pressures accompanying the recession of an ice sheet (e.g. Mackay, 1972, 1989; Mackay & Dallimore, 1992). Once the aggrading permafrost intercepts the sub-permafrost aquifer, sustained water flow to a stationary freezing front results in the formation of thick bodies of tabular ice. The observed tendency of massive ice in the western Canadian Arctic to form at the stratigraphic boundary between clay-rich materials and subjacent sands with much higher permeabilities provides a key line of evidence for this hypothesis with the resultant ground ice being referred to as 'intrasedimental ice'. With this in mind, the characteristics of massive ice described within coastal exposures at sites like Peninsula Point near Tuktoyaktuk are considered consistent with this mechanism, displaying for example a prominent sub-horizontal layering and a conformable contact with the overlying sediments. In addition, ice dykes extending upward from the top of the massive ice into the overburden demonstrate the existence – during massive-ice formation – of porewater pressures sufficiently high to hydrofracture the overlying frozen sediments (Figure 5.21, Mackay & Dallimore 1992). Rampton (1988, 1991) has argued that the high porewater pressures and massive-ice formation consequently relate to a glacially imposed hydraulic gradient driving subglacial meltwater to an aggrading proglacial permafrost table during deglaciation (see also Lacelle *et al.*, 2004).

An alternative explanation is that massive ice represents buried glacier ice that has been preserved within permafrost following ice retreat. The ability of glacier recession to result in the burial of stagnant ice within ice-marginal environments and for this ice to be preserved for centuries in non-permafrost conditions is well established (e.g. Everest & Bradwell, 2003) (Figure 5.12). This process is particularly common in polythermal glaciers with cold-based margins where ice-marginal compression can lead to thrusting, isolation and preservation of buried ice within controlled moraines (see Section 5.3.2). If glacier recession is associated with the aggradation and continued presence of permafrost, then the buried glacier ice can potentially be preserved for millions of years (e.g. Sugden *et al.*, 1985). This hypothesis is the preferred explanation amongst Russian researchers for the massive ice that is widespread in western Siberia and whose preservation is therefore considered the key product of 'incomplete' or 'retarded deglaciation' (e.g. Astakhov &

Figure 5.21 Upper contact of massive ice exposed at Peninsula Point, Canadian Northwest Territories. Note the prominent ice dykes to the left and right of the spade that project upwards into the overlying frozen diamicton. These are thought to relate to hydrofracturing with the dykes featuring small ice crystals, chilled margins and few bubbles, consistent with rapid freezing (Image: Richard Waller).

Isayeva 1988; Grosval'd *et al.,* 1986; Kaplyanskaya & Tarnogradskiy 1986; Astakhov *et al.,* 1996). Potential remnants of former ice sheets have also been described at various locations in the Canadian Arctic (e.g. Dyke & Savelle, 2000; Dyke & Evans, 2003; Lacelle *et al.,* 2007, 2018; Coulombe *et al.,* 2019; Evans *et al.,* 2021) as well as in Alaska (e.g. Kanevskiy *et al.,* 2008, 2013) and East Antarctica (e.g. Sugden *et al.,* 1985). Key lines of evidence for this mode of origin include the stratigraphic setting and structural characteristics of the massive ice (e.g. Ingólfsson & Lokrantz, 2003). In contrast to intrasedimental ice, massive ice of glacial origin generally features an unconformable upper contact (thaw unconformity) and is often entirely surrounded by diamicton, whose low permeability precludes an open system water supply to an aggrading freezing front. Consequently, such ice is frequently described within ice-cored moraine complexes (e.g. French & Harry 1988; Dyke & Savelle 2000; Lacelle *et al.,* 2007; Lakeman & England, 2012). In addition, the ice frequently contains both large-scale and small-scale tectonic structures such as thrust faults and recumbent folds (e.g. French & Harry 1990; Astakhov *et al.,* 1996) (Figure 5.22). Such features are widely considered indicative of shear deformation within the basal layers of an ice sheet and to be inconsistent with an origin through segregation following deglaciation, although Rampton (1988, 1991) has argued that such features might also be imparted on intrasedimental ice by glacier re-advance.

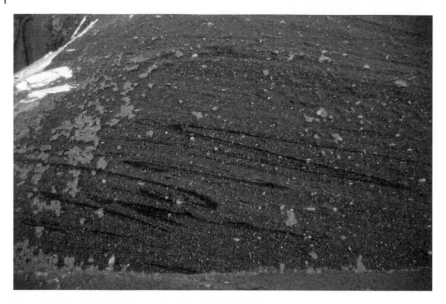

Figure 5.22 Debris-rich massive ice with folded clear ice layers exposed at Mason Bay, western Canadian Arctic. The face is approximately 3 m high (Image: Richard Waller).

Massive ice within glaciated permafrost terrains has therefore usually been interpreted either as buried glacier ice or as intrasedimental ice formed in subaerial permafrost regions. Distinguishing between these origins is fundamental to understanding the interactions between glaciers and permafrost, and the characteristics and behaviour of former ice sheets but has proven challenging in practice (Vtyurin & Glazovskiy 1986; French & Harry 1990). The development and application of a range of new techniques, including co-isotopic techniques used routinely in glaciology (e.g. Lorrain & Demeur, 1985) and the measurement of the molar ratios of gasses (comprising O_2, CO_2, N_2 and Ar) entrapped in bubbles (Lacelle *et al.*, 2007) has improved our ability to distinguish firnified glacier ice from intrasedimental ice. Discriminating debris-rich basal ice of glacial origin (as the ice type most likely to form and be preserved around former ice sheets) (Section 3.5.4) from intrasedimental ice formed by subaerial processes remains more problematic however, with both ice types forming through the same freezing processes operating in porous media (cf. Mackay 1989; Rempel 2007). The lack of a clear genetic distinction between basal glacier ice and massive intrasedimental ice and the associated similarities in their physical and chemical properties explains why distinguishing the two has proven so challenging. It seems more likely that any distinction between the two may relate to their spatial distribution and their stratigraphic setting. Therefore, the more useful distinctions amongst massive-ice occurrences in permafrost regions are likely to be between (i) subglacial or submarginal ice (e.g. Murton *et al.,* 2005), (ii) proglacial ice (e.g. Lacelle *et al.,* 2004) and (iii) non-glacial ice (e.g. Lawson 1986).

An illustration of the challenges associated with the interpretation of the origin and significance of massive ice is provided by Murton *et al.* (2004, 2005) who examined a series of exposures of massive ice and icy sediments at former ice-marginal sites of the Laurentide

Ice Sheet (LIS) located within the Tuktoyaktuk Coastlands of western Arctic Canada. The massive ice was observed to display particular features characteristic of both basal glacier ice and intrasedimental ice. Basal-ice features comprised: (i) ice facies and facies groupings similar to those from the basal ice layers of contemporary glaciers and ice sheets in Alaska, Greenland and Iceland; (ii) ice crystal fabrics similar to those from basal ice in Antarctica and ice-cored moraines on Axel Heiberg Island, Canada and (iii) a thaw or erosional unconformity along the top of the massive ice and icy sediments, buried by glacigenic or aeolian sediments. Intrasedimental ice consisted of pore ice and segregated ice formed within Pleistocene sands deposited prior to the area being overridden by the LIS.

The co-existence of basal and intrasedimental ice within massive ice and icy sediments suggests that they formed within the basal ice layer of the LIS. This layer is thought to have developed through the accretion of both new and existing ice with the formative mechanisms including (i) large-scale freeze-on of meltwater and sediment at the transition from warm- to cold-based ice; (ii) permafrost aggradation beneath thinning, stagnant basal ice; and (iii) pore-water expulsion in ice-free areas prior to glacial overriding. Stagnation of the ice sheet and melt-out of till from the ice surface downwards allowed burial and preservation of the basal ice layer on a regional scale in the Tuktoyaktuk Coastlands. At other sites close to but within the Laurentide ice limit, buried glacier ice beneath frozen glacitectonite that contains subglacially eroded clasts of ground ice indicates that near-surface permafrost did not degrade beneath some parts of the ice margin (cf. Mackay *et al.,* 1972; Mackay & Matthews 1983; Rampton 1988).

To integrate all of the studies of massive ice in the Tuktoyaktuk Coastlands, a two-stage model of ground-ice development has been proposed based on the cryostratigraphic distinction of two generations of ground ice: pre- and post-deformation ice (Murton, 2005). Pre-deformation ice was glacially deformed or eroded beneath the cold-based margin of the LIS during Marine Isotope Stage 2 (Murton *et al.,* 2007) and comprises (i) buried basal ice, (ii) massive segregated ice and (iii) ice clasts subglacially eroded from pre-existing ground ice. In contrast, post-deformation ice has not been glacially disturbed because it formed during or after deglaciation, and includes (iv) dykes and sills of intrusive ice, (v) massive segregated-intrusive ice, (vi) ice wedges and composite wedges, (vii) segregated ice and (viii) pool ice. The superimposition of post-deformation ice into permafrost containing older, pre-deformation ice indicates that substantial quantities of overpressurised water were injected into ice-marginal permafrost during or after deglaciation. The required external water source for the post-deformation intrusive ice was probably overpressurised subpermafrost groundwater in front of the retreating margin of the LIS (Section 4.2.5). Injection of this water into proglacial permafrost hydraulically fractured the permafrost and formed ice dykes (Figure 5.21), ice sills and massive segregated-intrusive ice. This model may have wider applications to the development of massive ice wherever ice sheets advanced and retreated across lowland regions of continuous permafrost.

Ongoing efforts to distinguish the origin of buried massive ice found within glaciated permafrost environments are crucial to understanding their wider glaciological and palaeoenvironmental significance. Intrasedimental ice masses for example have the potential to provide an insight into the hydrogeological processes and pathways associated with ice-marginal and proglacial environments (Section 4.2.4). Alternatively, if they represent the remnants of former Pleistocene ice sheets, then these products of 'incomplete' or 'retarded' deglaciation provide a palaeoglaciological archive of immense value. Grosval'd *et al.* (1986) estimate that approximately 10,000 km³ of relict glacier ice remain preserved within western Siberia alone. Whilst this is likely to represent an overestimate, even if only a fraction of this is related to Pleistocene ice sheets, its detailed investigation still has the potential to make a significant contribution to our knowledge and understanding of the configuration, physical properties and dynamic behaviour of continental ice sheets within high-latitude regions (Waller *et al.*, 2009a).

Research by Lacelle *et al.* (2018) provides one of the first targeted attempts to utilise proxy information uniquely preserved within buried massive ice to address outstanding questions within palaeoglaciological reconstructions. Whilst the mapping of moraines and other glacial landforms has provided a clear insight into the maximum extent of the Laurentide Ice Sheet during the Late Wisconsinan and its subsequent pattern of decay (e.g. Dyke & Prest, 1987), the reconstruction of its geometry and surface profile using glaciological and geophysical models with differing assumptions has led to the production of a diverse range of models with conflicting ice-sheet configurations. These in turn have important implications for the reconstruction of the ice-sheet's changing volume and its contribution to global eustatic sea level during deglaciation. The authors utilise a palaeo-altimitry method based around the measurement of stable isotopes ($\delta^{18}O$) previously used to reconstruct changes in the Holocene elevation of the Greenland Ice Sheet and Agassiz Ice Cap. Isotope data is collected from four massive ice exposures in the Eastern and Western Canadian Arctic that are interpreted as buried glacier ice and combined with pre-existing data from Late Wisconsinan ice found within the Barnes and Penny Ice Caps on Baffin Island to provide a more spatially extensive data set. With the ice being assumed to be derived from local accumulation centres and isotopic values being depressed by 5–7 per mil, the isotopic data are used to infer the elevation of the key ice domes and divides in which the ice formed and to suggest that the Laurentide Ice Sheet displayed an asymmetric profile at the LGM with a high-elevation central Keewatin Dome (c. 3200 m) and a much lower western plains ice divide (c. 1700–2100 m).

5.3.5 Icings, Pingos and Blowouts

With non-temperate glaciers terminating in permafrozen forelands being associated with a range of distinctive hydrological processes and flow pathways that differ from their temperate counterparts (Section 4.2), the formation of meltwater-related landforms provides an additional means of examining the landscape expression of glacier–permafrost interactions.

As noted previously in Section 4.2.2, proglacial icings are a hydrological feature commonly observed to form in the proglacial areas of both polythermal and entirely cold-based glaciers. They provide an excellent example of a transient landform indicative of glacier–permafrost interactions with the glacier providing the perennial water source (Wainstein *et al.*, 2014) and the permafrost environment the setting in which the discharged water can freeze to produce these layered ice accumulations that can reach several metres in thickness and several thousand square metres in extent (Bukowska-Jania & Szafraniec, 2005).

A large proglacial icing located at the margin of Fountain Glacier on Bylot Island provides a particularly relevant case study that Wainstein *et al.* (2008) explicitly relates to the existence of an unstable equilibrium between the glacier and the proglacial permafrost.

The icing extends up to 11 km down valley of the glacier margin although it is only the most proximal 1.2 km that has featured continuous ice cover since 1948 (Figure 4.4). GPR surveys of this perennial icing by Moorman & Michel (2000a) revealed a thickness of up to 12 m with significant spatial variations relating to the topography of the underlying valley floor. Whilst the icing was generally characterised by a tabular structure, it also displayed significant compositional and structural variations between the perennial and seasonal parts. These suggest the operation of a number of formative processes including the freezing of surface sheet flows and slush flows, the injection of water into the icing to produce ice lenses and the infilling and freezing of stream channels.

Returning to the geomorphological significance of the hydrological connections between glaciers and permafrost introduced in Section 4.2.5, it is the presence of the nearby polythermal glacier with a cold and impermeable margin that is considered pivotal to the existence, volume and persistence of the icing (Wainstein *et al.*, 2008, 2014). With the warm-based interior of the glacier delivering a year round supply of meltwater, an unfrozen proglacial talik through the ice-marginal permafrost maintains a hydrological connection between the subglacial environment and the glacier foreland with continued flow through the winter allowing the icing to regenerate. Glacier recession however can result in a hydrothermal disturbance that drives the system out of balance. Subsequent permafrost aggradation results in the closure of the talik and the up-valley migration of the spring, restoring the threshold distance between the glacier margin and the spring location and ensuring the replenishment of the icing. Such surface features are however inevitably going to be vulnerable to the impacts of climate change and are therefore likely to be transient in nature. In undertaking repeat GPR surveys, Wainstein *et al.* (2008) recorded an average thinning of 5.3 m and a total volume loss of >500,000 m^3 between 1993 and 2007. The burial of icings provides a potential mechanism through which remnants could be preserved for longer-time scales although Moorman & Michel (2000b) considers this unlikely in view of their formation within dynamic and largely erosional settings.

Open-system pingos are conspicuous permafrost landforms that have been used to demonstrate the broader hydrological interconnections between glaciers and permafrost at the landscape scale. In this case, the freezing of meltwater occurs below the surface to produce a core of ground ice and the distinctive conical landform. In identifying a spatial association between springs and open-system pingos, Liestøl (1977) provides an intriguing illustration of a distinctive process-form linkage indicative of glacier–permafrost interactions whereby the creation of one of the most distinctive permafrost landforms can be explicitly connected to a glacially derived water source (Figure 4.11). The author draws on a series of earlier studies to identify 25 pingos distributed throughout Spitsbergen. These display a diversity of forms and commonly occurred in clusters, suggesting that the location of the spring that provided the source for ground ice formation had migrated over time. Evidence of active springs within some of the pingos that displayed characteristic horseshoe-shape planforms provided evidence of their association with taliks, with high salinities being consistent with subpermafrost water sources. When observed in winter or early spring, some of the pingos were associated with large ice domes with active discharge contributing to the associated development of icings. In the case of two large examples observed in Reindalen, the inclusion of fractured sandstone beds within the crater walls of the pingos attests to the generation of very high artesian pressures whilst suggesting that surface–groundwater interactions are not necessarily restricted to areas of unconsolidated sediment (Liestøl, 1996).

Scholz & Baumann (1997) have also described a classic cone-shaped pingo measuring 60–70 m in diameter and 15–20 m in height located within a moraine complex in front of the Leverett Glacier in west Greenland (see Waller & Tuckwell, 2005) (Figure 5.23). This is considered a rather unusual geomorphological setting that illustrates potential connections between the formation of the pingo and the moraine complex in which it occurs, with both requiring artesian subpermafrost water pressures (Section 5.3.2). Geochemical analysis of spring water emanating from a crater at its crest indicated high levels of bromide, chloride and sulphide suggestive of a deep-seated origin as thermal fluids. Scheidegger *et al.* (2012) have more recently suggested that in areas of heavily consolidated sedimentary or crystalline bedrock terrains, fault zones can play an important role in providing the secondary permeability and focused hydraulic pathways necessary for the development and sustenance of groundwater flow and spring systems in permafrost environments more generally. Whilst these preliminary analyses of Scholz & Baumann (1997) were used to question the viability of a glacial water source, more detailed examination of the geochemical and isotopic signature of a pingo ice core in Grøndalen, Spitsbergen by Demidov *et al.* (2019) has confirmed the ability of subglacial meltwater to follow shallower sub-permafrost flow pathways, with fault structures again playing a key role in focusing the upwelling groundwater and promoting both spring and pingo development.

The generation of negative effective pressures close to the margins of glaciers and ice sheets terminating in permafrost may become sufficiently high to cause hydrofracturing of

Figure 5.23 Open-system pingo located close to the margin of the Leverett Glacier in western Greenland. Note the prominent staining to the right of the figure associated with the active spring draining from the crater at its crest (Image: Richard Waller).

the surficial, low permeability aquiclude and the creation of hydrodynamic blow out structures. This is particularly likely if the subglacial aquifer thins towards the margin leading to the development of strong upward flow vectors (Boulton *et al.*, 1995). In making the connection between this scenario and the formation of large push moraines, it is interesting to note that the over pressurisation of the aquifer required to mobilise the moraine complex is also expected to generate hydrofracturing and liquifaction close to its proximal extremity. Boulton & Caban (1995) suggest that large 'circular disintegration ridges or rimmed kettles' (p586) associated with the Dirt Hills push moraine in Saskatchewan, formerly interpreted as ice-stagnation features, might instead relate to the release of over pressurised groundwater. This process has been advocated by Evans *et al.* (2020) to explain the formation of a series of features including 'doughnuts', 'doughnut chains' and blow out features around the margins of a surging ice lobe in Alberta, although this is related primarily to an advance against an adverse bedrock slope rather than to ice-marginal permafrost. In considering the diverse nature and origin of these enigmatic features commonly found within the Canadian Prairies that are also referred to as prairie mounds, Mollard (2000) discusses both artesian groundwater discharge and the formation and decay of ground ice as potential mechanisms of relevance to glacier–permafrost interactions. Whilst such potential process-form connections are intriguing, with these features being likely to display 'manifold polygenetic origins' (p187), the establishment of any such a connection on the basis of their morphology alone remains problematic.

5.4 Sedimentological and Structural Evidence for Glacier–Permafrost Interactions

5.4.1 The Influence of Thermal Regime on the Sedimentology of Glacigenic Debris

Relatively little research attention has focused specifically on the influence of a glacier's thermal regime on the sedimentological characteristics of the material they transport and deposit. A rare example and seminal contribution provided by Boulton (1972) and discussed earlier in the specific context of glacial erosion (Section 3.4.2), also considers the influence of thermal regime on glacial sedimentation with respect to a conceptual model comprising four distinct basal boundary conditions. In zone C (characteristic of polythermal glaciers), active refreezing of meltwater to the glacier sole maintains the basal ice at the pressure melting point enabling the continued operation of basal sliding, abrasion and crushing as well as plucking. The resultant erosional products are subsequently entrained into the actively accreting debris-rich basal ice (see Section 3.5.4), the degradation of which subsequently results in the deposition of melt-out tills and flow tills. In zone D (characteristic of cold-based glaciers), the subglacial sediments are frozen and the glacier is assumed to adhere to the bed with an absence of basal motion precluding significant abrasion and crushing. Deposition in turn again occurs primarily in the form of melt-out tills and flow tills with the strong ice-bed coupling also promoting the plucking of large erratics (see next section).

This model illustrates and emphasises a specific and long-established connection between environment, process and form whereby glacier–permafrost interactions and distinctive basal thermal regimes enable both the formation of debris-rich basal ice (as discussed in Section 3.5.4) and the subsequent deposition of the entrained material via ablation to produce a distinctive type of glacigenic sediment. Employing early field observations in East Greenland to consider these processes of debris entrainment, transport and deposition, Chamberlin (1895) emphasised the remarkable stratification of the ice that was 'almost as distinctly bedded and laminated as sedimentary rock', featuring two distinct ice types including a 'lower laminated tract discolored by debris' (p203). Melting at the ice margin was subsequently observed to release this debris resulting in the formation of a talus slope (Section 3.6.2).

Detailed field observations of till deposition at the margins of contemporary glaciers in Svalbard led Boulton (1970b) to propose the term 'melt-out till' to describe diamicts released by the melting of debris-rich ice in confined and stable settings, in which case the sedimentological characteristics of the parent ice could be preserved (Boulton, 1970b). These till types are considered particularly prevalent in polar glacial environments with proglacial permafrost and the associated polythermal basal thermal regimes promoting the entrainment and transport of significant quantities of englacial material (Boulton, 1970a; Section 3.5.4). Deposition is thought to occur primarily through the ablation of ice from masses of buried ice that represent the stagnant remnants of former debris-rich basal ice layers and which occur within extensive tracts of ice-cored moraine (Section 5.3.2) (Figure 5.11). Ablation can occur from the top downwards and/or the bottom upwards with the former typically being considered the most rapid in view of the limited geothermal heat fluxes that drive the later (Section 5.6.3). Most importantly, with surface melting resulting in the formation of flow tills (Boulton, 1968), Boulton (1970b) argues that melt-out tills can only form beneath an overburden in situations where the meltwater can drain laterally and the overburden and slope are sufficient to preclude creep deformation.

The diagnostic characteristics of melt-out tills consequently relate to their ability to inherit the sedimentological and structural characteristics of the parent ice (Figure 5.24). With basal ice sequences containing potential proxy information on subglacial conditions, this raises the enticing possibility that the glaciological characteristics of former ice masses might be reconstructed from the study of glacial deposits and their associated geographical variability (e.g. Knight *et al.*, 2000; Adam & Knight, 2003; Cook *et al.*, 2011). Features diagnostic of melt-out have been argued to include the presence of unlithified, sorted and stratified sediments within the tills, strongly orientated clast fabrics in which the long axes are parallel to the direction of ice flow (Lawson, 1979), and a textural or lithological configuration that matches that of the debris-rich ice (Haldorsen & Shaw, 1982). Subsequent authors have however questioned the validity of these diagnostic attributes with an aggregation of clast-fabric data by Bennett *et al.* (1999) for example suggesting that the fabric strengths of melt-out tills are more variable than those reported within individual studies, such that this characteristic alone is unable to differentiate the mechanisms of till deposition. More fundamentally, Paul & Eyles (1990) have questioned the broader preservation potential of melt-out tills, arguing that the process of thaw consolidation will typically result in the generation of high porewater pressures and the overprinting or complete destruction of the aforementioned diagnostic characteristics.

One way in which this can be circumvented is through the removal of ice via sublimation rather than melting, with the resultant lack of water reducing the potential for syn and

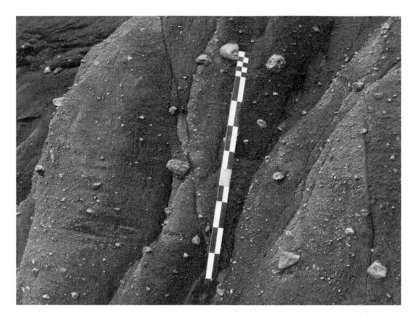

Figure 5.24 Exposure of supraglacial meltout till at Skeiðarárjökull, south-east Iceland, associated with deposition from the buried basal ice illustrated in figure 5.12. Note the subtle but distinctive sub-horizontal stratification which is considered one of the most distinctive features. Metre rule for scale (Image: Richard Waller).

post depositional modification. Following sedimentological research in the hyperarid polar environments of the Dry Valleys of Antarctica, Shaw (1977a) proposed an extended classification scheme for terrestrial tills in which deposition through sublimation is considered more conducive to the preservation of intricate primary structures present within the parent ice. The foliation and deformation structures evident within the sublimation tills are considered the result of the flow-related metamorphism of debris within the basal ice (Section 3.6.2). These are in turn sub-divided into two tectonic facies: a lower, poorly attenuated facies relating to debris-rich basal ice that has undergone limited deformation and an overlying highly attenuated facies that is more intensely deformed. Shaw (1977b) goes on to suggest that folding within this facies can cause localised thickening of the basal ice and the formation of transverse, asymmetric ridges within the resultant till deposits.

Research undertaken at the Russell Glacier, a polythermal glacier in western Greenland, has also explored the extent to which subglacial conditions and processes can be reconstructed from the study of sediments with the thick debris-rich basal ice layer characteristic of the ice margin again playing a crucial role (Knight, 1995). Having identified evidence for two distinct zones of debris entrainment within the basal ice layer (Sugden *et al.*, 1987; Knight, 1994) and an influential role played by ice-marginal deformation to create debris bands (Knight, 1989) and tectonic basal ice facies (Waller *et al.*, 2000), subsequent work sought to examine the extent to which features diagnostic of the basal ice and the related basal boundary conditions might be preserved within the ice-marginal sediments. Knight *et al.* (2000) suggest that whilst melt out can destroy sedimentary structures, there is a greater potential for distinctive sedimentary textures to be preserved with abundant fine-grained sediments characteristic of 'dispersed facies' basal ice being recognisable within intra-moraine sediment traps. Subsequent variations of the texture of ice-marginal moraine sediments can consequently be used to identify geographical variations in the basal ice layer and therefore to reconstruct the thermal and dynamic properties of the ice sheet. In studying the clast shape and roundness characteristics of the more debris-rich stratified facies basal ice and related debris bands, Adam & Knight (2003) identified a statistically significant increase in the angularity of clasts contained within the debris bands, providing an additional means through which geographical variations in the basal ice facies might be reconstructed.

A recent and highly relevant review provided by Hambrey & Glasser (2012) provides a systematic field-based evaluation of the extent to which the thermal and dynamic regimes of glaciers and ice sheets might be discriminated from a study of the associated glacigenic facies and environments. In so doing, they argue that a preoccupation with the nature and significance of subglacial sediment deformation (Section 3.3.3) has been at the expense of developing a fuller understanding of the influence of thermal regime on the processes and patterns of sediment production required to correctly interpret the sedimentary record of former glaciers. The authors present sedimentological data collected from 28 glaciers in 11 geographical regions that include temperate glaciers, cold glaciers and polythermal glaciers and demonstrate a diversity of lithofacies and lithofacies assemblages that vary according to both setting and thermal regime (Figure 5.25).

Focusing on the glacier types most commonly associated with interactions with permafrost, observations from Antarctica suggest that the glacigenic lithofacies characteristic of these very cold glacial environments contrast markedly from those associated with temperate and polythermal glaciers (Section 5.5.4). The diamictons commonly considered to be the inevitable product of subglacial transport for example are entirely absent, with the cold-based valley and outlet glaciers situated within terrestrial environments primarily depositing sand, gravel and glacier ice. This reflects the tendency of cold-based glaciers to mobilise and rework weathered bedrock or pre-existing sediments that are commonly permafrozen (Licht & Hemming, 2017). This contrasts with the sedimentologically distinctive glacigenic facies generated through a combination of erosion, entrainment and comminution within the basal traction zone in more temperate glacial environments. The lithofacies deposited directly from the ice are consequently dominated by sands and sandy boulder or cobble-gravels resulting from the reworking of pre-existing subglacial deposits and various aeolian, lacustrine and glaciofluvial deposits (Table 5.2). Some of the sands display well preserved primary sedimentary structures that have survived transport as frozen blocks that were entrained and transported *en masse* within the basal layers of the glacier (see following section for more detail on intraclasts).

This reworking of pre-existing sediments combined with the limited modification during transport means that the lithofacies associated with entirely cold-based glaciers typically lack the classic sedimentological signatures commonly attributed to glacial transport such as edge-rounded and striated clasts, and poorly sorted grain size distributions featuring abundant fines indicative of comminution (Boulton, 1978). This absence of the diagnostic sedimentological features commonly used to identify a glacial origin clearly has the potential to cause significant problems for those undertaking palaeoglaciological reconstructions and could easily lead to misinterpretations. In lacking the typical diagnostic features,

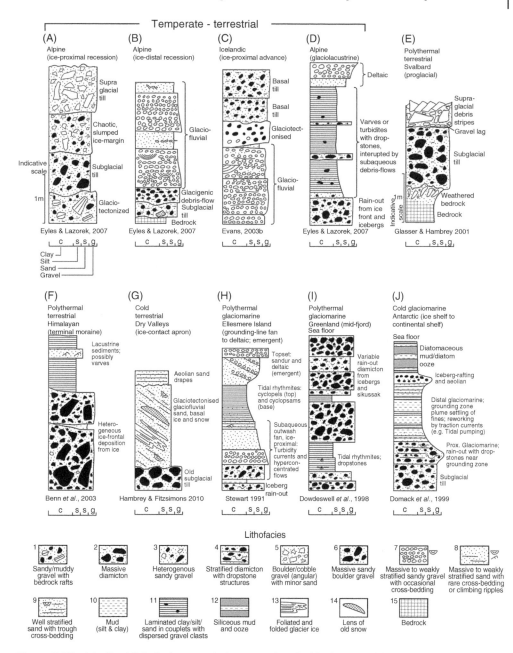

Figure 5.25 Idealised lithofacies associations associated with glaciers with different thermal regimes and settings described in the letters A to J (Hambrey & Glasser, 2012, p13). Grain size indicators: c = clay, s = silt, s = sand, g = gravel. See original paper for the citations referred to in the figure.

Table 5.2 Typical lithofacies, interpretation and relative abundance at cold glaciers. Key to relative abundance: ***** = dominant ← → * = rare.

Lithofacies	Interpretation	Abundance	% striated clasts	% faceted clasts	% clasts	Matrix (%)		Sorting coefficient	Sorting category (estimated)
Sandy boulder-gravel	Subglacial (reworked till)	****	0	0	50–80	95	5	n/d	Poorly sorted
Sandy cobble-gravel							0		
Gravelly sand	Subglacial	**	0	0	5–10	98	2	n/d	Poorly sorted
Sand	Subglacial	*****	0	0	0	100	0	n/d	Poorly sorted
Sand	Glaciofluvial	*****	0	0	0	100	0	n/d	Well sorted
Sand	Lacustrine	****	0	0	0	95	5	n/d	Well sorted
Sand	Aeolian	***	0	0	0	100	0	n/d	Well sorted
Siliceous mounds and laminae	Stromatolites and algal mats	*	0	0	0	n/d			
Foliated glacier ice	Basal ice remnants	**	0	0	0	n/d			
Firm	Snow-filled channel or depression	*	0	0	0	n/d			

sediments that are actually glacial in origin could easily be identified as being non-glacial. This compounds the associated difficulties of identifying geomorphological evidence of cold-based glaciation with areas of the High Arctic having previously been misinterpreted to be ice-free due to the absence of 'classic' glacial landforms (Section 5.2).

Sedimentological data obtained from polythermal glaciers in Svalbard, Canada and northern Sweden suggest that these glacier types typically produce glacigenic lithofacies that differ from those produced by entirely cold-based glaciers (Hambrey & Glasser, 2012). Polythermal glaciers produce a much greater diversity of lithofacies as a consequence of the operation of a more complex series of processes of glacial sediment entrainment and transport. In combination with periglacial activity and an increased influence of glaciofluvial processes, this typically leads to the creation of a more diverse suite of lithofacies. The depositional products are dominated by various types of diamicton that are texturally diverse with clasts ranging from rounded to angular (with sub-angular and sub-rounded forms being dominant). Clasts can also display features indicative of active basal processes such as striations and abrasional facets, suggesting a closer affinity with temperate glaciers than with cold-based glaciers. With Atkins (2002) having suggested that more variable sets of striae might be indicative of entirely cold-based glaciers (Section 5.3.3), the connection between basal thermal regime and these abrasional features clearly requires further investigation.

In appraising the various lithofacies typically deposited by polythermal glaciers, Hambrey & Glasser (2012) highlight the tendency of glacial geomorphologists to focus exclusively on the mineragenic components collectively referred to as glacial sediment. They make the important point that in volumetric terms, glacial deposition from these glaciers is dominated by the deposition of various types of ice (e.g. stagnant glacier ice or aufeis). In environments associated with glacier–permafrost interactions, this ice can be preserved for long time periods below a protective layer of debris so long as this exceeds the thickness of the active layer, thereby providing a key mechanism for the genesis of the extensive massive ice commonly found within formerly glaciated permafrost environments (Section 5.3.4). The presence of this abundant buried glacier ice has in turn rendered High Arctic environments particularly susceptible to ongoing climatic change that has recently started to destabilise these ice masses and the landscapes in which they occur (Section 5.6.2).

5.4.2   Rafts, Intraclasts and Hydrofractures

The combination of strong adhesion between cold-based glaciers and ice-rich permafrost substrates and the potential presence of discontinuities within the substrate relating to high porewater pressures at the base of the permafrost and/or the presence of favourably orientated fractures has led some to suggest that the quarrying of both lithified and unlithified materials is likely to be particularly effective when glaciers interact with permafrost (e.g. Boulton, 1972; Clayton & Moran, 1974; Section 3.4.2, Figure 3.6). Empirical evidence for this has been provided in a modern-day context by Fitzsimons *et al.* (1999) who observed blocks of sand and gravel entrained *en masse* into the basal ice layer of the cold-based Suess Glacier in Antarctica (Figure 3.9). More recently, Fitzsimons *et al.* (2023) have described the entrainment of thick rafts of sediment into the stratified facies basal ice of the Wright Lower Glacier resulting from the subglacial deformation of permafrost. These rafts were observed to contain sedimentary structures, algal layers and cryostrutures consistent with their original deposition within subaerial ice-marginal fluvial or lacustrine environments. Similar features have been observed in Pleistocene settings by both Astakhov *et al.* (1996) and Murton *et al.* (2004) who describe the occurrence of sand intraclasts within frozen deformation tills in west Siberia and the western Canadian Arctic respectively.

Lenses of unconsolidated sediment of various shapes and sizes have frequently been reported to occur within deformation tills (e.g. Ruszczyńska-Szenajch 1987, Menzies 1990, Hoffmann & Piotrowski 2001). These lenses typically range in size from a few centimetres to a few metres in diameter, usually feature sharp contacts with the surrounding till matrix and frequently contain coarse-grained materials with intact primary sediment structures such as cross-bedding (e.g. Menzies 1990) (Figure 5.26). Larger examples have been described by Stalker (1976) in the form of 'megablocks' of unconsolidated or poorly con-solidated pre-Quaternary deposits up to 20 m thick that are thought to have been trans-ported tens or maybe even hundreds of kilometres. In a focused review on the nature and origin of these features, Ruszczyńska-Szenajch (1987) defines a glacial raft as 'a fragment of unconsolidated or poorly cemented substratum with preserved lithologic character, detached from its primary bed and transported due to glacial activity, and which occurs within or immediately adjacent to the corresponding glaciogenic deposits' (p101). Two main genetic groups are proposed in the form of: 'glaciotectonic rafts' that have been detached and transported by purely mechanical means and, 'glacio-erosional rafts' that have been entrained by basal freeze-on and transported within the body of the glacier (i.e. a basal ice layer, Section 3.5.4), with 'rafts of composite origin' being used to describe those

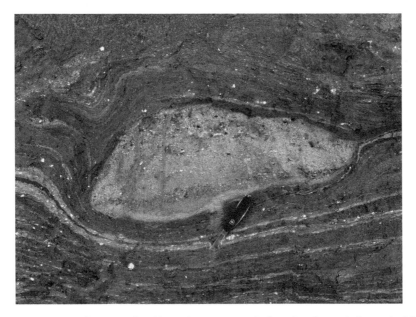

Figure 5.26 Coarse-grained intraclast composed of sand and gravel situated within a deformation till at Sheringham, north Norfolk. Note the tectonic laminations in the deformation till (composed of sheared chalk clasts) and the preservation of the sedimentary layering within the intraclast. Penknife for scale (Image: Richard Waller).

that have experienced both processes. In advocating the detachment of sedimentary layers through basal freeze-on and their subsequent transport as frozen masses, the association of glacio-erosional rafts with cold-based ice is clear. In considering their entrainment and transport pathways, their association with a permafrozen substrate and therefore with glacier–permafrost interactions is also explicitly acknowledged. This connection and the potential for sand intraclasts to provide sedimentological evidence of former glacier–permafrost interactions are considered in more detail in the next section.

The occurrence of large bedrock rafts within ice-thrust terrain and related landforms (e.g. Evans *et al.*, 2020) represents a particularly striking feature, the mobilisation and displacement of which some authors have attributed to the presence of permafrost. Moran *et al.* (1980) for example suggested that the process of basal adfreezing can lead to the entrainment of large bedrock rafts as well as the entrainment of discrete debris bands. They suggest that if the bedrock contains sufficient discontinuities, meltwater can be driven into the bedrock and towards the margin by pressure gradients to create potential décollement zones. In combination with basal freezing in ice-marginal locations, this may then be sufficient to reduce the shear strength of large segments of the bed facilitating the large-scale detachment and entrainment of bedrock rafts (see Section 5.4.2).

The chalk rafts contained within the glaciotectonised sequences in North Norfolk (UK), which can measure hundreds of meters in length and tens of meters in thickness, provide a well-studied example whose emplacement has been related to permafrost (Figure 5.27). Burke *et al.* (2009) noted that many of the rafts are overlain by a sand and gravel unit (Wroxham Crag Formation) that has been undeformed such that the primary

Figure 5.27 Large chalk raft entrained into the glacigenic sequence at Sidestrand in north Norfolk. Person for scale (Image: Richard Waller).

stratigraphic relationship is preserved. As this boundary would have provided an obvious plane of weakness, they conclude that its preservation during ice-marginal thrusting is best explained by the presence of a pre-glacial permafrost layer. This would have strengthened the bond between the two units and instead promoted shearing along the base of the permafrost layer. If this corresponded with a line of existing weakness within the chalk (e.g. a bedding plane), this would have facilitated the entrainment and displacement of the entire bedrock raft (chalk bedrock and overlying crag). It is worth noting here that the displacement of large rafts in sub-marginal environments provides a genetic association with thrust-block moraines (Section 5.3.2) that can be considered a geomorphological expression of the same general process, with Ruszczyńska-Szenajch (1987) suggesting that 'glaciotectonic end moraines' are typically rafts or assemblages of rafts.

Returning to the sedimentology of glacigenic deposits discussed in the previous section, Hiemstra *et al.* (2007) have related the mobilisation and subsequent comminution of bedrock rafts to a process-form continuum of till production. Combining sedimentological and structural data from a modern-day site in Antarctica and a Quaternary site in southwest Ireland, they identified an up-sequence transition that progressed from undeformed bedrock, to deformed bedrock, crushed and comminuted bedrock and ultimately matrix-supported tills. It is the detachment and mobilisation of bedrock rafts along shear planes at depths of up to 3 m that is considered key to till formation with their subsequent crushing and comminution producing the fine-grained till matrix. The explicit association of this process with basal freeze-on and strong ice-bed coupling and its observation in Antarctica suggests that this is most likely to occur beneath non-temperate ice masses and potentially in association with permafrozen bedrock in which ice-filled joints are known to provide potential lines of weakness (Section 4.4).

Hydrofractures and clastic dykes provide a final structural feature that have been related to the hydrological impacts of glacier–permafrost interactions (e.g. Kumpulainen, 1994). With proglacial permafrost impeding the upwelling of groundwater at the margin, artesian porewater is commonly forced to escape to the surface through unfrozen taliks associated with lakes and springs (Section 4.2.5). The development of very high groundwater potential gradients can cause the hydrofracturing of the permafrost and its more forceful expulsion to create the hydrodynamic blowouts described in Section 5.3.5. In their model of groundwater flow, Boulton & Caban (1995) predict that extensive proglacial permafrost can extend the conditions required for hydrofracturing tens of kilometres into the foreland, with large upward potential gradients promoting upward groundwater flow (Figure 5.16). Kumpulainen (1994) describes the formation of a fissure fills and pipes within delta deposits in Iceland that were subject to permafrost aggradation during deglaciation. A subsequent ice re-advance into the partially frozen feature pressurised the water within its unfrozen core to generate a network of vertical and sub-horizontal fissures within the surficial permafrost. Working on glacigenic sequences in eastern Ireland, Rijsdijk *et al.* (1999) describe a series of vertical gravel-filled clastic dykes up to 6 m in height and tens of centimetres in width that emanate from a gravel layer and cut through an over consolidated till unit, injecting the materials several metres into the overlying sediments to produce 'burst-out structures'. However, in this case, the authors ascribe the low permeability in the till to its overconsolidation rather than a frozen state with there being no independent evidence of permafrost. With this in mind, whilst the presence of permafrost seems likely to promote the development of hydrofractures and clastic dykes and to extend the areas in which they can form, their description within temperate glacial environments (e.g. Ravier *et al.*, 2015) indicates that permafrost is not necessarily a pre-requisite for their formation.

5.4.3 The Glacitectonic Deformation of Permafrost

The deforming-bed model, as originally described in detail within a seminal paper by Boulton & Jones (1979) envisaged warm-based glaciers overlying thawed beds. The combination of high basal melt rates and a low permeability bed was argued to result in high porewater pressures that counterbalanced ice overburden pressures. This resulted in a reduction in intergranular friction and the yield stress that in turn allowed the subglacial sediment to deform, even in response to relatively low basal driving stresses (Section 3.3.3). In the case of cold-based glaciers and permafrost, temperatures below the pressure melting point are commonly considered to preclude significant rates of subglacial sediment deformation for two reasons. First, the lack of basal melting prevents the saturation of subglacial sediments and the attainment of high porewater pressures required for deformation to occur (e.g. Alley, 1991). Second, the presence of interstitial ice within the pore spaces significantly enhances sediment cohesion, prevents dilation and thereby increases the sediment's shear strength (Williams & Smith, 1989).

The combination of experimental research revealing the complex rheological behaviour of ice–water–sediment mixtures at temperatures below the bulk freezing point and the role played by influence of interfacial water (Sections 2.2.2 and 2.5.1), coupled with the small but growing number of field-based studies describing active basal processes beneath non-temperate glaciers (Section 3.5.2), has led to a recognition that glaciers can actively deform beds composed of permafrost. Investigations of permafrost that has never thawed since it was overridden and deformed by Pleistocene ice sheets provide an important touchstone for interpreting glacitectonic sequences in regions where permafrost no longer exists. Previous work has highlighted evidence for the glacier-ice thrusting of permafrost near the northwestern limit of the Laurentide Ice Sheet in western Arctic Canada (Mathews & Mackay, 1960; Mackay & Mathews, 1964), and for a deformable bed of permafrost beneath the southern Kara Ice Sheet in western Siberia (Astakhov *et al.*, 1996). Murton *et al.* (2004) document the stratigraphy and glaciotectonic structures of permafrost deformed beneath the Laurentide ice margin in the Tuktoyaktuk Coastlands in western Arctic Canada. Several important inferences about glacier–permafrost interaction and the glacitectonic deformation of permafrost arise from these studies (Waller *et al.*, 2009a).

First, permafrost was deformed beneath the cold margins of both ice sheets. Ductile deformation of warm, ice-rich materials is inferred from recumbent and S-shaped folds, sand lenses and layers (Figure 5.28a) within glacially deformed permafrost in the Tuktoyaktuk Coastlands. Collectively, these structures indicate simple shear and extension – consistent with submarginal rather than proglacial glaciotectonic deformation – occurring within a frozen glaciotectonite. Significantly, the thickness of glacially deformed permafrost is substantial. Astakhov *et al.* (1996) reported thicknesses of tens of metres or more for glacially deformed sequences in the western Yamal Peninsula

(a)

(b)

Figure 5.28 Structures indicative of glacier–permafrost interactions at the northwest margin of the Laurentide Ice Sheet during Marine Isotope Stage 2 in the western Canadian Arctic (see text for details). (a) Sand lenses and fold noses between silty clay at North Head indicating the sub-marginal ductile deformation of ice-rich materials. (b) Raft of massive ice (~15 m long) within frozen glacitectonite at Liverpool Bay. (c) Frozen glacitectonite ~8 m thick overlying massive ice at Pullen Island (figure for scale).

(c)

Figure 5.28 (Continued)

and the lower Yenissei region whilst Murton *et al.* (2004) reported thicknesses of at least 5–20 m from the Tuktoyaktuk Coastlands.

Second, deformation was accompanied by the erosion of permafrost. Blocks of ice up to 15 m long and 1.5 m high, some folded, are dispersed throughout the glaciotectonite in the Tuktoyaktuk Coastlands (Figure 5.28b). Where bands in the ice blocks are truncated along the margins of the blocks, *in situ* formation of segregated or intrusive ice is discounted. Instead the ice was eroded from pre-existing ice and thus represent ice clasts. Submarginal erosion of ice also explains the occurrence of an angular unconformity (décollement surface) along the top of the underlying massive ice (Figure 5.28c). Had the unconformity resulted from downward thaw, then the ice clasts above it would have thawed. Preservation of the ice clasts and the underlying massive ice indicate that erosion occurred at subfreezing temperatures.

Third, basal ice from the Laurentide and Kara ice sheets is widely preserved in the permafrost of both glaciated regions as discussed previously in Section 5.3.4 (Figure 5.22). Burial mechanisms include: (i) glacial thrusting and shearing of frozen sediment (inferred where frozen glaciotectonite overlies basal ice); (ii) downward melt of ice previously exposed at the ground surface (inferred where supraglacial melt-out till overlies basal ice); (iii) deposition of aeolian, glaciofluvial or lacustrine sediments above basal ice (Murton *et al.*, 2005). The continuous presence of permafrost throughout the last glacial cycle has consequently resulted in 'incomplete deglaciation' and the preservation of both the buried ice (Astakhov & Isayeva, 1988; cf. Dyke & Evans, 2003) and the ice blocks mentioned above. Deglaciation of these regions will therefore not be completed until the permafrost preserving these ice masses has been degraded (see Section 5.6.3).

Fourth, injection of pressurised water into proglacial permafrost occurred during retreat of the Laurentide Ice Sheet. Superimposed on some glacially deformed permafrost are undeformed dykes (Figure 5.21) and sills of intrusive ice and some undeformed massive segregated-intrusive ice (Murton, 2005). Such ice clearly post-dates the time of glacial deformation. Because the underlying sedimentary sequence has remained frozen ever since it was deformed, water supplied by pore-water expulsion during permafrost aggradation is unlikely. Instead, an external source of overpressurised water is inferred, probably from glacial meltwater flowing beneath proglacial permafrost during the retreat of the Laurentide Ice Sheet (Section 5.3.4).

A recent reappraisal of the glacial landforms and sediments of the Smoking Hills area of the Canadian Northwest Territories has provided new insights into the unusual features and complex sequences associated with the interaction of polythermal ice sheets with pre-existing permafrost (Evans *et al.*, 2021). The authors describe the emplacement of complex glacitectonites and thrust stacks with thicknesses of up to at least 50 m comprising diamictons, intraclasts of local bedrock, glacifluvial and glacilacustrine sediments and various ice types including relict glacier ice, ice wedges and aufeis. One of the most intriguing features described comprises an exposure of buried glacier ice containing ice wedges that has been deformed down-valley in a direction parallel to flutings visible at the surface. This suggests the overriding glacier was capable of deforming the surficial materials whilst leaving the buried glacier ice largely intact.

These field descriptions of the products of glacier–permafrost interactions demonstrate the dynamic coupling of the cold-based margins of Pleistocene ice sheets, initially with subglacial permafrost and subsequently, during retreat of the active ice margin, with proglacial permafrost. The permafrost beneath the ice margins was probably warm and ice-rich such that it contained a significant amount of liquid water – a type of easily deformable permafrost termed 'plastic frozen ground' (Tsytovich, 1975) (Section 2.5). The combination of temperatures close to the bulk freezing point combined with the presence of clay-rich beds facilitated the mobilisation and shear deformation of the permafrost and the subglacial erosion and entrainment of frozen materials including blocks of massive ice (Figure 5.29). This suggests that the permafrost acted as a subglacial deforming layer tens of metres in thickness as described previously in western Siberia by Astakhov *et al.* (1996), although in the latter case the authors suggested the thickness of deformation reached up to 300 m. In addition, the strong ice-bed coupling associated with glacier–permafrost interactions and associated with erosion through plucking or block removal (Section 3.4.2) would have facilitated the transmission of basal driving stresses into the substrate (Schindler *et al.*, 1978 cited in Haeberli, 1981).

The geological evidence for glacier–permafrost interactions is relatively clear and incontrovertible in Arctic regions where the permafrost has remained intact and the structures indicative of glacitectonic deformation are in part composed of ice. Reconstructions of the extent of permafrost during glacial maxima suggest that the margins of continental ice sheets will almost certainly have overridden pre-existing permafrost in areas close to their maximum extents where it had several thousands of years to form and thicken in response to cold subaerial conditions. Modelling work on the extent and degradation of permafrost beneath the margin of the Green Bay lobe of the southern Laurentide Ice Sheet undertaken by Cutler *et al.* (2000) concluded that glacier–permafrost interactions would have persisted for 80–200 km up-glacier of the margin with the permafrost remaining intact for up to a few thousand years beneath the advancing ice. Even if this situation is considered likely, the

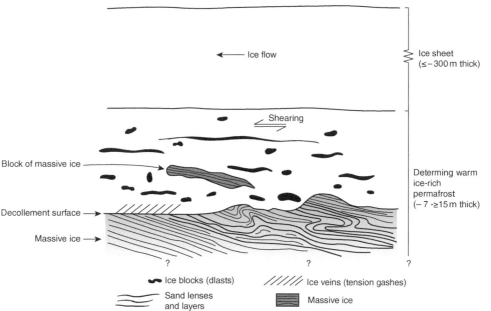

Figure 5.29 Schematic model of deforming, warm, ice-rich permafrost beneath the northwestern margin of the Laurentide Ice Sheet (Murton *et al.*, 2004, p411).

geological evidence indicative of the subglacial deformation of frozen ground has remained unclear. In reviewing the processes and products of glacitectonic deformation, Hart & Boulton (1991) concluded that,

> . . .*proglacial and subglacial deformation can act upon frozen and unfrozen sediments, but the means to differentiate between these states in ancient sediments requires more detailed research* (p342).

Following this invitation, Waller *et al.* (2009b, 2011) reconsidered the evidence of glacitectonic deformation exposed within the coastal cliffs of northern Norfolk in the UK, an area that was overridden by the British Ice Sheet in the Middle Pleistocene (MIS 12). These cliffs feature a complex sequence of glacial sediments that has been heavily deformed producing the so-called 'contorted drifts' that display extensive evidence of both subglacial and proglacial glacitectonic deformation (e.g. Hart & Boulton, 1991; Hart & Roberts, 1994). With initial theories of subglacial deforming layers suggesting that they are dependent upon the formation of high porewater pressures required to reduce the yield stress (Section 3.3.3), the evidence of glacitectonic deformation was interpreted as relating to subglacial sediment deformation within an unfrozen and water-saturated deforming layer beneath a warm-based, ice-sheet margin (e.g. Hart, 2007). Waller *et al.* (2009b, 2011) however argued that there are aspects of the sequence that are difficult to reconcile with these basal boundary conditions.

The first issue relates to the proposed thickness of the subglacial deforming layer. The glacially deformed sequences at sites such as West Runton are considered the product of the pervasive deformation of a subglacial deforming layer of up to 20 m in thickness (e.g. Phillips *et al.*, 2008). This interpretation is consistent with early observations of the process of subglacial sediment deformation in modern-day glacial environments, with seismic surveys beneath Ice Stream B (renamed the Whillans Ice Stream) in Antarctica for example identifying a subglacial layer 5–6 m thick characterised by low shear and compressional wave velocities thought to provide evidence for a saturated, dilatant and actively deforming till (Alley *et al.*, 1986; Blankenship *et al.*, 1987). However, most current observations and measurements from modern glaciers suggest that even under thick ice, subglacial deforming layers tend to be restricted in thickness to a few tens of centimetres, or 1–2 m at most (e.g. Iverson *et al.*, 1995; Engelhardt & Kamb, 1997; Fuller & Murray, 2002). This has resulted in an apparent paradox, with sedimentological investigations of geological sequences indicating subglacial deformation extending to depths of several metres, whilst most investigations beneath modern-day glaciers suggest that deformation is ordinarily restricted to a depth of <1 m. A noteworthy exception here are tills that have been subject to submarginal compression and stacking (Evans & Hiemstra, 2005).

This represents one element of a broader, ongoing debate regarding the most appropriate rheological model for tills, with viscous fluid models predicting thick, pervasively deforming layers and plastic-coulomb models being more consistent with thin deforming layers (Murray, 1997; Kavanaugh & Clarke, 2006). Whilst there are a number of potential explanations, on the basis of the thickness of the deformed sequence and the size and coherency of the structural features it contains (Figure 5.30), Waller *et al.* (2009b) argued that it is most likely to represent the product of deformation under different thermal conditions to those commonly observed beneath modern-day, warm-based glaciers. As noted earlier,

Figure 5.30 Pervasively deformed sequence of glacigenic sediments (Bacton Green till) exposed at West Runton, north Norfolk. Note the large number of sand and gravel intraclasts. The exposed section is approximately 15 m in height (Image: Richard Waller).

observations of subglacially deformed, ice-rich sediments beneath the margins of continental ice sheets in both the western Canadian Arctic and Western Siberia suggest that deforming-layer thicknesses in these situations commonly reach tens or even hundreds of metres in thickness (Astakhov *et al.*, 1996; Murton *et al.*, 2004; Evans *et al.*, 2021), consistent with the observed thickness of the deformed sequences found in north Norfolk.

A second line of evidence for glacier–permafrost interaction relates to the presence of distinctive intraclasts within the deformed sequence (Waller *et al.*, 2011) that were discussed more generally in the preceding section. Intraclasts composed primarily of sand have previously been identified as a distinctive feature within the glacitectonic sequences of north Norfolk (e.g. Hart, 2007; Lee & Phillips, 2008). At West Runton for example, they form part of a mélange facies comprising remobilised elements of a variety of tills as well as preglacial sands and gravels and chalk bedrock. The mélange is highly disrupted in appearance and locally contains numerous variably deformed lenses or intraclasts of poorly consolidated sand within a highly deformed sandy till matrix (Figure 5.26). The mélange facies is laterally extensive having also been recognised by Lee & Phillips (2008) at Bacton Green 15 km to the south east indicating that the processes leading to its formation occurred over a relatively wide area beneath the advancing ice sheet.

The intraclasts themselves vary in size from a few tens of centimetres to >10 m in length and display a variety of morphologies ranging from elongate slab-like bodies to more rounded 'eye' or 'augen'-shaped pods with variably developed 'tails', through to more highly deformed examples in which the sand intraclasts are folded and thrusted (Figure 5.31). The contacts between the intraclasts and the surrounding till are typically sharp and distinct, suggesting there has been little or no intermixing of sediment across this boundary. Of particular note is a marked contrast in the apparent intensity of deformation recorded by the sand intraclasts and the surrounding deformation till. The dominant structure within the till is a pervasive foliation delineated by thin chalk laminae that wraps around the intraclasts and is itself deformed by several generations of folds. In contrast to the host till, many of the intraclasts exhibit very little evidence of internal deformation and locally contain well-preserved primary sedimentary structures including graded bedding and parallel lamination. The contrast indicates that these sand bodies, rather than being weak as their unconsolidated nature would suggest, acted as a relatively rigid bodies during deformation. The deformation was instead partitioned into the weaker till with the presence of the sand intraclasts leading to localised mechanical instabilities within the actively deforming mélange matrix.

Previous work has appreciated that the presence of the intraclasts provides an important insight into subglacial processes and conditions but has differed in their interpretation of these features. The suggestion that the preservation of sorted sediments related to deposition within englacial channels flowing through bodies of stagnating debris-rich ice was advocated as early as the nineteenth century (e.g. Goodchild, 1875) and the intraclasts have subsequently been argued by some to be diagnostic of deposition through melt out (Haldorsen & Shaw, 1982; Shaw, 1982). With the wider sequence in which they occur being interpreted as the product of subglacial sediment deformation, others have argued that the intraclasts are created when a deforming-layer overrides a layer of sorted sediments and fragments of this layer are eroded and entrained. An increase in deforming layer thickness results in the migration of the base of the deforming layer into the underlying sands that

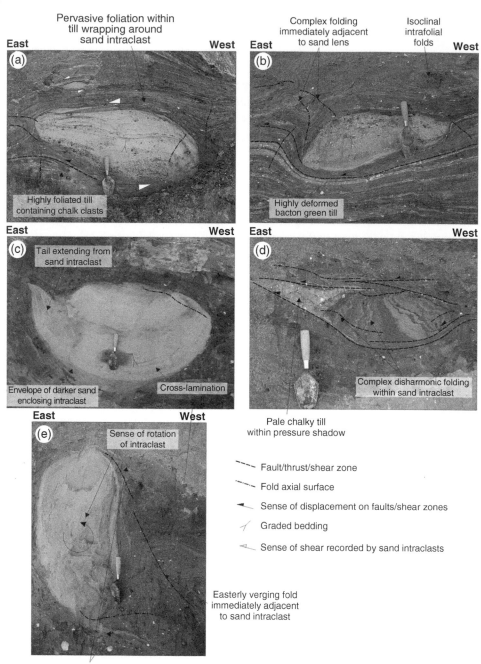

Figure 5.31 Morphology, orientation and key characteristics of a selection of intraclasts (a to e) exposed between West Runton and Sheringham, north Norfolk (Waller *et al.*, 2011, p3486/with permission from Elsevier).

are then entrained through a combination of folding, thrusting, attenuation and boudinage (e.g. Hart & Roberts, 1994; Lee & Phillips, 2008). If, as is traditionally the case, subglacial deformation is considered to be occurring under entirely unfrozen conditions, it is however difficult to envisage how sorted and cohesionless granular sediments can remain intact and act as more competent masses within what is argued to be a high strain environment. One possibility is that their competency relates to lower porewater pressures and higher levels of internal friction. However, if they are entirely enveloped by till displaying high porewater pressures, it is difficult to imagine how this distinction could develop and persist in an absence of available drainage pathways. Piotrowski *et al.* (2001) has consequently argued that their sharp contacts with the surrounding till are inconsistent with a deforming-bed hypothesis as deformation at high porewater pressures should lead to particle diffusion, mixing and smearing of the contacts. Menzies (1990) has also recognised the problem of the survivability of sand intraclasts within a rapidly deforming medium and concluded that they must have been frozen when they were entrained into and transported within a deforming layer. On account of a brecciated layer of till surrounding some of the intraclasts, Menzies (1990) suggested that whilst the till was 'thawed', the sand intraclasts remained 'frozen' during deformation such that on deposition, they acted as heat sinks. This led to the cryosuction of liquid water from the surrounding till and its desiccation and brecciation.

Waller *et al.* (2011) concur that cementation by pore ice provides the simplest way of explaining the preservation of the sand intraclasts but suggest what they argue to be a simpler interpretation for the examples observed at West Runton. The survival of undeformed intraclasts within an actively deforming till and the associated rheological heterogeneity are easily explained if deformation takes place at temperatures below, but close to the pressure melting point. The presence and quantity of liquid water, and therefore the mechanical properties of frozen ground at these relatively warm temperatures, are strongly grain-size dependent. Where such 'warm' permafrost develops in sediments of varied particle size, the permafrost is partially frozen and comprises a mixture of sediment, ice and liquid water (Section 2.5.1). Interfacial effects (Section 2.2.2) will be most pronounced in fine-grained materials resulting in the presence of significance amounts of 'pre-melted' liquid water. In contrast, the free water contained within sands and gravels will freeze rapidly when the temperature drops below the bulk freezing point resulting in their cementation by pore ice. Consequently, if such a sequence was overridden by ice, deformation would be expected to occur preferentially within the partially frozen fine-grained till matrix, whilst the ice-bonded sand intraclasts comprising relatively competent masses, should remain largely intact. This prediction is consistent with field observations of subglacially deformed permafrost in the western Canadian Arctic (Murton *et al.*, 2004) and western Siberia (Astakhov *et al.*, 1996) that feature streamlined and occasionally boudinaged sand lenses within a pervasively deforming till matrix. Waller *et al.* (2011) conclude therefore that the presence of unconsolidated intraclasts within highly deformed tills provides geological evidence for the deformation of partially frozen sediments that can potentially be used to identify the extent of glacier–permafrost interactions beneath former ice masses. If this is correct, then these features provide a sedimentological criterion that can complement and test geomorphological inverse models that provide the more commonly utilised approach for reconstructing the basal thermal regimes of former ice masses (Section 2.3.2).

Finally, a hypothesis of glacially deformed permafrost is consistent with the pre-glacial history of the area. There can be no doubt that cold permafrost subject to thermal contraction cracking was an integral component of the landscape of northern Norfolk prior to glaciation. Wedge-shaped structures occur within several horizons in the pre-glacial sediments (West, 1980) with the largest examples possessing convincing ice-wedge pseudomorph characteristics associated with permafrost (Whiteman, 2002). These features considered indicative of former continuous permafrost suggest that the ice sheet is likely to have initially advanced over frozen ground, probably warming it from cold to warm permafrost and in turn facilitating its deformation. Both Attig *et al.* (1989) and Clayton *et al.* (2001) have similarly argued that the presence of ice-wedge casts and relict polygons beyond the ice limit of the Laurentide Ice Sheet margin in Wisconsin indicated that the ice-sheet margin would have advanced over terrain characterised by permafrost.

5.5 Glacial Landsystem Models

5.5.1 Process-Form Models and Glacial Landsystems

Process-form models based around the development of genetic associations between specific glacial processes and the creation of distinctive forms, provide the cornerstone of empirical palaeoglaciological reconstructions. If these glacial processes are in turn associated with specific glaciological characteristics, then the presence of a distinctive landform or landform assemblage can be extended to infer the basal boundary conditions at their time of formation. Glacial striations for example are created through the process of glacial abrasion that is in turn a consequence of basal sliding. As such, they can be used to infer both the former presence of a glacier and the direction(s) of ice flow (Figure 5.32).

Figure 5.32 Cross-cutting striations on a bedrock outcrop located on the edge of Llyn Llydaw, Snowdonia, Wales. The two sets of striations illustrate the occurrence of basal sliding and the contrasting ice-flow directions of during the Last Glacial Maximum and the Younger Dryas (Image: Richard Waller).

In addition, with basal sliding widely being assumed to be restricted to warm-based glaciers with abundant basal meltwater, their presence is often used to infer this basal boundary condition (Section 3.4.2).

Rather than focusing solely on individual landforms, the connections between process and form that they provide are more commonly examined as the 'building blocks' within more integrated field investigations of the landform-sediment assemblages present within any glaciated landscape. These consider both the geomorphology of the landforms and where appropriate, their internal structure that can then be related to the local or regional stratigraphy. Glacial geomorphologists have enthusiastically adopted and utilised the landsystems concept in order to provide a more holistic framework for the evaluation of formerly glaciated terrains. This approach, originally developed to assess the agricultural potential of the vast Australian outback, involves the identification of areas with common terrain attributes that differ from those of the surrounding areas that can occur at a range of spatial scales (Evans, 2003). This approach has encouraged the combined study of surface geomorphology and subsurface materials with extensive investigations within a range of modern and ancient glacial environments, elucidating both the diversity of glacial landsystems and the associated glaciological and environmental controls (e.g. Kleman & Borgström, 1996; Evans, 2003; Hambrey & Glasser, 2012). This has enabled the development of more robust genetic associations that have provided the opportunity for more rigorous and insightful palaeoglaciological reconstructions. In emphasising the spatial dimension for example, it has also highlighted the significance of the geographical arrangement of the constituent landform-sediment assemblages within a broader landsystem. This can be central to the identification of glaciological parameters or events that cannot be identified through the presence or absence of features alone (Figure 5.33), thereby demonstrating the ongoing importance of field mapping (e.g. Chandler *et al.*, 2018).

The development of a glacial process-form model to interpret the glacial geomorphology of the interior of North America by Clayton & Moran (1974) is considered an early and seminal contribution to the landsystems literature within which the authors systematically develop detailed connections between environment, process and form and emphasise '. . .the need for mapping all glacial features of an area rather than considering a single landform as an isolated element' (p118). The paper was also ground breaking in its appreciation and explicit consideration of the role and significance of glacier–permafrost interactions, with the authors suggesting for example that erosion did not simply remain active beneath marginal 'frozen-bed zones', but that it was more effective than the abrasion associated with interior 'thawed-bed zones' (Section 3.4.2; Figure 3.7). In considering the development of the resultant glacial landscape, the authors identified four superimposed elements: (i) the pre-advance element comprising the landscape that pre-dated the most recent glacial advance, (ii) the subglacial element resulting from the erosion of or deposition upon this surface, (iii) the superglacial element involving the release of sediment from debris-rich ice and finally, (iv) the postglacial element comprising landforms incised into or deposited upon the glacial surfaces by postglacial processes. Any glaciated landscape was seen as including each of these elements, although their relative significance would vary according to the extent to which any of the elements had removed or masked the preceding elements. This in turn helps to account for the diverse nature of glacial landscapes more generally.

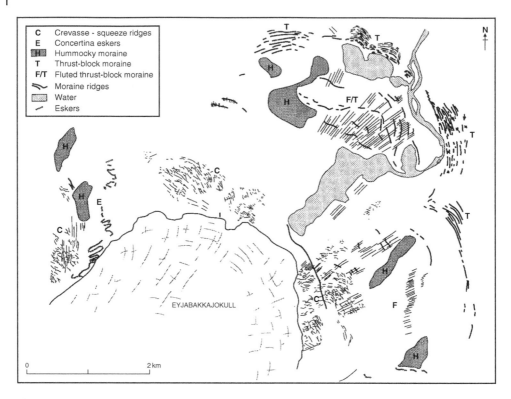

Figure 5.33 Map of the foreland of Eyjabakkajökull in south-eastern Iceland (Evans & Rea, 1999, p76). This illustrates the distinctive landform zonation associated with this surge-type glacier.

One of the most fundamental problems associated with field-based palaeoglaciological reconstructions is consequently the partial and fragmented nature of the available landscape evidence with successive phases of glacial activity commonly resulting in the destruction or overprinting of the evidence relating to earlier phases of landscape development. Whilst this constitutes a major challenge, Kleman & Borgström (1996) suggest that it also provides an opportunity with the development and application of 'inversion models' to these palimpsest landscapes allowing the extraction of information on earlier phases of glaciation, thereby introducing a temporal dimension to palaeoglaciological reconstructions (Kleman, 1994). These models also enable continental-scale syntheses of glacial geology than can be used to constrain the outputs of ice-sheet models based upon numerical modelling or patterns of isostatic rebound, whose utility is dependent upon our ability to fully understand the ways in which ice sheets interact with their beds (the 'genetic problem'). Of particular relevance is their ability to identify not only the processes and conditions that create new glacial landforms, but also the processes and conditions associated with the preservation of older landforms or strata thereby creating 'windows' back to earlier phases of landscape development (Section 5.2). In mapping the landscape impact of the Scandinavian Ice Sheet, Kleman & Borgström (1996) define three key geomorphological systems associated with the formation or preservation of landforms: (i) an inner 'dry-bed system' associated with frozen-bed conditions and a hiatus in landform development,

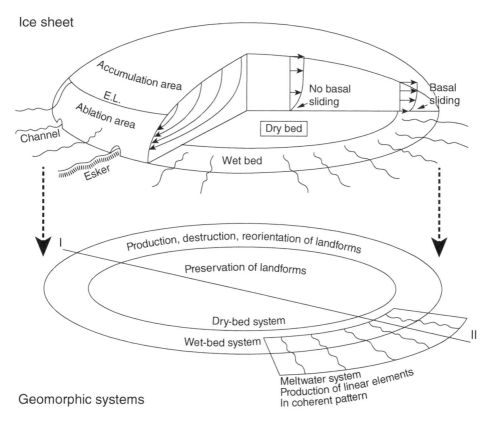

Figure 5.34 Schematic figure of the geomorphological systems associated with an ice sheet (Kleman & Borgström, 1996, p898). Note the connection between the basal boundary conditions and the associated geomorphological impacts. See text for details.

(ii) an outer 'wet-bed system' associated with basal sliding and the development of flow lineations and (iii) a 'marginal meltwater system' associated with glaciofluvial landforms such as eskers that are generated during deglaciation (Figure 5.34).

The application of inversion models consequently requires the systematic mapping of lineations (e.g. drumlins, flutes, striae), meltwater features (eskers, glacial lake traces and glaciofluvial channels) and transverse ice-marginal forms (e.g. ribbed moraines and end moraines) and the determination of the relative age of any cross-cutting features. The subsequent palaeoglaciological interpretation of this input data is based around a series of assumptions regarding the influence of basal thermal regime on process and form and the expected landscape signatures of glacier–permafrost interactions. These state that:

1) Basal sliding requires a thawed bed.
2) Lineations can only form if basal sliding occurs.
3) Lineations (drumlins, flutes, striae) are created in alignment to local flow and perpendicular to the ice-surface contours at the time of creation.
4) Frozen-bed conditions inhibit rearrangement of the subglacial landscape.

5) Regional deglaciation is always, except under polar desert conditions, accompanied by the creation of a spatially coherent systems of meltwater features such as channels, eskers and glacial lake shorelines. In the case of frozen-bed deglaciation, eskers may be lacking.

6) Eskers are formed in an inward-transgressive fashion close to the retreating ice front.

(Kleman & Borgström, 1996, p900–901)

Kleman & Glasser (2007) have subsequently employed these techniques and landform-process links to examine the subglacial landform record of both mid-latitude Pleistocene ice sheets and the modern-day Antarctic Ice Sheet (inferred from radar data) and to produce a conceptual model of the subglacial thermal organisation of entire ice sheets. This is particularly noteworthy in terms of the authors' emphasis of frozen-bed patches as an integral component of ice sheets that can occur at a range of spatial scales. In addition to dominating the central ice-dispersal zones and playing an influential role in landform creation in ice-marginal environments, the authors highlight their existence and significance as isolated, lenticular 'islands' co-existing with major ice streams within complex and dynamic 'ice-stream webs'. In these situations, they play a key role in determining lateral shear margins between fast and slow flowing ice that can extend for hundreds of kilometres. In controlling the locations of the ice-stream corridors in which fast ice flow occurs and isolating the peripheral drainage basins from the ice streams, the frozen-bed patches play a crucial role in determining the stability of the ice sheet (Section 4.3.4).

The following sections explore the diversity of glacial landsystems that have been described in a variety of different environmental settings, focusing in particular on the process-form connections that have been proposed to identify cold-based thermal regimes or glacier–permafrost interactions. In doing so, it is important to recognise that a number of recent process studies undertaken on non-temperate glaciers have provided clear evidence for the continued operation of subglacial processes at temperatures below the pressure melting point. This suggests that some of the assumptions regarding process and form that are routinely employed (e.g. lineations require warm-based or thawed bed conditions) are by no means universally applicable, highlighting an area of research where further work is still required (Section 6.3).

5.5.2   Pleistocene Ice Sheets

A significant amount of research examining the diversity of glacial landsystems has focused on the former Pleistocene Ice Sheets that covered significant parts of North America, Europe and Siberia during the Last Glacial Maximum. Some of the specific landscape features that have been used to infer the presence of cold-based ice and transitions from warm to cold-based ice have been described in previous sections (Sections 5.2 and 5.3.3). This section focuses on the related conceptual models that have been developed to describe and interpret the spatial diversity of glacial landforms and to infer the former presence of cold-based ice and permafrost.

A succession of studies have examined the nature and distribution of landforms associated with the southern margin of the Laurentide Ice Sheet in order to reconstruct its extent, dynamic behaviour and basal boundary conditions (e.g. Clayton & Moran, 1974; Moran *et al.*, 1980; Attig *et al.*, 1989; Clayton *et al.*, 2001). In focusing on the lower latitude southern margin of the ice sheet, the role of permafrost during the main phase of glaciation and during deglaciation would be expected to be comparatively limited. Nonetheless, a number of studies have explicitly inferred its presence and acknowledged its role in influencing the basal thermal regime around the ice margin and the associated landform record.

Once again, Clayton & Moran (1974) provided an early insight into the geomorphological products relating to permafrost, frozen-bed conditions and glacier–permafrost interactions in relation to the four key landscape elements introduced in the previous section. Considering these in turn, permafrost-related landforms have frequently been described in relation to the 'preadvance element' (i.e. the preglacial landscape), with the presence of patterned ground or ice-wedge polygons for example being invoked to infer the presence of permafrost prior to ice advance (see following section on Clayton *et al.*, 2001) and the preservation of preglacial features beneath cold, frozen or dry-based ice (e.g. Kleman & Borgström, 1990; Kleman, 1994; Kleman & Glasser, 2007). Within the 'subglacial element', frozen-bed conditions close to the terminus are related to the production of 'transverse compressional features' characterised by folds or thrust masses, essentially incorporating the genetic connection between subglacial permafrost, raft displacement and thrust-block moraine genesis discussed earlier (Sections 5.3.2 and 5.4.2). No specific associations between basal thermal regime, permafrost and landform development are provided in relation to the 'superglacial' or 'postglacial' elements. It is worth noting however that there is a substantial volume of other research that has established genetic associations in both cases and that could therefore be incorporated into their process-form model. The distinctive 'washboard moraines' they describe in relation to the superglacial element can for example be related both to the thick basal ice (Section 3.5.4) and controlled moraines (Section 5.3.2) associated with polythermal ice margins and submarginal permafrost, whilst permafrost can continue to play an important role in the postglacial evolution of glaciated landscapes (Section 5.6). Clayton & Moran (1974) finally combine these temporal process environments and the constituent landforms into distinctive 'suites' within an integrated landsystem model (Figure 5.35) before proposing ways in which the model might be tested through the systematic mapping of landforms and sediments.

A number of subsequent studies have taken on the challenge of systematically mapping the glacial geomorphology of specific regions with Attig *et al.* (1989) for example mapping the distribution of landforms in Wisconsin associated with the Laurentide Ice Sheet in order to reconstruct the extent of 'thawed-bed' and 'frozen-bed' conditions at its maximum late Wisconsinan (i.e. Last Glacial Maximum) extent (Figure 5.36). The authors identify an 'ice-marginal tunnel-channel zone' and an 'interior drumlin zone' that utilise explicit process-form connections to relate them to frozen bed and thawed bed conditions respectively. Within the former, frozen bed conditions are inferred through the presence of a zone of thick hummocky supraglacial sediment c. 20 km wide that is related to shearing, sediment entrainment and the stacking of basal ice associated with the transition from thawed to frozen bed conditions (Section 3.5.4). In addition, a series of tunnel channels that cut through the hummocky area are related to the episodic and high magnitude drainage of water impounded by the marginal thermal dam (Section 4.2.2). The presence of stacked thrust blocks are also suggestive of ice-marginal permafrost (Section 5.4.2) although these aren't included within the model as they are comparatively rare in Wisconsin. In contrast, as the description indicates, the thawed bed conditions are related primarily to the

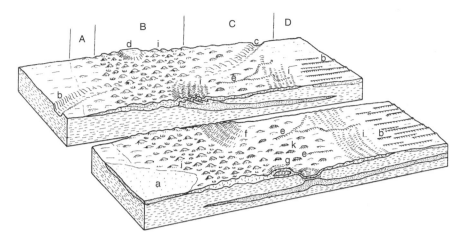

Figure 5.35 A typical sequence of glacial suites approximately 50 km wide from the southern Laurentide Ice Sheet comprising: A. Fringe suite, B. Marginal suite, C. Transitional suite and D. Inner suite (Clayton & Moran, 1974, p111). These feature a range of landforms including fluvial plains (a), meltwater channels (b), partly buried meltwater channels (c), ice-walled lake plains (d), eskers (e), transverse compressional features (f), thrust masses (g), longitudinal shear marks (h), washboard moraines (i), simple hummocks (j) and circular disintegration ridges (k). See original source for details.

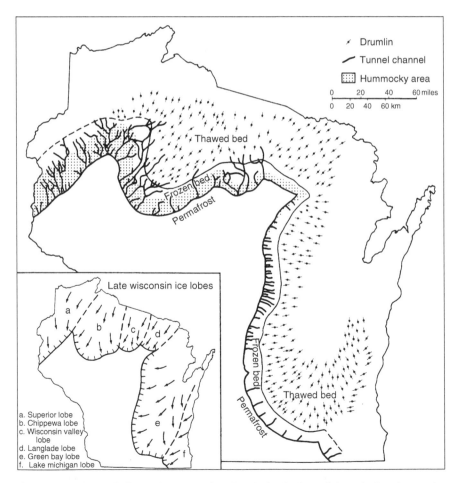

Figure 5.36 Map of Wisconsin illustrating the glacier-bed conditions during the maximum extent of the Late Wisconsin inferred from the distribution of tunnel channels, drumlins and areas of hummocky moraine (Attig *et al.*, 1989, p400). See text for details.

presence of an extensive drumlin field, the formation of which is considered indicative of active basal processes. The presence of eskers in turn corroborates the active basal melting required to drive these processes. A face value interpretation of this model therefore suggests that glacier–permafrost interactions did occur, but that their influence was limited to a narrow ice-marginal zone c. 20 km in width. The authors acknowledge however that frozen-bed conditions are likely to have been more extensive than the model suggests. They hypothesise that the irregular bed required for drumlin formation reflects spatial variations in erosion relating to a bed that was frozen in some places and wet in others with the subsequent streamlining that created these lineations occurring at a late stage in the landsystem's evolution. This highlights the potential for spatially extensive but temporally limited late phase warm-based conditions to dominate the landscape record such that a face value interpretation is likely to indicate the minimum extent of cold-based conditions (section 2.4.1; Kleman & Glasser, 2007).

A follow-up study on the glaciated landscapes of Wisconsin by Clayton *et al.* (2001) is particularly noteworthy in that it has the explicit aim of identifying evidence for permafrost and its consequent influence on glacial processes and landform development. Wide-ranging evidence is presented for the existence of thick permafrost during the last glaciation in the form of extensive ice-wedge casts and relict polygons (Figure 5.37), thick talus deposits, block streams, soliflueted sediment and collapse features relating to the melt-out of buried glacier ice and lake ice. In noting this evidence for permafrost within an area overridden by the southern Laurentide Ice Sheet, the authors acknowledge that 'the presence of permafrost in front of or under the glacier probably had a considerable influence on the behaviour of the glacier' (p180). In revisiting the model proposed previously by Attig *et al.* (1989) (Figure 5.36), this evidence is used to further emphasise the aforementioned role of ice-marginal permafrost in promoting the genesis of thick sequences of basal ice and influencing the subglacial drainage to facilitate the creation of tunnel channels. They also emphasise the temporal dimension, recognising that the nature of the glacier–permafrost interactions will have differed between its advance phase when it was overriding proglacial permafrost and its subsequent recession when permafrost was able to reform. Extensive palaeoglaciological reconstructions indicate that permafrost persisted from c. 25,000 years B.P. to c. 14,000 years B.P. in southern Wisconsin and to 10,000 years B.P. in the north of the state. During deglaciation for example, permafrost delayed the melt-out of buried ice and prevented the drainage of water from ice-walled lakes, the subsequent deglaciation of which resulted in the formation of a hummocky landscape containing ice-walled lake plains situated tens of metres above the current groundwater table (Figure 5.13). In conclusion, Clayton *et al.* (2001) consequently recommend that,

> *Although often overlooked, the physical evidence of permafrost should be an important element of paleoclimatic reconstructions and glacial-processes studies in the midcontinent region* (p187).

In examining the regional diversity of glaciated features associated with the ice-marginal areas of the southern Laurentide Ice Sheet, Colgan *et al.* (2003) identify four distinctive landsystems. Two of these (landsystems B and C) recognise the influence of permafrost and therefore help to assess the broader regional significance of glacier–permafrost

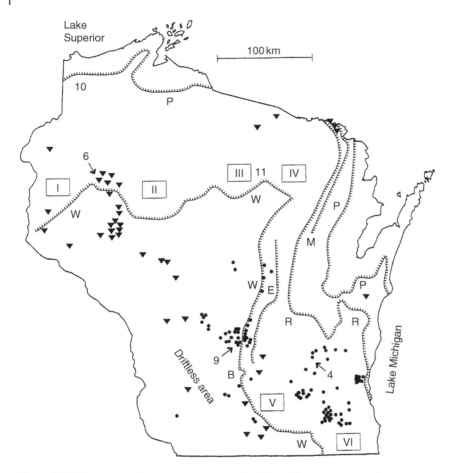

Figure 5.37 Location of ice-wedge polygons (dots) and ice-wedge casts (triangles) observed around the former limits of the Laurentide Ice Sheet in Wisconsin during various late Pleistocene glacial stages (W, E, R, M and P) (Clayton *et al.*, 2001, p174). The Roman numerals denote different ice lobes. See original source for details.

interactions. Landsystem B occurs in North Dakota, Minnesota, Wisconsin, northern Michigan (and a small area in southern Michigan), Pennsylvania, New York, New England. This is characterised by drumlins and a marginal zone of high-relief hummocky moraines (Figure 5.38) with glacier–permafrost interactions promoting the formation of the hummocky moraine (and associated supraglacial sediment) and potentially the tunnel channels as previously described by Attig *et al.* (1989). Landsystem C occurs in North and South Dakota, western Minnesota, Iowa and a small part of western Ohio and is characterised by low-relief aligned hummocks and ice-thrust masses. Whilst its key features are related primarily to surge events, ice-marginal permafrost may also have played an important role in facilitating the development of the ice-thrust features (Sections 3.6.2 and 5.3.2).

A re-appraisal of the complex topography found in the northern highlands of Wisconsin by Attig & Rawling III (2017) utilised newly available LiDAR imagery to aid the recognition

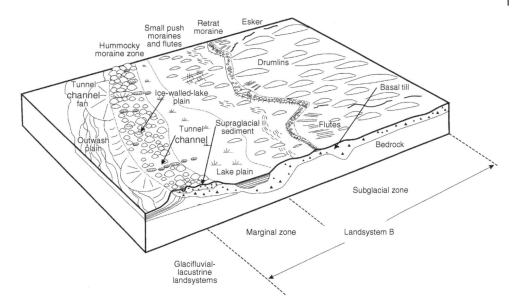

Figure 5.38 Schematic diagram illustrating the landform-sediment assemblages associated with the southern Laurentide Ice Sheet margin and 'Landsystem B' (Colgan *et al.*, 2003, p147). This comprises a marginal zone dominated by high-relief hummocky moraine and a subglacial zone dominated by drumlins and flutes. Glacier–permafrost interactions played an influential role in the formation of the marginal zone features. See text for details.

of hitherto unrecognised landforms within this extensively forested area. The marginal areas of the Wisconsin Valley and Langlade lobes were found to be dominated by features related to the burial and subsequent melt-out of buried ice. Whilst some areas featured high-relief hummocky moraine potentially related to the repeated ice advances and the stacking of debris-rich basal ice layers, the majority of the features were attributed to glacifluvial processes and the deposition of outwash onto stagnant marginal ice. The subsequent removal of the buried ice (that was in turn associated with the degradation of permafrost; see Section 5.6.3) led to the formation of a complex landscape dominated by abundant collapse depressions (up to 35 m deep) that obscured the ice-contact faces of the associated outwash heads and destroyed the original depositional surfaces of the outwash tracts. In places, the removal of this ice led to the re-appearance of drumlins, the relief of which remained visible beneath the drape of outwash sediment.

As expected, the presence of permafrost and its potential influence on glacial processes and products is much more pronounced in high latitude regions including the Canadian Arctic where permafrost remains extensive and has persisted for at least 40,000 years (Mackay *et al.*, 1972). With the geological evidence for glacier–permafrost interactions in the form of the glacitectonic deformation of pre-existing permafrost having been discussed within Section 5.4.3, this section examines the regional geomorphological signatures for these interactions in the context of the ice-marginal terrestrial landsystems found around the margins of the northern Laurentide and Innuitian Ice Sheets (Dyke & Evans, 2003). During the Last Glacial Maximum, the northern margin of the Laurentide

Ice Sheet coalesced with the Innuitian Ice Sheet (that formed over the Canadian Arctic Archipelago) and a series localised ice caps on northern Baffin Island and Somerset Island (Dyke & Evans, 2003). The Laurentide Ice Sheet had a profound impact on the landscape with dynamic ice streams producing extensive glacial lineations. In contrast, the Innuitian Ice Sheet had a much more subtle impact that has resulted in a long-standing debate regarding its actual existence, although clear evidence for glaciation has now been recognised in the form of lateral meltwater channels for example (Section 5.3.1). Subglacial permafrost is thought to have persisted throughout the last glacial cycle in some areas of this coalescent ice sheet as cold-based patches (cf. Kleman & Glasser, 2007) whilst it reformed in others during deglaciation as the ice thinned in a situation comparable with the modern-day changes in basal thermal regime occurring in Svalbard (Section 3.2).

Focusing initially on the glacial landsystems within the Canadian Arctic Archipelago, Dyke (1993, 1999) has identified a distinctive landscape zonation attributed to the basal thermal regimes and enduring ice-flow patterns occurring beneath the independent ice caps that formed and coalesced during the Late Wisconsinan. In each case, a central zone is dominated by extensive residuum-mantled, pre-Quaternary surfaces featuring tors, felsenmeer and subaerial fluvial networks that reflect prolonged periods of subaerial modification. During glacial periods these features were preserved beneath cold-based ice overlying subglacial permafrost. The circumjacent peripheral zone provides clearer evidence of active basal processes, subglacial melt and glacial erosion in the form of extensive lakes, sculpted bedrock and occasional end moraines and eskers. In combination, these features suggest a warm-based thermal regime where the heat generated by basal processes thawed the subglacial permafrost. The abrupt nature of the boundary between these zones suggests that the configuration and basal thermal regime of these ice caps was stable for tens of thousands of years.

Whilst the preservation of pre-Quaternary landscapes within the central zones led previous authors to conclude that these areas lack evidence of former glacial activity (see review by Hodgson, 1989), Dyke (1993, 1999) identified extensive networks of lateral meltwater channels incised into the weathered material. These features, which are commonly observed to be eroded by supraglacial meltwater draining along the impermeable margins of modern-day, cold-based glaciers (Section 5.3.1), invalidate the assumption made by earlier workers that cold-based glaciers in the High Arctic leave no geomorphic record. Moreover, Dyke also argued that the extensive network of lateral channels means that areas of cold-based ice often leave a more detailed record of recession than the areas of warm-based ice, potentially with an annual resolution. Finally, the extension of these channels into the warm-based peripheral zone suggests that the ice caps became entirely cold-based during ice retreat as the ice thinned and the flow decelerated.

In examining a far more extensive tract of glaciated terrain extending from Ellesmere Island to Keewatin, Dyke & Evans (2003) describe a landsystem comprising large concentric zones that feature characteristic subglacial and ice-marginal landform assemblages that are again strongly controlled by the basal thermal regime of the ice sheet. These are described as being more 'simple' in the north and more 'complex' in the south suggesting that colder climatic regimes associated with permafrost are associated with a more limited

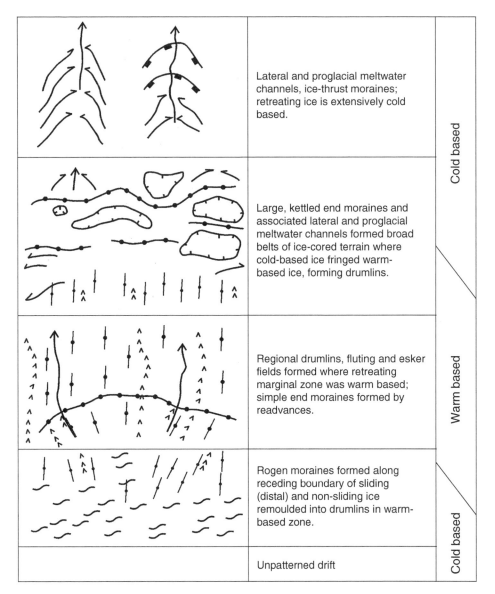

	Lateral and proglacial meltwater channels, ice-thrust moraines; retreating ice is extensively cold based.
	Large, kettled end moraines and associated lateral and proglacial meltwater channels formed broad belts of ice-cored terrain where cold-based ice fringed warm-based ice, forming drumlins.
	Regional drumlins, fluting and esker fields formed where retreating marginal zone was warm based; simple end moraines formed by readvances.
	Rogen moraines formed along receding boundary of sliding (distal) and non-sliding ice remoulded into drumlins in warm-based zone.
	Unpatterned drift

Figure 5.39 Schematic illustration of the landform zones associated with the northern Laurentide and Innuitian Ice Sheet margins (Dyke & Evans, 2003, p147). These display a distinctive north–south progression that is generally simpler in the north (top) and more complex in the south (bottom). See text for details.

diversity of landform assemblages. A schematic outline of these zones is illustrated in Figure 5.39 in relation to changes in the basal thermal regime. Moving from the north to the south these include:

- An outer zone characterised by lateral and proglacial meltwater channels, ice-thrust moraines and limited accumulations of glacigenic debris. This is related to the recession of predominantly cold-based glaciers with warm-based ice being limited to deep fjords during the glacial maxima.

- Broad belts of ice-cored moraine c. 10–40 km wide comprising ridged and hummocky moraine characterised by depressions and proglacial meltwater channels. Moving up-glacier, this zone typically transitions into areas featuring streamlined bedforms that suggests that these moraine belts formed at the transition from warm to cold-based ice. This promoted both the aggradation of thick sequences of basal ice (Section 3.5.4) and the accumulation of sediment required to form ice-cored or controlled moraine (Section 5.3.2). Subsequent partial melting of this buried ice (that is largely preserved by permafrost aggradation) resulted in the formation of large depressions (kettles) in between the moraine ridges.

- Extensive fields of glacial lineations, eskers and simple end moraines associated with punctuated recession of warm-based ice.

- Areas of Rogen moraine that occur on the up-glacier side of the fields of glacial lineations. As previously suggested by Kleman & Hattestrand (1999), these are believed to form at the boundary between non-sliding (cold-based) and sliding (warm-based) ice (Section 5.3.3). With the lineations being associated with warm-based conditions, they tend to form during the latter stages of glaciation when ice thicknesses are at their greatest. If the associated boundary between sliding and non-sliding ice propagates up ice during deglaciation, this results in the streamlining of the Rogen moraines that are only preserved close to the final position of the boundary (typically close to the final ice-dispersal centres).

- In examining the glacial landsystems found within northern Canada, Dyke & Evans (2003) explicitly identify what they consider to be the distinctive process-form relationships associated with ice-sheet margins terminating in permafrost environments that are by implication illustrative of glacier–permafrost interactions. These include:

  - **Glacially deformed permafrost:** Deep seated deformation relating to the dynamic coupling between the cold-based margin of the ice sheet and the pre-existing permafrost resulted in the formation of large thrust moraines (e.g. Herschel Island & Nicholson Peninsula) as well as the structures described in Section 5.4.3.

  - **Major belts of ice-cored moraine:** These occur throughout the Canadian Arctic with specific examples having been described in relation to the Bluenose Lake Moraine (St. Onge & McMartin, 1999), the Colville moraine on the Wollaston Peninsula on Victoria Island (Dyke & Savelle, 2000) and the Jesse moraine belt that extends for 350 km across both Banks Island and Victoria Island (Lakeman & England, 2012). These are characterised by significant volumes of massive ground ice that has been variously interpreted as representing relict glacier ice or intrasedimental ice (Section 5.3.4). Permafrost aggradation has promoted the preservation of this massive ice in areas where the thickness of supraglacial melt out till exceeds the active layer thickness. Thermokarst activity associated with the removal of surface materials or the formation of thaw lakes has led to the formation of extensive kettle-like basins with recent climate change resulting in these landforms forming the locus for landscape destabilisation and megaslump development (Section 5.6.3). The presence of widespread buried ice within these moraine belts has led to questions as to whether they represent 'pseudo-moraines' relating to the degradation of ice-rich terrain, hummocky moraine formed through regional stagnation, or ice-cored moraines associated with former active ice margins. Dyke & Evans (2003) argue that they constitute large, controlled moraines in view of the linearities evident in the moraine ridges and their association with distal outwash trains and fans. The recession of an ice-sheet margin that remains active (rather than stagnating) with periodic readvances can result in the formation of a series of ice-cored moraine belts separated by tracts of glacial lineations associated with the migration of the boundary between warm- and cold-based ice over time (Figure 5.9a).

- **Ice-shelf landforms:** These largely comprise moraine systems with low-to-negligible gradients, with any apparent gradient potentially representing the influence of postglacial isostatic uplift. These ice-shelf landforms are considered indicative of glacier–permafrost interaction with the sustenance of ice-shelves requiring cold-based ice with warm-based ice conversely lacking the tensile stress required to prevent calving.

In conclusion, it is worth noting that Dyke & Evans (2003) consider the landsystems associated with the Northern Laurentide and Innuitian ice sheets to be entirely compatible with those of the Southern Laurentide ice sheet described earlier. The key difference is that the continued presence of permafrost within the north throughout the postglacial period has resulted in their 'incomplete deglaciation' (Kaplyanskaya & Tarnogradskiy 1986) and the long-term preservation of detached zones of buried ice. In contrast, its complete melt-out in the temperate postglacial climates of the south has produced extensive tracts of hummocky moraine in which the ice-walled lake plains for example can be seen as being genetically related to the kettles currently observed in ice-cored moraine belts further north. This highlights the role of permafrost in retarding or delaying the paraglacial adjustment of glaciated landscapes, an important concept the implications of which are explored in more detail in Section 5.6.

5.5.3 Arctic Glacial Landsystems

A growing number of geomorphological investigations undertaken primarily in Svalbard and the Canadian High Arctic have provided a valuable insight into the diverse nature of the landsystems associated with non-temperate glaciers and glacier–permafrost interactions. The opportunities to relate fundamental glaciological characteristics (mass balance, velocity and basal thermal regime) to the operation of specific geomorphic processes and the creation of distinctive landform-sediment assemblages provides the opportunity to test and refine the genetic process-form models central to palaeoglaciological reconstructions (Section 5.5.1).

Ó Cofaigh (2006) identifies a series of important genetic associations between environment, process and form associated with the sub-polar glaciers (plateau icefields and piedmont lobes) of the Queen Elizabeth Islands in the Canadian High Arctic and the generation of a distinctive landsystems model (Figure 5.40). These highlight the significance and influence of the characteristic thermal transitions from interior warm-based ice to marginal frozen zones. This transition promotes the entrainment of debris through net basal adfreezing and apron entrainment, the formation of thick sequences of basal ice and the consequent formation of controlled moraine. It also facilitates the propagation of compressive stresses into permafrozen forelands leading to proglacial glacitectonic deformation

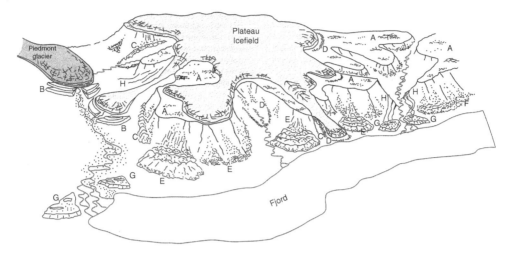

Figure 5.40 Schematic landsystems model for sub-polar glaciers in the Canadian High Arctic comprising: A blockfield/residuum, B thrust-block moraine, C ice-cored lateral moraine, D trimline moraine, E glacier-ice-cored protalus rock glacier, F proglacial protalus rock glacier, G raised, former ice-contact deltas and H lateral meltwater channels (Ó Cofaigh, 2006, p90).

and the formation of large thrust-block moraines comprising glacifluvial, glacilacustrine and emergent glacimarine sediments. Basal thermal regime and the occurrence of cold-based ice margins also promote the development of nested sequences of lateral meltwater channels, kame terraces and deltas that provide a record of the recent recession of these ice masses (Dyke, 1993). It can also result in variations in sediment flux with the margins of plateau ice fields and upland outlet glaciers displaying low debris turnovers and the formation of trimlines, boulder lines and rubble veneers rather than moraines and extensive till sheets. Distinctive types of paraglacial activity (Section 5.6) can also occur within these cold climatic regimes with the production of talus coupled with the preservation of buried ice during deglaciation resulting in the formation of rock glaciers (Evans, 1993).

Distinctive thermal zonations between geomorphologically active, warm-based ice and geomorphologically inactive cold-based ice have also been inferred to explain spectacular landscapes of 'selective linear erosion' in which steep-sided glacial troughs are juxtaposed with rolling plateaux that retain their preglacial form and display little evidence of glacial landscape modification (Figure 5.5). Following earlier work in the Cairngorms, further investigation of the plateaux surfaces on the Cumberland Peninsula of Baffin Island by Sugden & Watts (1977) illustrated the dominance of felsenmeer and tors and the influence of long-term subaerial weathering, although the presence of erratics demonstrated previous glacial activity. A consistent pattern of smoothing on the south-western sides of the tors suggested that this activity was associated with some erosion, which the authors related to the flow of debris-rich ice (Section 3.5.2). Subsequent cosmogenic exposure dating on these heavily weathered upland surfaces by Bierman *et al.* (1999) using ^{10}Be and ^{26}Al has suggested a total exposure and burial history of between 450 and 735 ka. Reconstructed maximum erosion rates of $<1\,\mathrm{m\,Ma^{-1}}$ suggest that any erosion that has occurred has been very limited in nature.

As part of their focused examination of glacier–permafrost interactions in Arctic and alpine mountain environments, Etzelmüller & Hagen (2005) provide a summary of the typical landsystems associated with terrestrial glacier margins on Svalbard. These are characterised by one or two prominent end moraines that enclose a glacier foreland characterised by small glacial mounds, areas of flow till and outwash plains. End moraines and areas of flow till are usually underlain by ice with ice-cored terminal or lateral moraines displaying evidence of creep to create rock glaciers with dynamics unrelated to the parent glacier. Some glaciers feature push moraines or composite ridges occurring beyond the ice-cored end moraines, which only occur below the Holocene marine limit. As these are no longer in direct content with the modern glaciers, they are considered relict features relating to proglacial glacitectonic deformation that occurred when the glaciers were at or close to their Little Ice Age limit with some potentially relating to surge events. These earlier phases of landsystem development are also considered responsible for the formation of areas featuring flutes and streamlined ridges within the glacier forelands, with the thicker ice enabling warm-based conditions and active basal motion.

Personal observations around the margin of Fountain Glacier, a polythermal glacier terminating in an area of continuous permafrost estimated to be between 200 and 400 m thick, illustrate the distinctive Arctic glacial landsystems associated with the glacier–permafrost interactions in the High Canadian Arctic. With a mean annual air temperature estimated to be −15 °C, these more continental environments are typically colder and drier than those found in Svalbard. Field observation showed the glacier margin to be characterised by a distinctive landform assemblage comprising boulder-dominated surfaces and moraine ridges (a number of which were ice-cored), areas of heavily weathered and shattered bedrock, incised lateral meltwater channels and outwash braidplains (Figures 5.6 and 5.41) as well as the large proglacial icing discussed in Section 4.2.2 (Figure 4.4). In addition to these newly created landforms, recession of part of the margin revealed an intact tundra surface complete with biologically viable vegetation that showed little if any glacial modification associated with its inundation during the Little Ice Age (Figures 5.1 and 5.2). Significant glacial sediment transfer occurred through a prominent debris-rich basal ice layer dominated by sub-angular and angular material of up to boulder size. The basal ice layer and ice margin as a whole displayed extensive evidence of glacitectonic deformation intimately associated with the structural glaciology of the glacier as a whole (Figure 3.11). Thrusting was observed to elevate basal material from the basal ice to the glacier surface where the material contributed to a supraglacial sediment flux. Fluvial sediment transfer was associated primarily with a series of ice-marginal meltwater systems producing the only lithofacies with a significant percentage of rounded clasts.

The glacial sediment-landform association at Fountain Glacier displays a number of distinctive elements illustrative of the specific geomorphic processes active in this environment. Bedrock shattering is predominant with there being little evidence of glacial abrasion, thereby leading to the development of glacigenic lithofacies with a surprising degree of angularity (Section 5.4.1). In spite of the cold climatic conditions and limited meltwater production, fluvial processes remain important geomorphologically, leading to the erosion of lateral meltwater channels, the deposition of small outwash surfaces and the generation of large proglacial icings. The most enigmatic element however relates not to the new

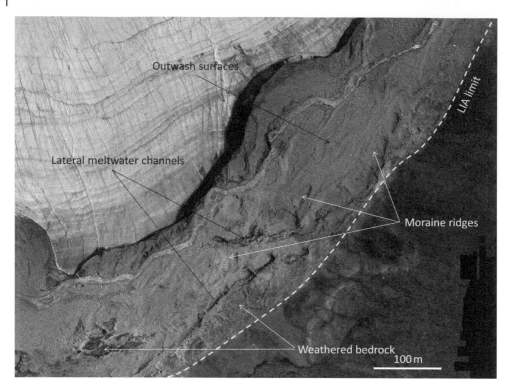

Figure 5.41 Drone image of part of the margin of Fountain Glacier, Bylot Island, illustrating the location of some of the key elements of the ice-marginal landsystem. See text for details (Image: Richard Waller).

features being actively created but to the ability of the glacier to preserve delicate vegetated tundra surfaces during ice advance (Section 5.2). The proximity of these intact surfaces with more heavily modified areas suggests a spatial variability in process that merits further investigation and explanation.

5.5.4 Cold Extremes – Antarctic Glacial Landsystems

Field studies of the behaviour and geomorphic impacts of glaciers in Antarctica provide an insight into the distinctive landforms and landscapes associated with glacier–permafrost interactions in some of the coldest places on Earth. These glaciers represent 'the extreme end member of the broad spectrum of glaciers with different thermal regimes' (Hambrey & Fitzsimons, 2010, p858) such that they are considered the closest terrestrial analogues for glaciers on Mars (Section 5.5.5). Nonetheless, research undertaken on the Meserve (Cuffey *et al.*, 1999, 2000) and Suess Glaciers (e.g. Fitzsimons, 1996; Fitzsimons *et al.*, 2001) in the McMurdo Dry Valleys as well as the Manhaul Bay Glacier in South Victoria Land (Atkins *et al.,* 2002, Davies *et al.,* 2009; Atkins, 2013) have demonstrated the potential for entirely cold-based glaciers with basal temperatures below -15 °C to remain geomorphologically active and to create a range of subtle yet distinctive features (Figure 5.17). Having considered the specific landforms indicative of active erosion and deposition in

Section 5.3.3, this section considers the ways in which these features and related sediment facies can occur in combination to describe the glacial landsystems characteristic of these cold extremes.

Fitzsimons (2003) has synthesised field observations made within East Antarctica and south Victoria Land to describe the ice-marginal terrestrial landsystems characteristic of polar continental glacier margins. In the context of Antarctica, these ice-marginal environments feature cold-based glaciers occurring within cold desert environments characterised by very low mean annual air temperatures (−10 to −20 °C) and precipitation totals that promote the formation of extensive permafrost (e.g. Bockheim *et al.*, 2007). Fitzsimons (2003) differentiates the ice-marginal landsystems into those associated with low-relief landscapes (such as the Vestfold Hills and Bunger Hills in East Antarctica) and those associated with high-relief landscapes (including the McMurdo Dry Valleys). Low-relief ice-marginal landsystems are typically associated with extensive areas of ice-cored moraine that can make it difficult to identify the actual ice margin. The ice-cored moraines are in turn associated with inner moraines (Section 5.3.2) that supply basal debris to the glacier surface that comprises massive, matrix-supported diamictons. This material can subsequently be re-mobilised to contribute to the formation of ice-contact fans and screes forming sharp-crested ridges up to 20 m high and 500 m long and comprising a range of lithofacies that reflect the influence of alluvial and colluvial reworking. Thrust-block moraines can also form where glacier margins flow across inlets or lakes (see Fitzsimons, 1996). Many of the high-relief ice-marginal landsystems found in the McMurdo Dry Valleys are characterised by either an absence of landforms or landforms that are very small. Glacifluvial features such as outwash surfaces are also largely absent due to the limited meltwater generation and glacigenic sediment supply. Some ice margins are however associated with well-developed end moraines with diverse characters that relate to the broad-ranging formative processes that can include apron entrainment (Section 3.6.1) and the displacement of blocks of frozen submarginal sediment (Fitzsimons *et al.*, 1999).

Fitzsimons (2003) concludes by stating that whilst our incomplete knowledge precludes the development of a detailed landsystem model, polar continental glacier margins are associated with the following distinctive elements:

- The production of limited volumes of sediment that results in the development of sedimentary landforms with limited volumes and preservation potentials.
- The development of constructional moraine types at stable ice margins, notably ice-contact fans and screes that are associated with debris supplies from inner moraines.
- The formation of thrust-block and push moraines where cold ice interacts with saturated and unfrozen sediment.
- The potential for active bed deformation of permafrozen sediments beneath dry-based glaciers that can lead to the creation of structural landforms (see Fitzsimons *et al.*, 2023).
- The limited development of glaciofluvial landforms.

In investigating the margins of a selection of glaciers in the Dry Valleys in southern Victoria Land and focusing on the Wright Lower Glacier in particular, Hambrey & Fitzsimons (2010) provide one of the few detailed case studies examining the glacial landsystem signatures of cold glaciers terminating in areas of permafrost. In doing so, they highlight the paucity of similar investigations and the lack of sedimentological information on the related glacigenic

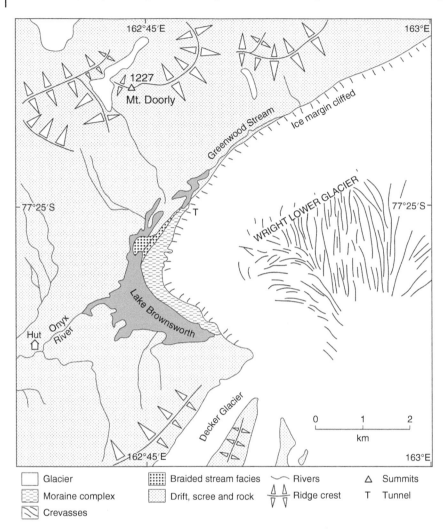

Figure 5.42 Geomorphological map of the snout and proglacial area of the Wright Lower Glacier in the Dry Valleys of Antarctica illustrating the principal landform-sediment associations (Hambrey & Fitzsimons, 2010, p862). See text for details.

lithofacies that is an essential prerequisite for the identification of cold glaciers in the geological record. Observations made around the ice margin and foreland of the Wright Lower Glacier describe a landsystem dominated by older drift deposits, modern moraine ridges, ice-contact debris aprons and an ice-marginal braid plain that flows into a delta complex within nearby Lake Brownworth (Figure 5.42). Moving from the glacier margin into the foreland, the glacier itself features a characteristic cliff-like margin up to 10 m high that displays evidence of active debris entrainment in the form of foliation-parallel planar structures containing sandy gravel and blocks of sand with preserved cross-laminations (Fitzsimons *et al.*, 2023). The immediate glacier margin features an extensive ice-contact apron up to 15 m high and 70–150 m in breadth. Dissection by sporadic supraglacial meltwater flows to produce a series of gullies reveals a core of glacitectonised ice and sediment in its proximal part and a drape of slope-parallel bedded sands on its distal flank as well as extensive evidence of aeolian activity to produce patchy surficial deposits of loose sands and aeolian ripples. The more distal part of the foreland is characterised by an ice-covered proglacial lake (Lake Brownworth) that contains a series of recessional moraines several metres in height comprising a sandy boulder-gravel drape over an ice core. These continue beyond the lake where they comprise similar sandy boulder-gravels containing tilted masses of laminated sands. Finally, in spite of the extreme cold, the landsystem displays evidence of fluvial activity both in the reworking of sediment from the ice-contact aprons to produce an alluvial fan and in the presence of an ice-margin-parallel braid plain with well-sorted sands and bars comprising pebble-cobble sandy gravel that terminates in a delta at the northern end of the proglacial lake.

In providing additional sedimentological detail on the constituent landform-sediment assemblages, the authors describe a sedimentary cycle that produces a more limited range of lithofacies than those typically associated with temperate and polythermal glaciers (Figure 5.43). Whilst these cold end member glaciers display abundant evidence of geomorphological activity capable of producing classic glacial landforms such as moraines as well as more distinctive features such as ice-contact aprons, the textural characteristics of the constituent sediments are fundamentally different from those associated with temperate and polythermal glaciers (section 5.4.1, Hambrey & Glasser, 2012). Diamicts and fine-grained sediments are largely absent due to the lack of abrasion with the lithofacies instead being dominated by sands and gravels that have been largely reworked from older sediments. The lack of modification and consequent absence of features diagnostic of glacial reworking provides an ongoing challenge for those seeking to recognise evidence of cold glaciers in the geological record. The depositional landforms also feature extensive buried ice in the form of glacier ice, lake ice and snow that contributes to their heights and which has the potential to persist for millennia in these environments (Section 5.3.4). Finally, in spite of the remarkable aridity, these glacial environments display evidence of limited fluvial activity in the form of ice-marginal braid plains, alluvial fans and lake sediments (Figure 5.44).

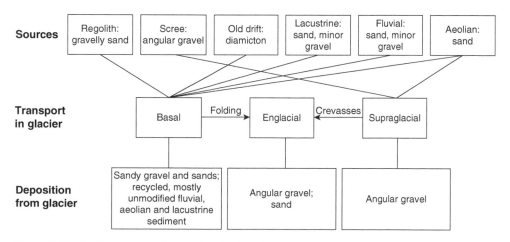

Figure 5.43 Sedimentary cycle associated with cold-based glaciers in the Dry Valleys of Antarctica (Hambrey & Fitzsimons, 2010, p878). See text for details.

Figure 5.44 Ice margin and foreland of the Suess Glacier located in South Victoria Land, Antarctica. Note the steep ice cliff, the ice-and-debris apron, the boulder covered ice-cored moraines and the braid plain feeding the proglacial lake (with permission from Sean Fitzsimons).

5.5.5 Martian Landsystems

Research examining the landsystem signatures of glaciers in general and glacier–permafrost interactions in particular is not restricted to our own planet and has recently been extended to Mars, where a number of research publications have identified a range of 'glacier-like forms' (Hubbard *et al.*, 2014) as well as extensive evidence for periglacial activity (e.g. Balme & Gallagher, 2009). The extreme environmental conditions found on Mars have resulted in much of this research focusing on the Antarctic Dry Valleys as the closest terrestrial analogue. With this most extreme of climates on Earth featuring glaciers that are exclusively cold-based situated within an environment characterised by continuous permafrost, then the glacial landsystems described on Mars are also likely to represent the product of 'extraterrestrial' glacier–permafrost interactions.

As discussed earlier (Section 3.5.4), one of the principal products of glacier–permafrost interactions is the formation of significant quantities of buried ground ice that can be preserved over very long time scales if glacier recession is accompanied by permafrost aggradation. With this in mind, various lines of geomorphological evidence have been put forward to infer the presence of buried (water) ice in the tropical and mid-latitude regions of Mars, which Scanlon *et al.* (2015) divides into three categories:

1) Surface textures related to the degradation and partial removal of ground ice (i.e. thermo-karst activity). These include features such as sublimation pits (Mustard *et al.*, 2001), scalloped depressions (Dundas *et al.*, 2015) and 'brain terrain' (e.g. Levy *et al.*, 2010).

2) Distinctive topographic profiles. A suite of related features including lobate debris aprons (LDA), lineated valley fills (LVF) and concentric crater fills (CCF) have been argued to represent debris-covered glaciers with residual cores of ice on account of their convex-upward marginal profiles.

3) Unusual crater morphologies indicative of impacts into ice-rich substrates.

Focusing on the third category, 'expanded secondary craters' comprise a specific landform and crater morphology used by Viola *et al.* (2015) to infer the potential extent of subsurface ice in the Arcadia Planitia in the northern mid-latitudes of Mars. Secondary craters are created by the ejecta formed during a planetary impact and are therefore widely distributed around the larger primary crater. The simultaneous creation of a series of secondary craters is thought to provide an insight into the local near surface ground conditions. Expanded secondary craters comprise specific examples that have developed a shallow extension that has been related to thermokarst activity induced by the impact that can trigger the degradation of subsurface ice and subsequent surface subsidence (Figure 5.45). In relating the expanded secondary craters to the presence of excess ice below the surface, the authors estimate that at least $6000\,\text{km}^3$ of ground ice is likely to remain preserved in uncratered parts of the Arcadia Planitia. This is in turn hypothesised to represent the remains of a large ice sheet that formed in the region over 20 million years ago and which has been partially preserved beneath a surface lag deposit. It is however important to note that as with equivalent studies on Earth (Section 5.3.4), the mode of origin of the ground ice remains subject to debate with the prospect of its resolution through detailed geochemical or isotopic analyses for example being even more remote.

Much of the research that has postulated the presence of cold-based glaciers has made explicit use of the well documented terrestrial analogues and process-form models that have formed the focus of this section, even if the terminology used by planetary geologists to describe the associated geomorphological features is very different. This research has similarly recognised the dangers inherent in basing any interpretations on singular landforms and the associated advantages of using a landsystem approach that considers the spatial arrangement of a range of related features (Evans, 2003). Head & Marchant (2003)

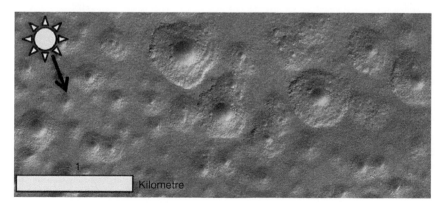

Figure 5.45 Examples of expanded secondary craters on Mars (located near 52.7°N, 216.3°E) (Viola *et al.*, 2015, p191/with permission from Elsevier).

for example describe a distinctive fan-shaped landsystem c. 180,000 km² in extent located on the western flank of the Arsia Mons volcano that they attribute to the degrading remnants of a cold-based mountain glacier. The landsystem comprises three distinct 'facies' including (i) an outermost 'ridged facies' characterised by distinct ridges tens of kilometres in length and hundreds of metres to kilometres apart, (ii) a 'knobby-terrain facies' characterised by a more chaotic terrain with hills several kilometres in diameter that sometimes form distinct chains and (iii) a 'smooth facies' featuring arcuate ridges tens of metres high occurring on broad lobes hundreds of metres thick that appears to overlie facies ii. In interpreting these facies, the outer ridges are thought to represent 'drop moraines' associated with the deposition of supraglacial material at a receding ice margin whilst the knobby facies constitutes an area of hummocky moraine associated with the sublimation and downwasting of glacier ice. Finally, the smooth facies is interpreted as a rock glacier associated with the final waning stages of the glacier that is potentially still active. In explicitly relating each of these facies to landforms observed in the Dry Valleys of Antarctica, they are argued to collectively delineate the remnants of a cold-based glacier on the volcano's western flank. Follow-up research by Scanlon *et al.* (2015) focusing on the ridged and smooth facies provides further detail on this Martian glacial landsystem (Figure 5.46) and the potential relevance of glacier–permafrost interactions. The linear arrangement of mounds and depressions within 'pit-and-knob' terrain for example is explicitly related to

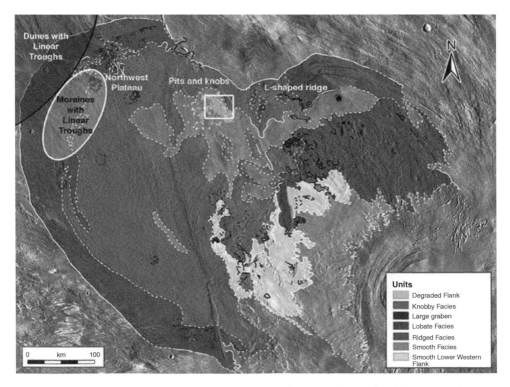

Figure 5.46 Geomorphological map of the Arsia Mons fan-shaped deposit (FSD) on Mars (Scanlon *et al.*, 2015, p146/with permission from Elsevier). See text for details.

the controlled moraine that Evans (2009) considers diagnostic of both polythermal basal thermal regimes and the processes of debris entrainment and concentration that occur at the glacier–permafrost interface (Section 5.3.2). Therefore, in incorporating both ice-marginal moraine ridges and an extensive zone of controlled moraine, this glacial landsystem can be viewed as broadly comparable to those considered indicative of frozen-bed conditions and glacier–permafrost interactions at the margins of the Laurentide Ice Sheet (Section 5.5.2).

Other research contributions have inferred the presence of cold-based glaciers and glacier–permafrost interactions within the context of more unusual Martian landsystem settings that lack clear terrestrial analogues. Guidat *et al.* (2015) for example propose the existence of a 3 Ga old polythermal ice sheet in the Isidis Planitia, a large impact crater located close to the Martian equator. In systematically mapping the component parts of the distinctive and controversial 'thumbprint terrain' apparent on the crater floor, the authors adopt an inverse modelling approach (Section 5.5.1) to interpret a suite of landforms with a distinctive spatial organisation that they consider to be glacial in origin. They utilise process-form connections to infer the basal thermal regime and to suggest that the parent ice mass was polythermal in nature with a cold-based interior, a warm or wet-based periphery and finally a cold-based fringe (Figure 5.47) – a thermal regime and distribution of cold-based conditions (and therefore subglacial permafrost) similar to ice sheets on Earth (Section 2.4.1). The cold-based interior is related to a series of isolated cones and cone fields that are thought to represent dirt cones forming on degrading cold-based ice. In contrast, peripheral wet-based conditions are inferred on the basis of features including sinuous ridges (20–200 m wide and hundreds of metres to kilometres in length) interpreted as esker systems on account of their branching nature and broader radial pattern. In addition, the arcuate ridges (300–600 m wide, 300 m–40 km in length and 20–60 m high) that collectively define the whorl-shaped patterns characteristic of the thumbprint terrain are argued to represent ribbed moraines that are also argued to reflect wet-based conditions.

This brief review of some of the published literature on the glacial landsystems of Mars demonstrates both the wider applicability of the detailed process-form linkages developed in terrestrial glacial environments and more specifically, the relevance of glacier–permafrost interactions and their landsystems signatures. However, it also serves as a reminder of the ongoing assumptions and uncertainties regarding the connections between basal thermal boundary conditions and subglacial processes that can lead to the disputed and potentially inaccurate usage of geomorphological evidence within palaeoglaciological reconstructions more broadly (Section 3.5). In relating the arcuate ridges to ribbed moraines, Guidat *et al.* (2015) argue that these key features within the landsystem described above are indicative of the reshaping of subglacial sediment beneath wet-based ice. Kleman & Hattestrand (1999) in contrast relate the same landforms in a terrestrial setting to the brittle fracture of frozen subglacial sediment, suggesting they are instead diagnostic of thermal transitions between frozen and thawed-bed conditions (Section 5.3.3). Whilst planetary geologists refer extensively to the work of terrestrial glacial geomorphologists, an obvious concern here is the limited nature of collaborations between these research communities outside of the specific confines of the Dry Valleys of Antarctica. Undertaken

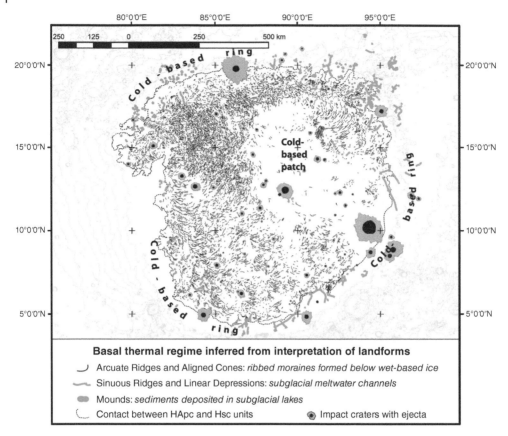

Figure 5.47 Interpretive map of the glacial landsystem in Isidis Planitis on Mars (Guidat *et al.*, 2015, p264). The key identifies the principal landforms and their glacigenic interpretations. HApc and Hps relate to different geological units.

more broadly, such collaborations clearly have the potential to enhance analyses and to refine interpretations of the products of glacier–permafrost interactions on both Earth and Mars.

5.6 Glaciers, Permafrost and Paraglaciation

5.6.1 Introduction

The term 'paraglacial' was introduced by Church & Ryder (1972) as a core concept within their consideration of the impacts of glaciation on fluvial processes and sediment yields. With glaciers typically depositing large and unstable sediment accumulations that can subsequently be reworked by a range of denudational processes, they used the term to define 'nonglacial processes that are directly conditioned by glaciation' (p3059). In doing so, they specifically drew a comparison with the more established term 'periglacial' that in modern-day usage does not require the presence of glaciers.

Focusing initially on the influence of glaciations on fluvial sediment yields, Church & Ryder (1972) demonstrated that glaciations cause a perturbation of 'normal' fluvial conditions and a significant increase in denudation and sedimentation rates such that sediment yields bear no relation to the sediment production provided by weathering processes. In using examples where paraglacial denudation was both ongoing (central Baffin Island) and complete (south-central British Columbia), the authors identified a distinct 'paraglacial period' lasting for a few millennia. During this time, sediment yields were estimated to be at least an order of magnitude higher than those associated with 'normal' conditions, with measured values on Baffin Island reaching 610–1170 metric tons km^{-2}, equivalent to denudation rates of 230–440 mm yr^{-1}. Whilst the authors recognised the potential influence of a range of complicating factors including geology, climate, vegetation and isostatic uplift, they provided a tentative schematic model of the pattern of sedimentation during the paraglacial period that featured a peak during the initial phase of deglaciation followed by an exponential decay back to 'normal' conditions (Figure 5.48). Interestingly, with denudation rates on Baffin Island during full glacial conditions having previously been estimated to be 25–90 mm per 1000 years, this suggested that sediment yields during the initial paraglacial phase could be an order of magnitude higher than during the glacial period itself.

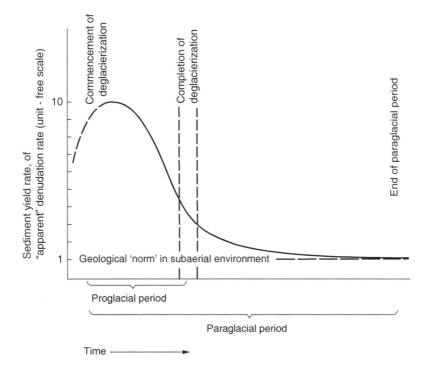

Figure 5.48 Schematic figure of the sediment yield during the paraglacial period. Note the prominent peak associated with the commencement of deglacierisation and the continuation of the paraglacial period beyond the proglacial period and the removal of glaciers from the catchments (Church & Ryder, 1972, p3069).

A subsequent, detailed review of paraglacial geomorphology undertaken by Ballantyne (2002) has broadened the geomorphic scope of the concept beyond the exclusively fluvial focus considered by the original authors, resulting in a revision of the working definition of the term 'paraglacial' to encapsulate:

> *...nonglacial earth-surface processes, sediment accumulations, landforms, landsystems and landscapes that are directly conditioned by glaciation and deglaciation* (p1938).

Explicit use is made of the landsystems concept (Section 5.5.1) to identify six specific paraglacial landsystems (rock slopes, drift-mantled slopes, glacier forelands and alluvial, lacustrine and coastal systems) associated with unstable or metastable sediment stores that are reworked by a variety of processes as the landscapes adjust to non-glacial conditions. 'Primary' sediment stores (such as talus cones, alluvial fans and valley fills) are created through the reworking of glacigenic sediment sources, providing what are often only temporary sediment stores that can subsequently be reworked by 'secondary' paraglacial systems. Each landsystem can consequently be conceptualised as an interrupted sediment cascade in which a range of processes operate over different spatial and temporal scales to transfer sediments from sources, through primary and secondary stores and ultimately to sinks (Figure 5.49).

Contrasting models of sediment transfer consider the influence of both catchment size and sediment exhaustion over time. In relation to catchment size, a peak in specific sediment yield is envisaged to diminish in amplitude and to be increasingly delayed in occurrence as the 'wave' of reworked sediment derived from the small upland catchments progresses through the system. Alternatively, sediment transfer can be related to an 'exhaustion model' in which the sediment yield decreases exponentially over time as the available sediment stores are depleted. A range of extrinsic factors can however perturb systems that had previously stabilised resulting in the remobilisation of material and secondary peaks in sediment yields. These can include base level changes, climate change, extreme climatic events and anthropogenic activity such as overgrazing. Ultimately, the combination of these intrinsic and extrinsic controls operating at varying spatial and temporal scales produces sediment cascades and paraglacial landscapes of 'great geomorphological variety and evolutionary complexity, in which individual landforms develop even as others decay over timescales ranging from a few years to many millennia' (Ballantyne, 2002, p2004).

This section focuses on the influence of permafrost upon the processes and timescales of landscape adjustment that accompany deglaciation in specific relation to two of the landsystems settings identified by Ballantyne (2002). First, Section 5.6.2 considers the combined influence of glacier recession and permafrost degradation on the destabilisation of mountain slopes in high mountain regions. Second, Section 5.6.3 examines the influence of permafrost on the preservation and subsequent melt-out of massive ground ice (Section 5.3.4), with recent permafrost degradation having been observed to lead to a rapid expansion in the areas affected by thermokarst activity and the development of large 'megaslumps' within formerly glaciated permafrost regions.

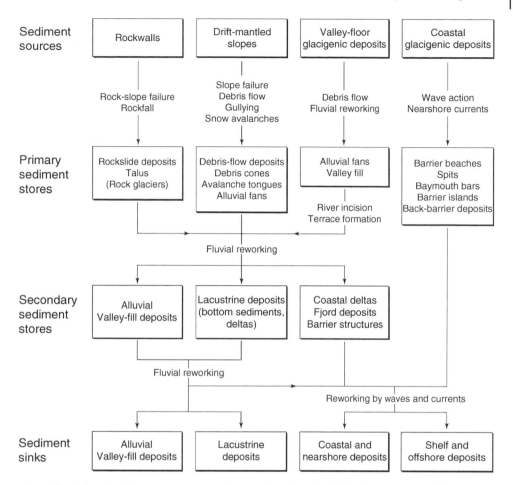

| Sediment sources | Rockwalls | Drift-mantled slopes | Valley-floor glacigenic deposits | Coastal glacigenic deposits |

Rock-slope failure
Rockfall

Slope failure
Debris flow
Gullying
Snow avalanches

Debris flow
Fluvial reworking

Wave action
Nearshore currents

| Primary sediment stores | Rockslide deposits Talus (Rock glaciers) | Debris-flow deposits Debris cones Avalanche tongues Alluvial fans | Alluvial fans Valley fill | Barrier beaches Spits Baymouth bars Barrier islands Back-barrier deposits |

River incision
Terrace formation

Fluvial reworking

| Secondary sediment stores | Alluvial Valley-fill deposits | Lacustrine deposits (bottom sediments, deltas) | Coastal deltas Fjord deposits Barrier structures |

Fluvial reworking

Reworking by waves and currents

| Sediment sinks | Alluvial Valley-fill deposits | Lacustrine deposits | Coastal and nearshore deposits | Shelf and offshore deposits |

Figure 5.49 Simplified sediment cascade illustrating the principal primary and secondary stores and sediment transfer processes. Relatively minor processes such as wind action are excluded (Ballantyne, 2002, p2004).

In considering the impact of extrinsic controls (climate change) on the paraglacial adjustment of landsystems containing both glaciers and permafrost, this section highlights the connectivity of these key components of the cryosphere and the need therefore to explicitly consider the nature and significance of glacier–permafrost interactions. Both Haeberli (2005) and Etzelmüller & Hagen (2005) emphasise the thermal and hydrological connections between these two key components in the context of the potential geomorphological consequences of ongoing climate change. An important dynamic to note here relates to their differing reaction times with glaciers responding over much shorter timescales (years to decades) than permafrost (hundreds to thousands of years). The resulting complex responses can consequently produce seemingly counter intuitive system responses to external drivers with a *warming* climate and associated glacier recession potentially resulting in the *aggradation* of permafrost and a *cooling* of the glacier's thermal regime (see Sections 4.2.4, 6.5). Before looking at the specific circumstances within the selected case studies, it is important therefore to appreciate the,

> ...*close interaction between the glacial and periglacial system, affecting all chains of the sediment cascade system within landform evolution. In the framework of understanding geomorphological processes and response of the system to climate change, glaciers and permafrost must be regarded as an integrated part of the analyses.*
>
> (Etzelmüller & Hagen, 2005, p23).

5.6.2 Rock Slope Failure

The adjustment of glacially steepened slopes caused by glacier recession and their exposure and 'debuttressing' is considered by Ballantyne (2002) to be 'one of the most important geomorphological consequences of deglaciation in mountain environments' (p1938). This can occur most dramatically in the form of large-scale and catastrophic rock slides or rock avalanches but can also occur as large-scale but much slower rock slope failures, and also as high frequency and low magnitude rockfall events. In each case, the influence of glaciers is of central importance with glaciation and glacial erosion (particularly in areas of focused ice flow) initially creating higher and steeper slopes and a general increase in the stress field within the rock mass. Subsequent deglaciation and the downwasting of the ice within the valley that previously supported the slopes leads to debuttressing and a relaxation of the residual stresses within the rock mass that can induce failure of a variety of different types depending on the stress conditions, lithologies and joint patterns.

Adjustment can occur soon after deglaciation with numerous recent examples of catastrophic slope failures having been attributed at least in part to the dramatic reduction in ice volume that has occurred since the Little Ice Age. The most likely cause of a catastrophic rock slope failure on Mount Fletcher in New Zealand in September 1992 for example was a 250 m reduction in the thickness of the Maud Glacier since the middle part of the nineteeth century that unloaded the toe of a steep and structurally defective slope (McSavaney, 2002). Alternatively, the release of residual stresses and the accompanying slope adjustment can be delayed such that it takes place over a paraglacial period lasting for several millennia. Widespread slope failure is thought to have occurred in a number of formerly glaciated mountain ranges during the Late Pleistocene or Early Holocene following the recession of ice after the Last Glacial Maximum. Whilst the reliability of any such paraglacial process association between deglaciation and regional slope adjustment remains uncertain due to a lack of sufficient dating control (Ballantyne, 2002), the use of cosmogenic isotope dating to date a major rockslide on the Isle of Skye by Ballantyne *et al.* (1998) confirms the potential for large-scale events to occur several thousands of years after deglaciation.

In contrast to the widespread research interest afforded to the role of glaciers in causing paraglacial rock slope failures, the potential influence of permafrost and subsurface ice has until recently received comparatively little interest. As recently as the late 1990s, Haeberli *et al.* (1997) noted that 'the influence of permafrost on the destabilisation of rock walls [...] has remained until now a virtually untouched field of research' (p408). In reality, the formation and degradation of subsurface ice within ice-rich permafrost can act as a criti-cally important preparatory factor in the failure of slopes composed of both bedrock and loose sediments such as moraines and ice-rich screes. These processes can occur at a range of different temporal and spatial scales, ranging from diurnal fluctuations associated with freeze-thaw weathering, to the seasonal formation and degradation of segregated ice within the active layer, and longer-term changes in the extent, thickness and thermal character of permafrost. The associated thermal and hydrological changes occur within mountain landsystems in which the glaciers and permafrost commonly display complex yet funda-mental interconnections (Haeberli *et al.*, 1997). Hanging glaciers on steep mountain slopes for example are associated with deep-seated thermal anomalies that can generate high shear stresses in the marginal ice cliffs, thereby creating the potential for failure. In addi-tion, the exposure of shaded rock walls caused by the downwasting of glaciers can result in permafrost aggradation, the formation and propagation of ice-filled fractures and an eleva-tion in porewater pressures in the subjacent unfrozen materials. Conversely, permafrost degradation and active layer deepening on steep slopes with loose sediments can result in a loss of cohesion within the water-saturated near surface materials that can be particularly dangerous if situated on top of steeply inclined permafrost tables.

The immense Kolka-Karmadon rock/ice slide that took place in the Caucasus Mountains on the 20th September 2002 provides a recent illustration of the need to consider glaciers and permafrost as coupled systems when assessing the impact of climate change on the stability of rock slopes in high mountain areas and the associated risks to society (Haeberli *et al.*, 2004). The slide started on a steep mountainside at an elevation of 4300–3500 m and involved both the upper 40 m of the rock mass and a similar thickness of overlying snow, firn and glacier ice. The detachment therefore occurred below the surface and is thought to have related to the development of a zone of warm permafrost below the surface subject to high and very variable water pressures. The most extraordinary aspect of the event how-ever was the subsequent erosion and incorporation of much of the Kolka Glacier, with the impact of the initial slide causing the detachment of c. 60–75% of the glacier's mass to a depth of 40–60 m. The incorporation and melting of this ice resulted in the formation of a debris flow that attained a mean velocity of 180–280 km h^{-1}. Finally, a constriction in the valley resulted in the development of a mud flow that continued for 15 km resulting in the destruction of houses and industrial infrastructure and an estimated 120 fatalities.

In considering the impact of ongoing and future climate change on high-mountain land-scapes characterised by a combination of glaciers and permafrost, Haeberli *et al.* (2017) highlight the development, behaviour and potential risks posed by transitional landscapes comprising exposed bedrock, unstable debris accumulations, sparse vegetation, degrading permafrost and extensive lakes. They suggest that these landscapes are characterised by pronounced landscape disequilibria that reflect the complex interactions of glaciers and permafrost and more specifically, temporal differences in their response to climate change. Valley glaciers respond relative rapidly with their recession and downwasting resulting in the exposure of bedrock, the debuttressing of slopes and the formation of lakes within overdeepenings. In contrast, permafrost degradation occurs over much longer timescales leading to a future transition to landscapes dominated by subsurface ice (i.e. ice-rich per-mafrost) rather than surface ice (i.e. glaciers) and a combination of paraglacial and

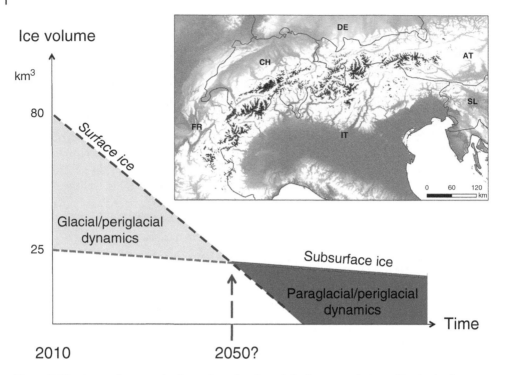

Figure 5.50 Approximate and schematic estimation of the future evolution of ice in the European Alps. Note how the rate of removal of surface ice is far more rapid than for subsurface ice, resulting in the latter becoming dominant at some stage in the mid-twenty-first century (Haeberli *et al.*, 2017, p408/with permission from Elsevier).

periglacial processes (Figure 5.50). This in turn results in the development of complex process chains that reflect the broader response of the mountain landscapes to climate change with glacier recession leading to the exposure of unstable rock and sediment slopes and the formation of hanging glaciers, all of which are prone to failure.

It is the formation of extensive lakes at the foot of steep and unstable slopes that are considered a particularly hazardous component of these process chains with the potential to act as a 'hazard multiplier' (Haeberli *et al.*, 2016, 2017). Their formation means that 'mass flows will no longer stop on a remote glacier but can now possibly reach a mobile water body, transmit their energy to it and create an impact wave which may, in critical cases, overflow the lake outlet, triggering far-reaching flood waves or debris flows in the valley below' (Haeberli *et al.*, 2017, p408). An illustration of this effect is provided by a rock/ice avalanche that occurred on Mount Hualcán in the Cordillera Blanca of Peru in April 2010 (Vilímek *et al.*, 2015). The detachment of part of a hanging glacier coupled with bedrock material traversed a 190 m bedrock step before impacting a lake and producing a displacement wave that overflowed the bedrock threshold by c. 20 m and created a debris flow on the steep slope below. On reaching the steep lower parts of the Rio Chucchún, this triggered a secondary debris flow that reached the town of Carhuaz some 15 km from the original onset zone, resulting in the destruction of houses, bridges and a water treatment plant. In this way, lakes can cause a significant extension of the hazard zones and consequently a significant increase in the risk posed to mountain communities. In the absence of costly mitigation schemes, the cessation of this hazard multiplier will only occur when the hazardous lakes drain or infill. Whilst the enhanced sediment fluxes associated with paraglacial periods have the potential to result in rapid infilling, observations in the European Alps of glacier recession and lake formation since the Little Ice Age suggest that larger water bodies can persist for centuries if not longer (Haeberli *et al.*, 2017).

5.6.3 Thermokarst, Dead Ice and Megaslumps

The potential response of ice-rich and thermally sensitive landscapes to external perturbations including both climate change and human activity have provided a long-term research focus for permafrost researchers who have coined the term 'thermokarst' to refer to the processes and products related to permafrost degradation. The term was originally developed to describe the hummocky terrains created through the degradation of ice-rich permafrost in northern Siberia (Ermolaev, 1932; cited by French, 2017), although its use has been broadened to incorporate the processes of ground ice thawing that can result in thaw subsidence and landscape instability. The formation of sub-surface voids and features indicative of surface subsidence such as sinkholes have resulted in these degrading ice-rich terrains being considered analogous to classic karst landscapes. This connection can be seen in the use of the terms thermokarst and 'glacial karst' to identify types of pseudokarst in which mass loss and void formation relates to the melting of ice rather than the dissolution of rock. Clayton (1964) for example describes a diverse range of features characteristic of karst regions forming on the stagnant and debris-covered margin of the Martin River Glacier in Alaska including sinkholes, caves, sinking streams, blind valleys and springs (Figure 5.51). These glacial environments can therefore be viewed as a type of karst topography that evolves in a similar way to the Davisian limestone karst cycle associated with the stages of youth (initial ice exposure), maturity (sinkhole formation and sub-surface drainage) and old age (return to surface drainage). This can ultimately result in the formation of large tracts of hummocky moraine (as seen around the southern margin of the Laurentide Ice Sheet) that display a range of features reminiscent of those found in karst regions including uvullas, poljes and hums.

Process-based studies undertaken in modern-day glacial environments have provided more detailed insights into the specific processes involved in the removal of buried ice from deglaciated forelands, their rates of operation and the controlling factors (Schomacker, 2008). Climate is clearly a primary control on the rates at which glacier forelands can be 'de-iced' with expected rates being particularly rapid in temperate glacial environments where permafrost is absent. Krüger & Kjær (2000) for example examined the processes, rates and geomorphic evolution of ice-cored moraines at the margin of Kötlujökull in southern Iceland under humid sub-polar conditions characterised by a mean annual air temperature of 2.6 °C and an estimated annual precipitation of 5000 mm. In areas of dead ice associated with extensive buried ice, de-icing was associated with two main processes: the backwasting of exposed ice walls (Figure 5.11) and the downwasting of buried ice associated with the ablation of both its upper and lower contacts. The combination of these processes contributed to a total surface lowering of $1.4\,\mathrm{m\,a^{-1}}$ although

Figure 5.51 Sediment-rich stream emerging from a cave in an extensive area of dead ice at Skeiðarárjökull, south-east Iceland. Note the wall of exposed debris-rich ice (c. 3 m high) and the capping layer of sediment released by supraglacial meltout (Image: Richard Waller).

rates of up to $2.5\,\mathrm{m\,a^{-1}}$ had previously been recorded when exposed ice was more extensive and surface streams were active. Where the de-icing process was more advanced however, the thicker surficial sediment cover (0.5–2 m) and absence of exposed ice resulted in much slower surface lowering rates of $0.8\,\mathrm{m\,a^{-1}}$, with areas of partially ice-cored moraine with the thickest sediment covers (1–3 m thick) displaying the lowest rates of $0.3\,\mathrm{m\,a^{-1}}$. These measurements therefore highlight the evolutionary nature of the de-icing process and the temporal reduction in the rate of de-icing as the volume of ice diminishes and the thickness of the insulating sediment cover increases (Figure 5.52).

In considering the time period over which buried ice might persist and contribute to wider paraglacial landscape adjustment, Krüger & Kjær (2000) estimate that a period of 50 years should be sufficient to completely remove a 40 m thick layer of stagnant, debris-rich ice from an Icelandic glacier foreland. In using geophysical surveys to identify the continued presence of buried glacier ice within the deglaciated forelands of other glaciers in southern Iceland, Everest & Bradwell (2003) similarly conclude that it has survived for at least 50 years at Virkisjökull and Hrútárjökull. However, the continued presence of debris-rich ice within an end moraine complex at Skeiðarárjökull thought to date back to the late eighteeth century suggests that buried ice has the potential to persist for centuries, even within relatively warm and wet climatic regimes (Figure 5.12). This longer-term preservation may relate to the elevation of the buried ice above the surrounding outwash plain and its consequent isolation from the regional groundwater systems that Krüger & Kjær (2000) consider responsible for rapid rates of thermal erosion and ice loss, particularly during periods of ice advance.

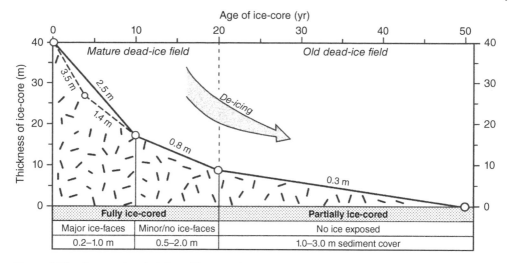

Figure 5.52 Progressive de-icing of ice-cored terrain as observed at the margin of Kötlujökull, southern Iceland (Krüger & Kjær, 2000, p745).

Subsequent work on a sequence of ice-cored moraines at Brúarjökull has helped to clarify the influence of climate on the timescales of preservation of buried ice, the rates of landscape adjustment and the geomorphological impacts within a surging-glacier landsystem (Schomacker *et al.*, 2006; Schomacker & Kjær, 2007). Being located in the Icelandic highlands, the foreland of Brúarjökull is associated with a mean annual air temperature of −0.3 °C and a mean annual precipitation total of 643 mm, conditions sufficiently cold and arid for sporadic permafrost to occur. Estimations of the rates of downwasting using a combination of photogrammetry and field measurement revealed values of between 7 and 17.7 cm a^{-1} with the highest values occurring in younger terrains with more limited debris cover and also in areas affected by active groundwater fluxes (Schomacker & Kjær, 2007). These values are roughly an order of magnitude lower than the average rates calculated for Kötlujökull in equivalent 'fully ice-cored terrains' (Krüger & Kjær, 2000, p737), reflecting a climatically controlled limitation in the time window for the sub-surface ablation of buried ice to a 4 to 7 week period in the late summer. With the annual rate of foreland de-icing being so limited, successive surge events have reworked the buried ice deposited by preceding events such that the foreland contains multiple generations of ice cores. This is argued to have created a distinctive geomorphic signature with surge advances over ice-cored moraine resulting in the formation of at least one ice-cored drumlin (Schomacker *et al.*, 2006). In the context of paraglacial landscape adjustment, this can however to be considered a 'transitional-state landform' that will transform into an area of hummocky moraine once the ice core has melted.

Extensive research undertaken in the permafrozen and partially deglaciated forelands of glaciers in Svalbard that commonly feature extensive tracts of ice-cored moraine, provide important insights into the changing nature and influence of glacier–permafrost interactions within the context of a rapidly warming climate (e.g. Lyså & Lønne, 2001; Sletten *et al.*, 2001; Schomacker & Kjær, 2008). Etzelmüller (2000) provides a particularly relevant contribution that explicitly considers the influence of permafrost and thermokarst processes in the thermal erosion of deglaciated forelands at four sites with contrasting thermal regimes and surge histories (Erikbreen, austre and vestre Brøggerbreen, vestre Lovénbreen and Finsterwalderbreen), all of which terminate in areas of continuous permafrost. Measurements of surface lowering derived from aerial photography and photogrammetry yield average rates of $<0.1\,\mathrm{m\,a^{-1}}$ (1938–1970) and $0.2\,\mathrm{m\,a^{-1}}$ at Erikbreen, similar in magnitude to those recorded at Brúarjökull in Iceland. The highest rates of surface lowering (with an average rate of $0.5\,\mathrm{m\,a^{-1}}$) are associated with the lateral recession of exposed ice cliffs that has the potential to entirely de-ice the ice-cored moraine within approximately 200 years. An important point to note here is that with permafrost allowing the long-term preservation of ice covered by sediment layers with thicknesses exceeding that of the active layer depth, it is the processes capable of removing this protective sediment cover and exposing the ice that are key to the de-icing of permafrozen glacier forelands. Etzelmüller (2000) also emphasises the influential role of thermal erosion associated with subaerial drainage systems that can act in combination with subaerial thermokarst processes to cause rapid ice loss, the destabilisation of these 'sediment magazines' and the evacuation of significant quantities of material from the catchment (Section 4.4). These processes are considered 'to be an important part of paraglacial activity in high-arctic environments dominated by permafrost' (Etzelmüller, 2000, p355) leading to the relatively rapid removal of ice deposited in glacier forelands in spite of the presence of permafrost (Ballantyne, 2002, p1995).

Recent technological advances enabling the acquisition of higher resolution imagery within polar regions have allowed the quantification of the rates of ground ice loss in the ice-rich forelands of glaciers in Svalbard at higher spatial and temporal resolutions (e.g. Tonkin *et al.*, 2016; Midgley *et al.*, 2018). Schomacker & Kjær (2008) for example utilise a combination of field measurement, high-resolution aerial photography and QuickBird 2 satellite imagery to quantify ice loss in the marginal dead-ice zone of Holmströmbreen, recording downwasting rates of $0.9\,\mathrm{m\,a^{-1}}$ (between 1984 and 2004) that are significantly higher than those recorded by Etzelmüller (2000) and of a similar magnitude to those recorded at Kötlujökull. The authors suggest that this reflects the accumulation and impact of surficial meltwater within the dead ice zone, the drainage of which is precluded by the presence of permafrost and the absence of glacier karst. The absence of drainage promotes mass movement activity that prevents the establishment of a stable debris cover and continuously exposes new ice. In addition, the formation of an ice-walled lake (Lake Emmy) has accelerated the rate of ice wall backwasting around its shores such that it has increased in size at an exponential rate. This suggests that the accumulation of meltwater provides an additional mechanism that can counteract the preserving influence of permafrost, resulting in rates of ice loss similar to those associated with temperate environments.

In assessing the role played by climate and permafrost in the paraglacial adjustment of areas of dead ice, Schomacker (2008) provides a useful review that synthesises the backwasting rates recorded at 14 sites in Iceland, Svalbard, the Yukon Territories (Canada), the Himalayas, New Zealand and Antarctica. It is worth noting here that downwasting rates were excluded from the analysis, with the local influence of varied debris cover thicknesses, thermal conductivities and water contents adding a degree of complexity that precludes the generation of any clear association with climate. The reported backwasting rates displayed the lowest values in the cold and arid climates of Antarctica and the highest

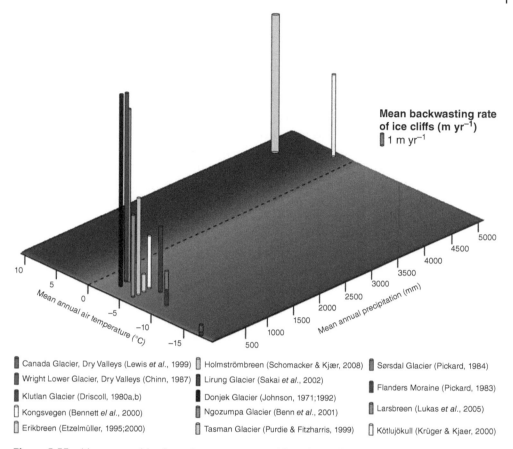

Mean backwasting rate of ice cliffs (m yr⁻¹)

☐ 1 m yr⁻¹

▌ Canada Glacier, Dry Valleys (Lewis *et al.*, 1999)	☐ Holmströmbreen (Schomacker & Kjær, 2008)	▌ Sørsdal Glacier (Pickard, 1984)
▌ Wright Lower Glacier, Dry Valleys (Chinn, 1987)	▌ Lirung Glacier (Sakai *et al.*, 2002)	
▌ Klutlan Glacier (Driscoll, 1980a,b)	▐ Donjek Glacier (Johnson, 1971;1992)	▌ Flanders Moraine (Pickard, 1983)
☐ Kongsvegen (Bennett *el al.*, 2000)	▌ Ngozumpa Glacier (Benn *et al.*, 2001)	▌ Larsbreen (Lukas *et al.*, 2005)
☐ Erikbreen (Etzelmüller, 1995;2000)	☐ Tasman Glacier (Purdie & Fitzharris, 1999)	☐ Kötlujökull (Krüger & Kjaer, 2000)

Figure 5.53 Mean annual backwasting rates reported from ice cliffs in 14 dead-ice areas at glaciers around the world presented in relation to mean annual air temperature and precipitation (Schomacker, 2008, p108). The column representing the Tasman Glacier (light grey) has been reduced in height by half.

values in New Zealand (humid, temperate climate) and the Himalayas (high insolation) (Figure 5.53). Subsequent T-tests demonstrate the highest correlations (R^2 values) between backwasting rates and mean annual air temperature (0.65) and the sum of positive degree days (0.49). These values are however lower than those recorded in glacier mass balance studies suggesting that local factors such as ice cliff aspect and the proximity to water bodies are more influential in areas of dead ice. The correlations with mean annual and mean summer precipitation are very low (0.17 and 0.30 respectively) suggesting these climatic parameters are of comparatively limited influence.

Observations of the evolution of ice-cored terrain at the margins of Midtre Lovenbreen by Midgley *et al.* (2018) spanning a decade provide a more detailed insight into the influence of local site-specific characteristics that can result in a complex paraglacial response and significant spatial and temporal variations in the rates of ice loss. Variations in the debris content of the ablating ice coupled with topographic variability were observed to lead to marked variations in the thickness of the debris cover with the latter also influencing the distribution of seasonal snowpatches. In combination with the aforementioned role played by glaciofluvial drainage and meltwater accumulation, these controls resulted in a highly complex paraglacial response characterised by significant spatial variations in the rates of surface charge. In a related study at Rieperbreen, Lyså & Lonne (2001) describe the recession and downwasting of the glacier margin by 900 m and 50–60 m, respectively, during the period 1936–1990 that has resulted in the evolutionary development of ice-cored lateral and frontal moraine complexes. They identify aspect as an additional local characteristic influencing the degradation of the marginal ice with high levels of insolation on the northern side of the glacier leading to enhanced ablation and the development of a pronounced transverse asymmetry. In addition to creating a larger northern lateral moraine 60 m high, this has also influenced the glacier's hydrology with the primary supraglacial drainage systems being diverted to the northern lateral margin to create a lateral meltwater channel that has migrated up-glacier over time as the asymmetry has become more pronounced.

Whilst research in modern-day glacial environments has focused on rates of ice loss and largely concluded that the removal of dead ice is likely to occur over a timescale of decades to centuries, the preservation of significant amounts of ground ice within extensive moraine complexes relating to Pleistocene Ice Sheets within polar regions suggests that the process of deglaciation in its literal sense (i.e. the complete removal of the glacier ice from the landscape) can be a much slower and more episodic process. The discovery, investigation and dating of buried ice with the Beacon Valley in East Antarctica by Sugden et al. (1985) demonstrates the geological timescales over which buried ice can be preserved within extreme environments under certain conditions. In radiometrically dating the volcanic ash deposited in contraction cracks within an overlying glacial diamict, they suggest that the ice is of Miocene age having survived for a minimum of 8.1 million years. Subsequent cosmogenic isotope dating of dolerite erratics of the till surface produced a conservative minimum age of 2.3 Ma (Schäfer et al., 2000), supporting the remarkable original conclusion that in the most extreme cold climates, buried ice can survive for over a million years.

Field research in western Siberia and Arctic Canada has revealed extensive tracts of buried ice within permafrost regions that are now generally regarded as being glacial in origin and in excess of 10,000 years old (e.g. Kaplyanskaya & Tarnogradskiy, 1986; Astakhov & Isayeva, 1988; French & Harry, 1990; St. Onge & McMartin, 1999; Murton et al., 2005; Lakeman & England, 2012). The deposition of this buried ice was associated with the deglaciation of these regions following the Last Glacial Maximum with the accompanying aggradation of permafrost preserving the relict ground ice and delaying the typical paraglacial processes of landscape adjustment such that these landscapes have been described as being 'imprisoned by permafrost' (Kokelj et al., 2017). The lack of geomorphic activity and absence of exposures within the deglaciated terrains means that the widespread extent of buried glacier ice within the Canadian Arctic for example hasn't been appreciated until relatively recently (Dyke & Savelle, 2000). Geomorphological mapping and field studies within this area indicate that the buried ice is primarily associated with large swathes of ice-cored moraine that reflect former ice-marginal positions of the northwestern Laurentide Ice Sheet (see Section 5.3.2). Dyke & Savelle (2000) for example describe the retention of glacier ice cores within major end moraine complexes of Younger Dryas age (11–10 ^{14}C ka B.P.) located on the Wollaston Peninsula, Victoria Island. These are seen as analogous to the hummocky moraine complexes associated with the southern margin of the Laurentide Ice Sheet (e.g. Ham & Attig, 1996; section 5.5.2), although in this case the continued influence of permafrost has allowed the buried ice to persist, largely preserving the original ice-cored moraines that are of 'sufficient size and roughness that they look like mountain ranges when viewed from a distance' (Dyke & Savelle, 2000, p607). Ice-cored moraines are consequently seen as the most common type of moraines within the region with the presence of extensive kettles and large ice-wedge polygons providing geomorphic evidence for the presence of ice cores where exposures are absent. Of particular interest here is the authors' suggestion that the depth of the kettles provides a possible insight into the nature of the buried ice. They argue that the deep lakes reflective of more advanced thermokarst relate to areas of ice with limited debris, whilst the ridges with shallow thermokarst relate to areas with more debris-rich ice. This provides an intriguing possibility that the patterns of thermokarst might be used to delineate variations in the debris content of former ice margins.

Whilst these ice-cored moraine belts have been explicitly related to the hummocky moraine associated with the Laurentide Ice Sheet's southern margin, examination of these genetic precursors has challenged the widespread belief that they provide evidence of regional ice stagnation with Dyke & Savelle (2000) suggesting the ice-cored moraines on the Wollaston Peninsula instead represent the product of continuous, sequential end moraine development by active ice. Subsequent field investigation in the western Canadian Arctic of the Jesse moraine belt, a large ice-cored moraine complex that extends for 350 km across Banks and Victoria Island has helped to elucidate the palaeoglaciological scenario and processes associated with their emplacement (Lakeman & England, 2012). The combination of kettle lakes, debris-rich ground ice and sub-parallel, arcuate end moraines occurring in association with networks of lateral meltwater channels led the authors to suggest that the ice-cored moraine complex is the product of controlled moraine deposition at the cold-based margin of a polythermal glacier located within the Prince of Wales Strait. The long-term preservation of the ice is thereby promoted both by the accretion of debris-rich ice through basal adfreezing (Section 3.5.4) and by the thrusting of sediment that generates significant volumes of supraglacial sediment that facilitates the abandonment of the ice-cored moraine complex during deglaciation. The associated requirement for a polythermal basal thermal regime is in turn argued to provide evidence for an expansion of warm-based conditions that led to a change in the ice sheet's configuration and dynamic behaviour during deglaciation.

Thermokarst activity induced by rapid climate change within the Arctic has revealed the extent of buried ice, something that was largely unknown when these landscapes remained thaw stable. A recent increase in the incidence of summer storm events in particular has intensified rates of fluvial incision, denuding the surface overburden and exposing widespread buried ice associated with the 'incomplete deglaciation' of these high-latitude regions (Section 5.3.4) (Kokelj *et al.*, 2015). This has led to the formation and rapid expansion of large retrogressive thaw slumps or 'mega slumps' that commonly exceed 20 ha in area, resulting in the mobilisation of up to a million m^3 of previously frozen material that can reconfigure local drainage networks (Kokelj *et al.*, 2017; Lewkowicz & Way, 2019). This rejuvenation of geomorphic activity and the recent mobilisation of material from the primary sediment stores (buried ice and frozen glacigenic sediment) can be viewed as a recommencement of paraglacial activity after a hiatus of several thousand years caused by deglacial permafrost aggradation. The slumps themselves in turn represent secondary paraglacial systems that can be reworked over longer timescales to elevate suspended sediment fluxes within the wider drainage basins as predicted in Ballantyne (2002, p1995). In examining the hydrological impacts of thaw slump activity on the lower Mackenzie and Peel River catchments of northwestern Canada, Kokelj *et al.* (2013) recorded suspended sediment and solute concentrations in affected streams that were several orders of magnitude higher than those in streams that were unaffected by slump activity. These dramatic changes in the headwaters were in turn associated with catchment-scale (10^3 km^2) changes in the form of diurnal fluctuations in turbidity (driven by ablation rates) and a significant increase in solute fluxes.

In seeking to determine the geographical extent of these changes, Kokelj *et al.* (2017) employ regional-scale geomorphological mapping of a c. 1.3 million km^2 area of northwestern Canada using satellite imagery to yield what is seen as a conservative estimate of 136,000 km^2 for the area affected by slump activity. Interestingly, rather than occurring in the lower latitude and more thaw susceptible regions of the study area, the areas displaying the highest slump densities were found to correspond with the ice-marginal landsystems associated with active recession of the Laurentide Ice Sheet as previously outlined by Dyke & Savelle (2000) and Lakeman & England (2012) (Figure 5.54). This led the authors to conclude that the 'glacial legacy has an overriding influence on permafrost terrain sensitivity across northwestern Canada' (Kokelj *et al.*, 2017, p373). In assessing the comparative influence of landscape factors and climate on thaw slump dynamics within four formerly glaciated areas of ice-rich permafrost (the Jesse Moraine, the Tuktoyaktuk Coastlands, the Bluenose Moraine and the Peel Plateau), Segal *et al.* (2016) similarly noted that it was the coldest site (the Jesse Moraine) that displayed the highest modern density of thaw slumps (50.1 slumps per 100 km^2), the greatest proportional area affected by disturbance (49 ha per 100 km^2) and the greatest recent increase in the total active slump area (407% increase between 1960 and 2004). This seemingly counterintuitive finding highlights the central role played by summer rainfall intensification that can increase gullying, bank erosion and landslide activity to expose the buried ice and increase sediment removal from the slump scar zone to continually expose ice within the headwall (Kokelj *et al.*, 2015). This re-emphasises the aforementioned assertion of Etzelmüller (2000) that it is the processes of sediment removal that are key to the de-icing of ice-cored moraine complexes in permafrost environments. In addition, the climate history is also considered to have played an important role with those sites featuring deep thaw unconformities developed during the Holocene thermal maximum (such as the Tuktoyaktuk Coastlands) being much more resilient to recent climate-change driven thermokarst activity than those with shallow active layers (such as the Jesse Moraine).

Research on the hummocky moraine found extensively around the margins of the southern Laurentide Ice Sheet (Section 5.3.2) demonstrates the geomorphological end product of a process that is still ongoing in the modern-day Arctic regions with the ice-cored moraines representing genetic precursors at an earlier stage in their paraglacial landscape evolution. It is worth noting however that even in these lower latitude settings, permafrost is thought to have played an important role in delaying the processes of de-icing with Florin & Wright Jr (1969) using radiocarbon dating of wood fragments from former kettle lakes in Minnesota

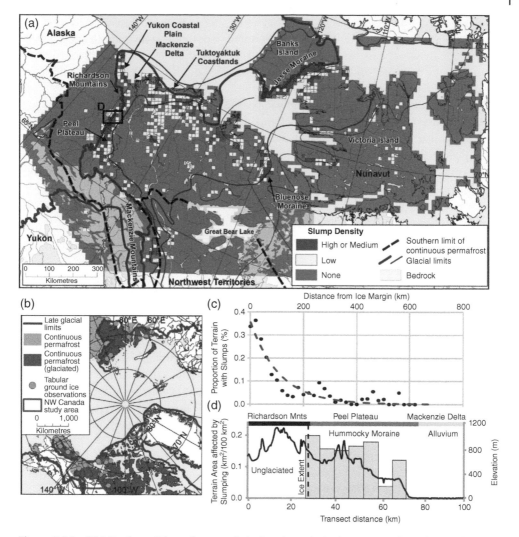

Figure 5.54 Distribution of thaw slump and glaciated terrain in the western Canadian Arctic (Kokelj *et al.*, 2017, p372). (a) slump-affected terrain and positions of the Laurentide Ice Sheet from c. 18–11 ka B.P. (b) Circumpolar map showing published observations of thaw slumps and thick ground ice. (c) proportion of terrain with slumps expressed in relation to the distance from the late Wisconsinan ice margin. (d) Topography and slump-affected area along a west-east corridor in the Mackenzie Valley (see a).

to argue that the parent stagnant ice survived for several thousand years after the disappearance of active ice. Research into the evolution of late glacial landscapes in north-central Wisconsin suggests that the dynamics of the permafrost played a key role in defining two distinct phases of geomorphic activity and landform development that reflect temporal variations in the stability of the buried ice (Ham & Attig, 1996; Attig & Rawling III, 2017). In the earlier stable phase, the cold climatic regime coupled with a surficial cover of sediment inhibited melting with the geomorphic activity being associated with the development of lakes and ice-walled river systems similar to those described in modern-day Svalbard. Subsequent climate warming resulted in permafrost degradation and a transition to an unstable phase associated with the melting of the buried ice and the formation of hummocky moraines and ice-walled lake plains that delineate the maximum limits of the Wisconsinan ice lobes.

This raises questions that might initially appear to be somewhat semantic but which are of fundamental importance to those undertaking palaeoglaciological and palaeoenvironmental reconstructions in areas associated with glacier–permafrost interactions; namely when does deposition occur and what are the key drivers? As noted by Dyke & Savelle (2000), the deposition of the final products of deglaciation (melt-out tills and hummocky moraine) may occur thousands of years after the disappearance of the parent ice sheet. In such a case, their deposition 'would be key to interpreting the permafrost dissipation but would be irrelevant to interpreting conditions or processes operating at the time of deglaciation more than 10,000 years earlier' (Dyke & Savelle, 2000, p616).

5.7 Summary

- In marked contrast to their warm-based counterparts, the cold-based ice typically associated with glacier–permafrost interactions has traditionally been considered to protect and preserve pre-existing landforms and landscapes from subaerial processes. In lacking the panoply of classic 'textbook' features ascribed to glaciation, this 'glacial protection' has created significant challenges for geomorphologists seeking to differentiate between areas that were unglaciated from those that were glaciated by cold-based ice.
- Ongoing research within both modern and ancient glacial environments has however identified a suite of glacial landforms commonly associated with cold-based glaciers and glacier–permafrost interactions. These can include:
  - Lateral meltwater channels, whose formation is promoted by the reduced permeability and limited crevassing of cold ice masses. In forming around the glacier margins, they can play a vital role in palaeoglaciological reconstructions by enabling the reconstruction of patterns of ice recession. Whilst these features are not exclusive to non-temperate glaciers, the formation of laterally extensive 'flights' of channels provides one of the most distinctive landforms associated with cold ice masses.
  - Specific types of end moraine including ice-cored moraines and various types of push moraine. In relation to ice-cored moraines, glacier–permafrost interactions are considered central to the ice-marginal processes of debris entrainment that enable the detachment of the buried ice and its longer-term preservation. As such, they are one of the most common moraine types found in the Arctic that also provides a key mechanism for the formation and preservation of massive ground ice. Permafrost is also widely agreed to facilitate the formation of push moraine complexes through its influence on the rheology of proglacial sediments and the associated hydrogeological conditions. However, there creation at the margins of surging temperate glaciers again suggests that the presence of permafrost is not necessarily a pre-condition for their formation.

- A range of erosional and depositional subglacial features including small-scale striations, scrapes, grooves and debris trains and modified tors that have been mobilised by cold-based ice. Ribbed moraines have also been related to the thermal transition between cold- and warm-based ice although this potential connection with subglacial permafrost remains more controversial.
- Massive ground ice that has been related to glacier–permafrost interactions both through its connection with glacial groundwater fluxes (to produce intrasedimental ice) and with the formation and preservation of debris-rich glacier ice. If ground ice of glacial origin can be reliably identified, it could provide a valuable palaeoglaciological archive with direct evidence of the characterisitics of former glaciers.
- Meltwater-related features including proglacial icings, open-system pingos and blow-out structures related to the hydrological and hydrogeological coupling of polythermal glaciers and proglacial permafrost environments that can in combination facilitate the development of artesian porewater pressures.

- Research focusing explicitly on the influence of basal thermal regime on the sedimentological and structural characteristics of glacigenic sediments remains relatively limited. Nevertheless, the published literature provides an insight into the sedimentological and structural signatures likely to be associated with glacier–permafrost interactions:
  - In facilitating the entrainment of sediment and the formation of thick sequences of debris-rich basal ice, glacier–permafrost interactions could be associated with the formation of extensive melt-out tills as the depositional end product. However, the immediate and longer-term preservation potential of this till type remains subject to debate with ablation and meltwater production commonly leading to extensive post depositional re-working of the sediment. The removal of the ice by sublimation provides a greater potential for preservation although the associated formation of sublimation tills is limited to cold and arid environments.
  - Reviews into the sedimentology of subglacial glacigenic sediments indicate that cold-based glaciers typically re-work pre-existing sediments with limited modification occurring during transport. As such, the related sediments are commonly dominated by coarse fractions and lack the abundant fines associated with abrasion and comminution. The constituent clasts also lack the typical forms with angular material being more prevalent. Striated surfaces can however still be found with a greater variability in striation orientations providing a potentially diagnostic criterion for cold-based conditions. Various types of ice can also be considered some of the most volumetrically significant depositional lithofacies with deglacial permafrost aggradation leading to their potential long-term preservation.
  - Structurally, the inclusion and preservation of coarse-grained intraclasts as competent masses within pervasively deformed tills are potentially diagnostic of glacier–permafrost interactions with the associated rheological heterogeneity being consistent with expected variations in the pre-melted water content. Glacier–permafrost interactions are also likely to promote the entrainment of large bedrock rafts and the formation of hydrofractures but in both cases these features are unlikely to be diagnostic of permafrost conditions.

- In elucidating the connections between environment, process and form, process-form models and the expanding studies of glacial landsystems have helped to clarify the landform-sediment assemblages associated with cold-based glaciers and glacier–permafrost interactions at a range of spatial scales. In acting to preserve pre-glacial surfaces, cold-based ice can for example provide temporal 'windows' and access to earlier phases of landscape development.

- The detailed examination of former Pleistocene ice-sheet margins suggests that permafrost has played and in some places continues to play an influential role in the development and evolution of a diverse range of glacial landsystems. In mid-latitude regions where permafrost is no longer present, key landsystem elements related to permafrost include extensive tracts of hummocky moraine (through the promotion of thick supraglacial sediment cover), tunnel channels (through the focused drainage of subglacial drainage through thermal dams) and ice-walled lake plains (through the preservation of stagnant ice). In high-latitude regions where permafrost is still present, key landsystem elements include pre-glacial landforms (preserved beneath cold ice), lateral meltwater channels (reflective of cold-based ice), extensive tracts of ice-cored moraine (where permafrost has preserved the buried glacier ice), ice-shelf landforms (where cold ice has promoted ice-shelf preservation) and glacially deformed permafrost.

- Systematic observation of the glacial landsystems associated with Arctic regions featuring both glaciers and permafrost have emphasised the importance of thermal transitions and their influence on large-scale patterns of glacial erosion and the formation of landscapes of 'selective linear erosion'. They have also illustrated the diversity of ice-marginal processes and products relating to glacier–permafrost interactions that include the formation of ice-cored and push moraines, intense bedrock weathering and the entrainment of angular materials, the continued operation of glacifluvial processes and the widespread deposition and preservation of ice that can be subject of permafrost creep.

- Research in Antarctica has demonstrated that geomorphic activity can remain active in the most extreme glacial environments on Earth. Whilst the very low rates of glacial sediment transfer typically produce landforms that are small, subtle and consequently of limited preservation potential, they are nonetheless capable of creating distinctive landsystems comprising small-scale erosional features, ice-cored and push moraines and ice-marginal aprons. Meltwater-related processes are generally limited by the extreme aridity and the short-melt seasons although they can still generate lateral meltwater channels, braid plans and alluvial fans. The absence of glacial abrasion and comminution results in glacigenic sediments that reflect the reworking of pre-existing sediments and lack classic diamictic textures.

- The observation and investigation of a range of distinctive landforms on Mars including distinctive crater morphologies, depressions, large lobate fans and 'thumbprint' moraines suggest that the influence of glacier–permafrost interactions on the landsystem development may not be limited to Earth.

• In view of the intrinsic thermal and hydrological connections associated with glaciers and permafrost in high-latitude and high-altitude environments, the dynamic permafrost changes that can accompany deglaciation and their influence on the paraglacial processes of landscape adjustment provide a highly dynamic process environment in which glacier–permafrost interactions can play a pivotal role.

- The deglaciation of high mountain areas and the combined influence of glacier recession, slope debutressing and permafrost degradation can instigate large-scale rock failures. These events constitute a significant and growing threat to resident populations, particularly where the deglacial process chains are associated with glacial over-deepenings and the formation of extensive lakes.

– Within high latitude environments, the aggradation of permafrost associated with ice-sheet recession can delay the onset and progress of paraglacial landscape adjustment. In limiting ablation rates, the formation of surficial debris layers can lead to the long-term preservation of buried glacier ice within ice-cored moraine complexes in particular, leading to their 'incomplete deglaciation' millennia after the disappearance of surface ice sheets. This has rendered these environments highly susceptible to thermo-karst activity with the recent degradation of this protective carapace by coastal erosion and fluvial incision leading to the formation and dramatic expansion of large retrogressive thaw slumps. This marks a renewed phase of deglaciation and paraglacial adjustment triggered by rapid climate change.

References

Adam, W.G. & Knight, P.G., 2003. Identification of basal layer debris in ice-marginal moraines, Russell Glacier, West Greenland. *Quaternary Science Reviews*, 22, 1407–1414.

Agassiz, L., 1841. On glaciers, and the evidence of there once having existed in Scotland, Ireland and England. *Proceedings of the Geological Society of London*, 3, 327–332.

Alley, R.B., 1991. Deforming-bed origin for southern Laurentide till sheets? *Journal of Glaciology*, 37, 67–76.

Alley, R.B., Blankenship, D.D., Bentley, C.R. & Rooney, S.T., 1986. Deformation of till beneath ice stream B, West Antarctica. *Nature*, 322, 57–59.

Astakhov, V.I. & Isayeva, L.L., 1988. The 'Ice Hill': an example of 'retarded deglaciation' in Siberia. *Quaternary Science Reviews*, 7, 29–40.

Astakhov, V.I., Kaplyanskaya, F.A. & Tarnogradsky, V.D., 1996. Pleistocene permafrost of West Siberia as a deformable glacier bed. *Permafrost and Periglacial Processes*, 7, 165–191.

Atkins, C.B., 2013. Geomorphological evidence of cold-based glacier activity in South Victoria Land, Antarctica. *Geological Society, London, Special Publications*, 381, 299–318.

Atkins, C.B. & Dickinson, W.W., 2007. Landscape modification by meltwater channels at margins of cold-based glaciers, Dry Valleys, Antarctica. *Boreas*, 36, 47–55.

Atkins, C.B., Barrett, P.J. & Hicock, S.R., 2002. Cold glaciers erode and deposit: evidence from Allan Hills, Antarctica. *Geology*, 30, 659–662.

Attig, J.W. & Clayton, L., 1993. Stratigraphy and origin of an area of hummocky glacial topography, northern Wisconsin, USA. *Quaternary International*, 18, 61–67.

Attig, J.W. & Rawling, J.E. III, 2017. Influence of persistent buried ice on late glacial landscape development in part of Wisconsin's Northern Highlands. *Quaternary Glaciation of the Great Lakes Region: Process, Landforms, Sediments, and Chronology: Geological Society of America Special Papers*, 530, 103–114.

Attig, J.W., Mickelson, D.M. & Clayton, L., 1989. Late Wisconsin landform distribution and glacier-bed conditions in Wisconsin. *Sedimentary Geology*, 62, 399–405.

Ballantyne, C.K., 1997. Periglacial trimlines in the Scottish Highlands. *Quaternary International*, 38, 119–136.

Ballantyne, C.K., 2002. Paraglacial geomorphology. *Quaternary Science Reviews*, 21, 1935–2017.

Ballantyne, C.K., Stone, J.O. & Fifield, L.K., 1998. Cosmogenic Cl-36 dating of postglacial landsliding at The Storr, Isle of Skye, Scotland. *The Holocene*, 8, 347–351.

Balme, M.R. & Gallagher, C., 2009. An equatorial periglacial landscape on Mars. *Earth and Planetary Science Letters*, 285, 1–15.

Beget, J., 1987. Low profile of the northwest Laurentide ice sheet. *Arctic and Alpine Research*, 19, 81–88.

Benn, D.I., Fowler, A.C., Hewitt, I. & Sevestre, H., 2019a. A general theory of glacier surges. *Journal of Glaciology*, 65, 701–716.

Benn, D.I., Jones, R.L., Luckman, A., *et al.*, 2019b. Mass and enthalpy budget evolution during the surge of a polythermal glacier: a test of theory. *Journal of Glaciology*, 65, 717–731.

Bennett, M.R., 2001. The morphology, structural evolution and significance of push moraines. *Earth-Science Reviews*, 53, 197–236.

Bennett, M.R., Waller, R.I., Glasser, N.F., Hambrey, M.J. & Huddart, D., 1999. Glacigenic clast fabrics: genetic fingerprint or wishful thinking? *Journal of Quaternary Science*, 14, 125–135.

Bennett, M.R., Huddart, D. & Waller, R.I., 2005. The interaction of a surging glacier with a seasonally frozen foreland: Hagafellsjökull-Eystri, Iceland. *Geological Society, London, Special Publications*, 242, 51–62.

Bergsma, B.M., Svoboda, J. & Freedman, B., 1984. Entombed plant communities released by a retreating glacier at central Ellesmere Island, Canada. *Arctic*, 49–52.

Bierman, P.R., Marsella, K.A., Patterson, C., Davis, P.T. & Caffee, M., 1999. Mid-Pleistocene cosmogenic minimum-age limits for pre-Wisconsinan glacial surfaces in southwestern Minnesota and southern Baffin Island: a multiple nuclide approach. *Geomorphology*, 27, 25–39.

Bierman, P.R., Corbett, L.B., Graly, J.A., *et al.*, 2014. Preservation of a preglacial landscape under the center of the Greenland Ice Sheet. *Science*, 344, 402–405.

Blankenship, D.D., Bentley, C.R., Rooney, S.T. & Alley, R.B., 1987. Till beneath Ice Stream B: 1. Properties derived from seismic travel times. *Journal of Geophysical Research: Solid Earth*, 92(B9), 8903–8911.

Bockheim, J.G., Campbell, I.B. & McLeod, M., 2007. Permafrost distribution and active-layer depths in the McMurdo Dry Valleys, Antarctica. *Permafrost and Periglacial Processes*, 18, 217–227.

Bonney, T.G., 1871. On the Formation of "Cirques," and their bearing upon Theories attributing the Excavation of Alpine Valleys mainly to the Action of Glaciers. *Quarterly Journal of the Geological Society*, 27, 312–324.

Boulton, G.S., 1968. Flow tills and related deposits on some Vestspitsbergen glaciers. *Journal of Glaciology*, 7, 91–412.

Boulton, G.S., 1970a. On the origin and transport of englacial debris in Svalbard glaciers. *Journal of Glaciology*, 9, 213–229.

Boulton, G.S., 1970b. On the deposition of subglacial and melt-out tills at the margins of certain Svalbard glaciers. *Journal of Glaciology*, 9, 231–245.

Boulton, G.S., 1972. The role of thermal regime in glacial sedimentation. In: Price, R.J. & Sugden, D.E. (Eds.), *Polar Geomorphology*. Institute of British Geographers Special Publication No. 4, 1–19.

Boulton, G.S., 1974. Processes and patterns of glacial erosion. In: Coates, D.R. (Ed.), *Glacial Geomorphology*. A proceedings volume of the Fifth Annual Geomorphology Symposia series. Springer, New York, 41–87.

Boulton, G.S., 1978. Boulder shapes and grain-size distributions of debris as indicators of transport paths through a glacier and till genesis. *Sedimentology*, 25, 773–799.

Boulton, G.S. & Caban, P., 1995. Groundwater flow beneath ice sheets: part II—its impact on glacier tectonic structures and moraine formation. *Quaternary Science Reviews*, 14, 563–587.

Boulton, G.S. & Jones, A.S., 1979. Stability of temperate ice caps and ice sheets resting on beds of deformable sediment. *Journal of Glaciology*, 24, 29–43.

Boulton, G.S., Caban, P.E. & Van Gijssel, K., 1995. Groundwater flow beneath ice sheets: part I—large scale patterns. *Quaternary Science Reviews*, 14, 545–562.

Boulton, G.S., Van der Meer, J.J.M., Beets, D.J., Hart, J.K. & Ruegg, G.H.J., 1999. The sedimentary and structural evolution of a recent push moraine complex: Holmstrømbreen, Spitsbergen. *Quaternary Science Reviews*, 18, 339–371.

Bukowska-Jania, E. & Szafraniec, J., 2005. Distribution and morphometric characteristics of icing fields in Svalbard. *Polar Research*, 24, 41–53.

Burke, H., Phillips, E., Lee, J.R. & Wilkinson, I.P., 2009. Imbricate thrust stack model for the formation of glaciotectonic rafts: an example from the Middle Pleistocene of north Norfolk, UK. *Boreas*, 38, 620–637.

Callanan, M., 2014. Out of the Ice: Glacial Archaeology in Central Norway. Norges teknisk-naturvitenskapelige universitet, Det humanistiske fakultet, Institutt for historiske studier. Available at: https://ntnuopen.ntnu.no/ntnu-xmlui/handle/11250/229726. Date accessed: 13 January 2020.

Chamberlin, T.C., 1895. Recent glacial studies in Greenland. *Bulletin of the Geological Society of America*, 6, 199–220.

Chandler, B.M., Lovell, H., Boston, C.M., *et al.*, 2018. Glacial geomorphological mapping: A review of approaches and frameworks for best practice. *Earth-Science Reviews*, 185, 806–846.

Church, M. & Ryder, J.M., 1972. Paraglacial sedimentation: a consideration of fluvial processes conditioned by glaciation. *Geological Society of America Bulletin*, 83, 3059–3072.

Clark, C.D., Hughes, A.L., Greenwood, S.L., Jordan, C. & Sejrup, H.P., 2012. Pattern and timing of retreat of the last British-Irish Ice Sheet. *Quaternary Science Reviews*, 44, 112–146.

Clayton, L., 1964. Karst topography on stagnant glaciers. *Journal of Glaciology*, 5, 107–112.

Clayton, L. & Moran, S.R., 1974. A glacial process-form model. In: Coates, D.R. (Ed.), *Glacial Geomorphology*. A proceedings volume of the Fifth Annual Geomorphology Symposia series. Springer, Dordrecht, Netherlands, 89–119.

Clayton, L., Attig, J.W. & Mickelson, D.M., 2001. Effects of late Pleistocene permafrost on the landscape of Wisconsin, USA. *Boreas*, 30, 173–188.

Clayton, L., Attig, J.W., Ham, N.R., *et al.*, 2008. Ice-walled-lake plains: implications for the origin of hummocky glacial topography in middle North America. *Geomorphology*, 97, 237–248.

Colgan, P.M., Mickelson, D.M. & Cutler, P.M., 2003. Ice-marginal terrestrial landsystems: southern Laurentide ice sheet margin. In: Evans, D.J.A. (Ed.), *Glacial Landsystems*. Arnold, London, 111–142.

Cook, S.J., Graham, D.J., Swift, D.A., Midgley, N.G. & Adam, W.G., 2011. Sedimentary signatures of basal ice formation and their preservation in ice-marginal sediments. *Geomorphology*, 125, 122–131.

Coulombe, S., Fortier, D., Lacelle, D., Kanevskiy, M. & Shur, Y., 2019. Origin, burial and preservation of late Pleistocene-age glacier ice in Arctic permafrost (Bylot Island, NU, Canada). *The Cryosphere*, 13, 97–111.

Cuffey, K.M., Conway, H., Hallet, B., Gades, A.M. & Raymond, C.F., 1999. Interfacial water in polar glaciers and glacier sliding at −17°C. *Geophysical Research Letters*, 26, 751–754.

Cuffey, K.M., Conway, H., Gades, A.M., *et al.*, 2000. Entrainment at cold glacier beds. *Geology*, 28, 351–354.

Cutler, P.M., MacAyeal, D.R., Mickelson, D.M., Parizek, B.R. & Colgan, P.M., 2000. A numerical investigation of ice-lobe–permafrost interaction around the southern Laurentide ice sheet. *Journal of Glaciology*, 46, 311–325.

Dahl, E., 1955. Biogeographic and geologic indications of unglaciated areas in Scandinavia during the glacial ages. *Geological Society of America Bulletin*, 66, 1499–1520.

Davies, M.L., Atkins, C.B., van der Meer, J.J., Barrett, P.J. & Hicock, S.R., 2009. Evidence for cold-based glacial activity in the Allan Hills, Antarctica. *Quaternary Science Reviews*, 28, 3124–3137.

Davis, P.T., Briner, J.P., Coulthard, R.D., Finkel, R.W. & Miller, G.H., 2006. Preservation of Arctic landscapes overridden by cold-based ice sheets. *Quaternary Research*, 65, 156–163.

Demidov, N., Wetterich, S., Verkulich, S., *et al.*, 2019. Geochemical signatures of pingo ice and its origin in Grøndalen, west Spitsbergen. *The Cryosphere*, 13, 3155–3169.

Dixon, E.J., Callanan, M.E., Hafner, A. & Hare, P.G., 2014. The emergence of glacial archaeology. *Journal of Glacial Archaeology*, 1, 1–9.

Drewry, D.J., 1986. *Glacial Geologic Processes*. Edward Arnold.

Dundas, C.M., Byrne, S. & McEwen, A.S., 2015. Modelling the development of martian sublimation thermokarst landforms. *Icarus*, 262, 154–169.

Dunlop, P., Clark, C.D. & Hindmarsh, R.C., 2008. Bed ribbing instability explanation: testing a numerical model of ribbed moraine formation arising from coupled flow of ice and subglacial sediment. *Journal of Geophysical Research: Earth Surface*, *113*, F03005, doi:10.1029/2007JF000954.

Dyke, A.S., 1993. Landscapes of cold-centred Late Wisconsinan ice caps, Arctic Canada. *Progress in Physical Geography*, 17, 223–247.

Dyke, A.S., 1999. Last glacial maximum and deglaciation of Devon Island, Arctic Canada: support for an Innuitian Ice Sheet. *Quaternary Science Reviews*, *18*, 393–420.

Dyke, A.S. & Evans, D.J.A., 2003. Ice-marginal terrestrial landsystems: northern Laurentide and Innuitian Ice Sheet margins. In: Evans, D.J.A. (Ed.), *Glacial Landsystems*. Arnold, London, 143–165.

Dyke, A.S. & Prest, V., 1987. Late Wisconsinan and Holocene history of the Laurentide ice sheet. *Géographie physique et Quaternaire*, 41, 237–263.

Dyke, A.S. & Savelle, J.M., 2000. Major end moraines of Younger Dryas age on Wollaston Peninsula, Victoria Island, Canadian Arctic: implications for paleoclimate and for formation of hummocky moraine. *Canadian Journal of Earth Sciences*, 37, 601–619.

Engelhardt, H. & Kamb, B., 1997. Basal hydraulic system of a West Antarctic ice stream: constraints from borehole observations. *Journal of Glaciology*, 43, 207–230.

Ermolaev, M.M., 1932. Geology and geomorphology of the Bol'shoy Liakhovky Island. In: *Polar geophysical station at the Bol'shoy Liakovsky Island*. USSR Academy of Sciences, Leningrad, 147–228. (In Russian)

Etzelmüller, B., 2000. Quantification of thermo-erosion in pro-glacial areas - examples from Svalbard. *Zeitschrift für Geomorphologie*, 44(3), 343–361.

Etzelmüller, B. & Hagen, J.O., 2005. Glacier-permafrost interaction in Arctic and alpine mountain environments with examples from southern Norway and Svalbard. In: Harris, C. & Murton, J.B. (Eds.), *Cryospheric Systems: Glaciers & Permafrost*. Geological Society, London, Special Publications, 242, 11–27.

Etzelmüller, B., Hagen, J.O., Vatne, G., Ødegård, R.S. & Sollid, J.L., 1996. Glacier debris accumulation and sediment deformation influenced by permafrost: examples from Svalbard. *Annals of Glaciology*, 22, 53–62.

Evans, D.J.A., 1991. Canadian landform examples 19 - High Arctic thrust block moraines. *The Canadian Geographer / Le Géographe Canadien*, 35, 93–97.

Evans, D.J.A., 1993. High-latitude rock glaciers: A case study of forms and processes in the Canadian arctic. *Permafrost and Periglacial Processes*, 4, 17–35.

Evans, D.J.A. (Ed.), 2003. *Glacial Landsystems*. Arnold.

Evans, D.J.A., 2009. Controlled moraines: origins, characteristics and palaeoglaciological implications. *Quaternary Science Reviews*, 28, 183–208.

Evans, D.J.A. & Hiemstra, J.F., 2005. Till deposition by glacier submarginal, incremental thickening. *Earth Surface Processes & Landforms*, 30, 1633–1662.

Evans, D.J.A. & Rea, B.R., 1999. Geomorphology and sedimentology of surging glaciers: a land-systems approach. *Annals of Glaciology*, 28, 75–82.

Evans, D.J.A., Atkinson, N. & Phillips, E., 2020. Glacial geomorphology of the Neutral Hills Uplands, southeast Alberta, Canada: the process-form imprints of dynamic ice streams and surging ice lobes. *Geomorphology*, 350, 106910.

Evans, D.J.A., Smith, I.R., Gosse, J.C. & Galloway, J.M., 2021. Glacial landforms and sediments (landsystem) of the Smoking Hills area, Northwest Territories, Canada: Implications for regional Pliocene–Pleistocene Laurentide Ice Sheet dynamics. *Quaternary Science Reviews*, 262, 106958.

Everest, J. & Bradwell, T., 2003. Buried glacier ice in southern Iceland its wider significance. *Geomorphology*, 52, 347–358.

Fabel, D., Stroeven, A.P., Harbor, J., *et al.*, 2002. Landscape preservation under Fennoscandian ice sheets determined from in situ produced 10Be and 26Al. *Earth and Planetary Science Letters*, 201, 397–406.

Falconer, G., 1966. Preservation of vegetation and patterned ground under a thin ice body in northern Baffin Island, N.W.T. *Geographical Bulletin*, 8, 194–200.

Finstad, E. & Pilø, L.H., 2019. *Secrets of the Ice*. secretsoftheice.com. Date accessed: 13 January 2020.

Fitzsimons, S.J., 1996. Formation of thrust-block moraines at the margins of dry-based glaciers, south Victoria Land, Antarctica. *Annals of Glaciology*, 22, 68–74.

Fitzsimons, S.J., 2003. Ice-marginal landsystems: polar continental margins. In: Evans, D.J.A. (Ed.), *Glacial Landsystems*. Blackwell Publishing, Oxford, 89–110.

Fitzsimons, S.J., McManus, K.J. & Lorrain, R.D., 1999. Structure and strength of basal ice and substrate of a dry-based glacier: evidence for substrate deformation at sub-freezing temperatures. *Annals of Glaciology*, 28, 236–240.

Fitzsimons, S., Sharp, M. & Lorrain, R., 2023. Subglacial deformation of permafrost beneath Wright Lower Glacier, Antarctica as revealed by direct observations and measurements from a subglacial tunnel. *Geomorphology*, 440, doi:10.1016/j.geomorph.2023.108886.

Florin, M. & Wright, H.E. Jr., 1969. Diatom evidence for the persistence of glacial ice in Minnesota. *Bulletin of the Geological Society of America*, 80, 695–704.

French, H.M., 2017. *The Periglacial Environment* (4th ed.). John Wiley & Sons.

French, H.M. & Harry, D.G., 1988. Nature and origin of ground ice, Sandhills Moraine, southwest Banks Island, western Canadian Arctic. *Journal of Quaternary Science*, 3, 19–30.

French, H.M. & Harry, D.G., 1990. Observations on buried glacier ice and massive segregated ice, western Arctic coast, Canada. *Permafrost and Periglacial Processes*, 1, 31–43.

Fuller, S. & Murray, T., 2002. Sedimentological investigations in the forefield of an Icelandic surge-type glacier: implications for the surge mechanism. *Quaternary Science Reviews*, 21, 1503–1520.

Garwood, E.J., 1910. Features of Alpine scenery due to glacial protection. *The Geographical Journal*, 36(3), 310–336.

Geikie, J., 1878. On the glacial phenomena of the Long Island, or Outer Hebrides. Second Paper. *Quarterly Journal of the Geological Society*, 34, 819–870.

Gellatly, A.F., Gordon, J.E., Whalley, W.B. & Hansom, J.D., 1988. Thermal regime and geomorphology of plateau ice caps in northern Norway: observations and implications. *Geology*, *16*(11), 983–986.

Goldthwait, R.P., 1951. Development of end moraines in east-central Baffin Island. *The Journal of Geology*, 59, 567–577.

Goldthwait, R.P., 1960. Study of Ice-cliff in Nunatarsuag, Greenland. *U.S. Army Snow, Ice and Permafrost Research Establishment (SIPRE) Technical Paper 39*. 106pp.

Goodchild, J.G., 1875. The glacial phenomena of the Eden Valley and the western part of the Yorkshire-Dale district. *Quarterly Journal of the Geological Society*, 31, 55–99.

Gordon, J.E., 1995. Early development of the glacial theory: Louis Agassiz and the Scottish connection. *Geology Today*, *11*, 64–68.

Gravenor, C.P. & Kupsch, W.O., 1959. Ice-disintegration features in western Canada. *The Journal of Geology*, 67, 48–64.

Greenwood, S.L., Clark, C.D. & Hughes, A.L., 2007. Formalising an inversion methodology for reconstructing ice-sheet retreat patterns from meltwater channels: application to the British Ice Sheet. *Journal of Quaternary Science*, *22*, 637–645.

Grosval'd, M.G., Vtyurin, B.I., Sukhodrovskiy, V.L. & Shishorina, Z.G., 1986. Underground ice in western Siberia: origin and geological significance. *Polar Geography*, 10, 173–183.

Guidat, T., Pochat, S., Bourgeois, O. & Souček, O., 2015. Landform assemblage in Isidis Planitia, Mars: Evidence for a 3 Ga old polythermal ice sheet. *Earth & Planetary Science Letters*, 411, 253–267.

Haeberli, W., 1979. Holocene push-moraines in alpine permafrost. *Geografiska Annaler: Series A*, 61, 43–48.

Haeberli, W., 1981. Ice motion on deformable sediments (correspondence). *Journal of Glaciology*, 27, 365–366.

Haeberli, W., 2005. Investigating glacier-permafrost relationships in high-mountain areas: historical background, selected examples and research needs. In: Harris, C. & Murton, J.B. (Eds.), *Cryospheric Systems*. Geological Society of London, Special Publications, 242, London, 29–37.

Haeberli, W., Wegmann, M. & Mühle, D., 1997. Slope stability problems related to glacier shrinkage and permafrost degradation in the Alps. *Eclogae Geologicae Helvetiae*, 90, 407–414.

Haeberli, W., Huggel, C., Kääb, A., *et al.*, 2004. The Kolka-Karmadon rock/ice slide of 20 September 2002: an extraordinary event of historical dimensions in North Ossetia, Russian Caucasus. *Journal of Glaciology*, 50, 533–546.

Haeberli, W., Buetler, M., Huggel, C., *et al.*, 2016. New lakes in deglaciating high-mountain regions–opportunities and risks. *Climatic change*, 139, 201–214.

Haeberli, W., Schaub, Y. & Huggel, C., 2017. Increasing risks related to landslides from degrading permafrost into new lakes in de-glaciating mountain ranges. *Geomorphology*, *293*, 405–417.

Haldorsen, S. & Shaw, J., 1982. The problem of recognizing melt-out till. *Boreas*, 11, 261–277.

Hall, A.M. & Phillips, W.M., 2006. Glacial modification of granite tors in the Cairngorms, Scotland. *Journal of Quaternary Science*, 21, 811–830.

Ham, N.R. & Attig, J.W., 1996. Ice wastage and landscape evolution along the southern margin of the Laurentide Ice Sheet, north-central Wisconsin. *Boreas*, 25, 171–186.

Hambrey, M.J. & Fitzsimons, S.J., 2010. Development of sediment–landform associations at cold glacier margins, Dry Valleys, Antarctica. *Sedimentology*, 57, 857–882.

Hambrey, M.J. & Glasser, N.F., 2012. Discriminating glacier thermal and dynamic regimes in the sedimentary record. *Sedimentary Geology*, 251, 1–33.

Hambrey, M.J. & Huddart, D., 1995. Englacial and proglacial glaciotectonic processes at the snout of a thermally complex glacier in Svalbard. *Journal of Quaternary Science*, 10, 313–326.

Harris, S.A., 1986. *The Permafrost Environment*. Croon Helm.

Hart, J.K., 2007. An investigation of subglacial shear zone processes from Weybourne, Norfolk, UK. *Quaternary Science Reviews*, 26, 2354–2374.

Hart, J.K. & Boulton, G.S., 1991. The interrelation of glaciotectonic and glaciodepositional processes within the glacial environment. *Quaternary Science Reviews*, 10, 335–350.

Hart, J.K. & Roberts, D.H., 1994. Criteria to distinguish between subglacial glaciotectonic and glaciomarine sedimentation, I. Deformation styles and sedimentology. *Sedimentary Geology*, 91, 191–213.

Hart, J.K. & Watts, R.J., 1997. A comparison of the styles of deformation associated with two recent push moraines, south Van Keulenfjorden, Svalbard. *Earth Surface Processes and Landforms*, 22, 1089–1107.

Head, J.W. & Marchant, D.R., 2003. Cold-based mountain glaciers on Mars: Western Arsia Mons. *Geology*, 31, 641–644.

Hiemstra, J.F., Evans, D.J.A. & Cofaigh, C.Ó., 2007. The role of glacitectonic rafting and comminution in the production of subglacial tills: examples from southwest Ireland and Antarctica. *Boreas*, 36, 386–399.

Hodgson, D.A., 1989. Quaternary geology of the Queen Elizabeth Islands. In: Fulton, R.J. (Ed.), *Quaternary geology of Canada and Greenland* (Vol. 1). Geological Survey of Canada, Geology of Canada, Ottawa, 443–478.

Hoffmann, K. & Piotrowski, J.A., 2001. Till melange at Amsdorf, central Germany: sediment erosion, transport and deposition in a complex, soft-bedded subglacial system. *Sedimentary Geology*, 140, 215–234.

Holdsworth, G., 1974. *Meserve Glacier Wright Valley, Antarctica: Part I. Basal Processes*. Research Foundation and the Institute of Polar Studies, Report No. 37, The Ohio State University Research Foundation.

Hubbard, B., Souness, C. & Brough, S., 2014. Glacier-like forms on Mars. *The Cryosphere*, 8, 2047–2061.

Ingólfsson, O. & Lokrantz, H., 2003. Massive ground ice body of glacial origin at Yugorski Peninsula, arctic Russia. *Permafrost and Periglacial Processes*, 14, 199–215.

Iverson, N.R., Hanson, B., Hooke, R.L. & Jansson, P., 1995. Flow mechanism of glaciers on soft beds. *Science*, 267, 80–81.

Kälin, M., 1971. *The active push moraine of the Thompson Glacier, Axel Heiberg Island, Canadian Arctic Archipelago, Canada.* McGill University, Axel Heiberg Island Research Reports, Glaciology 4

Kanevskiy, M., Jorgenson, M.T., Shur, Y. & Dillon, M., 2008. Buried glacial basal ice along the Beaufort Sea Coast, Northern Alaska. American Geophysical Union, Fall Meeting 2008, abstract id. C11D-0531.

Kanevskiy, M., Shur, Y., Jorgenson, M.T., *et al.*, 2013. Ground ice in the upper permafrost of the Beaufort Sea coast of Alaska. *Cold Regions Science and Technology*, 85, 56–70.

Kaplyanskaya, F.A. & Tarnogradskiy, V.D., 1986. Remnants of the Pleistocene ice sheets in the permafrost zone as an object for paleoglaciological research. *Polar Geography*, 10, 257–266.

Kavanaugh, J.L. & Clarke, G.K., 2006. Discrimination of the flow law for subglacial sediment using in situ measurements and an interpretation model. *Journal of Geophysical Research: Earth Surface*, 111(F1), doi:10.1029/2005JF000346.

Kleman, J., 1992. The palimpsest glacial landscape in northwestern Sweden: Late Weichselian deglaciation landforms and traces of older west-centered ice sheets. *Geografiska Annaler: Series A, Physical Geography*, 74, 305–325.

Kleman, J., 1994. Preservation of landforms under ice sheets and ice caps. *Geomorphology*, 9, 19–32.

Kleman, J. & Borgström, I., 1990. The boulder fields of Mt. Fulufjället, west-central Sweden: Late Weichselian boulder blankets and interstadial periglacial phenomena. *Geografiska Annaler: Series A, Physical Geography*, 72, 63–78.

Kleman, J. & Borgström, I., 1994. Glacial land forms indicative of a partly frozen bed. *Journal of Glaciology*, 40, 255–264.

Kleman, J. & Borgström, I., 1996. Reconstruction of palaeo-ice sheets: the use of geomorphological data. *Earth Surface Processes and Landforms*, 21, 893–909.

Kleman, J. & Glasser, N.F., 2007. The subglacial thermal organisation (STO) of ice sheets. *Quaternary Science Reviews*, 26, 585–597.

Kleman, J. & Hättestrand, C., 1999. Frozen-bed Fennoscandian and Laurentide ice sheets during the Last Glacial Maximum. *Nature*, 402, 63–66.

Kleman, J., Hättestrand, C. & Clarhäll, A., 1999. Zooming in on frozen-bed patches: scale-dependent controls on Fennoscandian ice sheet basal thermal zonation. *Annals of Glaciology*, 28, 189–194.

Klemen, J., Hättestrand, C., Borgström, I. & Stroeven, A., 1997. Fennoscandian palaeoglaciology reconstructed using a glacial geological inversion model. *Journal of Glaciology*, 43, 283–299.

Knight, P.G., 1989. Stacking of basal debris layers without bulk freezing-on: isotopic evidence from West Greenland. *Journal of Glaciology*, 35, 214–216.

Knight, P.G., 1994. Two-facies interpretation of the basal layer of the Greenland ice sheet contributes to a unified model of basal ice formation. *Geology*, 22, 971–974.

Knight, P.G., 1995. Debris structures in basal ice exposed at the margin of the Greenland ice sheet. *Boreas*, 24, 11–12.

Knight, P.G., Patterson, C.J., Waller, R.I., Jones, A.P. & Robinson, Z.P., 2000. Preservation of basal-ice sediment texture in ice-sheet moraines. *Quaternary Science Reviews*, 19, 1255–1258.

Kokelj, S.V., Lacelle, D., Lantz, T.C., *et al.*, 2013. Thawing of massive ground ice in mega slumps drives increases in stream sediment and solute flux across a range of watershed scales. *Journal of Geophysical Research: Earth Surface*, 118, 681–692.

Kokelj, S.V., Tunnicliffe, J., Lacelle, D., *et al.*, 2015. Increased precipitation drives mega slump development and destabilization of ice-rich permafrost terrain, northwestern Canada. *Global and Planetary Change*, 129, 56–68.

Kokelj, S.V., Lantz, T.C., Tunnicliffe, T., Segal, R. & Lacelle, D., 2017. Climate-driven thaw of permafrost preserved glacial landscapes, northwestern Canada. *Geology*, 45(4), 371–374.

Krüger, J., 1993. Moraine-ridge formation along a stationary ice front in Iceland. *Boreas*, 22, 101–109.

Krüger, J., 1996. Moraine ridges formed from subglacial frozen-on sediment slabs and their differentiation from push moraines. *Boreas*, 25, 57–64.

Krüger, J. & Kjær, K.H., 2000. De-icing progression of ice-cored moraines in a humid, subpolar climate, Kötlujökull, Iceland. *The Holocene*, 10(6), 737–747.

Kumpulainen, R.A., 1994. Fissure-fill and tunnel-fill sediments: expressions of permafrost and increased hydrostatic pressure. *Journal of Quaternary Science*, 9, 59–72.

Kupsch, W.O., 1962. Ice-thrust ridges in western Canada. *The Journal of Geology*, 70, 582–594.

La Farge, C., Williams, K.H. & England, J.H., 2013. Regeneration of Little Ice Age bryophytes emerging from a polar glacier with implications of totipotency in extreme environments. *Proceedings of the National Academy of Sciences*, 110, 9839–9844.

Lacelle, D., Bjornson, J., Lauriol, B., Clark, I.D. & Troutet, Y., 2004. Segregated-intrusive ice of subglacial meltwater origin in retrogressive thaw flow headwalls, Richardson Mountains, NWT, Canada. *Quaternary Science Reviews*, 23, 681–696.

Lacelle, D., Lauriol, B., Clark, I.D., Cardyn, R. & Zdanowicz, C., 2007. Nature and origin of a Pleistocene-age massive ground-ice body exposed in the Chapman Lake moraine complex, central Yukon Territory, Canada. *Quaternary Research*, 68, 249–260.

Lacelle, D., Fisher, D.A., Coulombe, S., Fortier, D. & Frappier, R., 2018. Buried remnants of the Laurentide Ice Sheet and connections to its surface elevation. *Nature Scientific Reports*, doi:10.1038/s41598-018-31166-2.

Lakeman, T.R. & England, J.H., 2012. Paleoglaciological insights from the age and morphology of the Jesse moraine belt, western Canadian Arctic. *Quaternary Science Reviews*, 47, 82–100.

Lawson, D.E., 1979. A comparison of the pebble orientations in ice and deposits of the Matanuska Glacier, Alaska. *Journal of Geology*, 87, 629–645.

Lawson, D.E., 1986. Response of permafrost terrain to disturbance: a synthesis of observations from northern Alaska, USA. *Arctic and Alpine Research*, 18, 1–17.

Lee, J.R. & Phillips, E.R., 2008. Progressive soft sediment deformation within a subglacial shear zone—a hybrid mosaic–pervasive deformation model for Middle Pleistocene glaciotectonised sediments from eastern England. *Quaternary Science Reviews*, 27, 1350–1362.

Levy, J.S., Head, J.W. & Marchant, D.R., 2010. Concentric crater fill in Utopia Planitia: history and interaction between glacial "brain terrain" and periglacial mantle processes. *Icarus*, 202, 462–476.

Lewkowicz, A.G. & Way, R.G., 2019. Extremes of summer climate trigger thousands of thermokarst landslides in a high arctic environment. *Nature Communications*, 10, 1–11.

Licht, K.J. & Hemming, S.R., 2017. Analysis of Antarctic glacigenic sediment provenance through geochemical and petrologic applications. *Quaternary Science Reviews*, 164, 1–24.

Liestøl, O., 1977. Pingos, springs and permafrost in Spitsbergen. *Norsk Polarinstitutt Årbok*, *1975*, 7–29.

Liestøl, O., 1996. Open-system pingos in Spitsbergen. *Norsk Geografisk Tidsskrift*, 50, 81–84.

Linton, D.L., 1950. Unglaciated enclaves in glaciated regions. *Journal of Glaciology*, 1, 451–452.

Lorrain, R.D. & Demeur, P., 1985. Isotopic evidence for relic Pleistocene glacier ice on Victoria Island, Canadian Arctic archipelago. *Arctic and Alpine Research*, 17, 89–98.

Lyså, A. & Lønne, I., 2001. Moraine development at a small High-Arctic valley glacier: Rieperbreen, Svalbard. *Journal of Quaternary Science*, 16, 519–529.

Mackay, J.R., 1956. Deformation by glacier-ice at Nicholson Peninsula, N.W.T., Canada. *Arctic*, 9, 218–228.

Mackay, J.R., 1959. Glacier ice-thrust features of the Yukon Coast. *Geographical Bulletin*, 13, 5–21.

Mackay, J.R., 1972. The world of underground ice. *Annals of the Association of American Geographers*, 62, 1–22.

Mackay, J.R., 1989. Massive ice: some field criteria for the identification of ice types. *Current Research, Part G, Geological Survey of Canada, Paper*, 5–11.

Mackay, J.R. & Dallimore, S.R., 1992. Massive ice of the Tuktoyaktuk area, western Arctic coast, Canada. *Canadian Journal of Earth Sciences*, 29, 1235–1249.

Mackay, J.R. & Mathews, W.H., 1964. The role of permafrost in ice-thrusting. *The Journal of Geology*, 72, 378–380.

Mackay, J.R. & Matthews, J.V., 1983. Pleistocene ice and sand wedges, Hooper Island, Northwest Territories. *Canadian Journal of Earth Sciences*, 20, 1087–1097.

Mackay, J.R., Rampton, V.N. & Fyles, J.G., 1972. Relic Pleistocene permafrost, western arctic, Canada. *Science*, 176, 1321–1323.

Mannerfelt, C.M., 1945. Marginal drainage channels as indicators of the gradients of Quaternary ice caps. *Geografiska Annaler*, 31, 19.

Margreth, A., Dyke, A.S., Gosse, J.C. & Telka, A.M., 2014. Neoglacial ice expansion and late Holocene cold-based ice cap dynamics on Cumberland Peninsula, Baffin Island, Arctic Canada. *Quaternary Science Reviews*, *91*, 242–256.

Mathews, W.H. & Mackay, J.R., 1960. Deformation of soils by glacier ice and the influence of pore pressures and permafrost. *Transactions of the Royal Society of Canada*, 54, 27–36.

Matthews, J.A., McCarroll, D. & Shakesby, R.A., 1995. Contemporary terminal-moraine ridge formation at a temperate glacier: Styggedalsbreen, Jotunheimen, southern Norway. *Boreas*, 24, 129–139.

McCarroll, D., 2016. Trimline trauma: the wider implications of a paradigm shift in recognising and interpreting glacial limits. *Scottish Geographical Journal*, 132, 130–139.

McSavaney, M.J., 2002. Recent rockfalls and rock avalanches in Mount Cook National Park, New Zealand. *Reviews in Engineering Geology*, 15, 35–70.

Menzies, J., 1990. Sand intraclasts within a diamicton mélange, southern Niagara Peninsula, Ontario, Canada. *Journal of Quaternary Science*, 5, 189–206.

Midgley, N.G., Tonkin, T.N., Graham, D.J. & Cook, S.J., 2018. Evolution of high-Arctic glacial landforms during deglaciation. *Geomorphology*, 311, 63–75.

Mollard, J.D., 2000. Ice-shaped ring-forms in Western Canada: their airphoto expressions and manifold polygenetic origins. *Quaternary International*, 68, 187–198.

Möller, P. & Dowling, T.P., 2018. Equifinality in glacial geomorphology: instability theory examined via ribbed moraine and drumlins in Sweden. *GFF*, 140, 106–135.

Moore, P.L., Iverson, N.R., Brugger, K.A., *et al.*, 2011. Effect of a cold margin on ice flow at the terminus of Storglaciären, Sweden: implications for sediment transport. *Journal of Glaciology*, 57, 77–87.

Moorman, B.J. & Michel, F.A., 2000a. Glacial hydrological system characterization using ground-penetrating radar. *Hydrological Processes*, 14, 2645–2667.

Moorman, B.J. & Michel, F.A., 2000b. The burial of ice in the proglacial environment on Bylot Island, Arctic Canada. *Permafrost and Periglacial Processes*, 11, 161–175.

Moran, S.R., Clayton, L., Hooke, R.L., Fenton, M.M. & Andriashek, L.D., 1980. Glacier-bed landforms of the prairie region of North America. *Journal of Glaciology*, 25, 457–476.

Motyka, R.J. & Echelmeyer, K.A., 2003. Taku Glacier (Alaska, USA) on the move again: active deformation of proglacial sediments. *Journal of Glaciology*, 49, 50–58.

Muir, J., 1988. *Travels in Alaska*. Sierra Club Books.

Murray, T., 1997. Assessing the paradigm shift: deformable glacier beds. *Quaternary Science Reviews*, 16, 995–1016.

Murton, J.B., 2005. Ground-ice stratigraphy and formation at North Head, Tuktoyaktuk Coastlands, western Arctic Canada: a product of glacier–permafrost interactions. *Permafrost and Periglacial Processes*, 16, 31–50.

Murton, J.B., Waller, R.I., Hart, J.K., *et al.*, 2004. Stratigraphy and glaciotectonic structures of permafrost deformed beneath the northwest margin of the Laurentide ice sheet, Tuktoyaktuk Coastlands, Canada. *Journal of Glaciology*, 50, 399–412.

Murton, J.B., Whiteman, C.A., Waller, R.I., *et al.*, 2005. Basal ice facies and supraglacial melt-out till of the Laurentide Ice Sheet, Tuktoyaktuk Coastlands, western Arctic Canada. *Quaternary Science Reviews*, 24, 681–708.

Murton, J.B., Frechen, M. & Maddy, D., 2007. Luminescence dating of mid-to Late Wisconsinan aeolian sand as a constraint on the last advance of the Laurentide Ice Sheet across the Tuktoyaktuk Coastlands, western Arctic Canada. *Canadian Journal of Earth Sciences*, 44, 857–869.

Mustard, J.F., Cooper, C.D. & Rifkin, M.K., 2001. Evidence for recent climate change on Mars from the identification of youthful near-surface ground ice. *Nature*, 412, 411–414.

Nesje, A., Pilø, L.H., Finstad, E., *et al.*, 2012. The climatic significance of artefacts related to prehistoric reindeer hunting exposed at melting ice patches in southern Norway. *The Holocene*, 22(4), 485–496.

Ó Cofaigh, C., 2006. The subpolar glacier landsystem of the Canadian High Arctic. In: Knight, P.G. (Ed.), *Glacier Science & Environmental Change*. Blackwell Publishing, 89–91.

Paul, M.A. & Eyles, N., 1990. Constraints on the preservation of diamict facies (melt-out tills) at the margins of stagnant glaciers. *Quaternary Science Reviews*, 9, 51–69.

Phillips, E., Lee, J.R. & Burke, H., 2008. Progressive proglacial to subglacial deformation and syntectonic sedimentation at the margins of the Mid-Pleistocene British Ice Sheet: evidence from north Norfolk, UK. *Quaternary Science Reviews*, 27, 1848–1871.

Piotrowski, J.A., Mickelson, D.M., Tulaczyk, S., Krzyszkowski, D. & Junge, F.W., 2001. Were deforming subglacial beds beneath past ice sheets really widespread? *Quaternary International*, 86, 139–150.

Rampton, V.N., 1988. *Quaternary Geology of the Tuktoyaktuk Coastlands, Northwest Territories, Memoir 423*. Geological Survey of Canada.

Rampton, V.N., 1991. Observations on buried glacier ice and massive segregated ice, western Arctic coast, Canada: discussion. *Permafrost and Periglacial Processes*, 2, 163–165.

Ravier, E., Buoncristiani, J.F., Menzies, J., Guiraud, M. & Portier, E., 2015. Clastic injection dynamics during ice front oscillations: a case example from Sólheimajökull (Iceland). *Sedimentary Geology*, 323, 92–109.

Reinhardy, B.T.I., Booth, A.D., Hughes, A.L.C., *et al.*, 2019. Pervasive cold ice within a temperate glacier - implications for glacier thermal regimes, sediment transport and foreland geomorphology. *The Cryosphere*, 13, 827–843.

Rempel, A.W., 2007. Formation of ice lenses and frost heave. *Journal of Geophysical Research*, 112(F2), doi:10.1029/2006JF000525.

Rijsdijk, K.F., Owen, G., Warren, W.P., McCarroll, D. & van der Meer, J.J., 1999. Clastic dykes in over-consolidated tills: evidence for subglacial hydrofracturing at Killiney Bay, eastern Ireland. *Sedimentary Geology*, 129, 111–126.

Rootes, C.M. & Clark, C.D., 2020. Glacial trimlines to identify former ice margins and subglacial thermal boundaries: A review and classification scheme for trimline expression. *Earth-Science Reviews*, 210, 103355, doi:10.1016/j.earscirev.2020.103355.

Ruszczyńska-Szenajch, H., 1987. The origin of glacial rafts: detachment, transport, deposition. *Boreas*, 16, 101–112.

Rutten, M.G., 1960. Ice-pushed ridges, permafrost and drainage. *American Journal of Science*, 258, 293–297.

Scanlon, K.E., Head, J.W. & Marchant, D.R., 2015. Remnant buried ice in the equatorial regions of Mars: morphological indicators associated with the Arsia Mons tropical mountain glacier deposits. *Planetary & Space Science*, 111, 144–154.

Schäfer, J.M., Baur, H., Denton, G.H., *et al.*, 2000. The oldest ice on Earth in Beacon Valley, Antarctica: new evidence from surface exposure dating. *Earth & Planetary Science Letters*, 179(1), 91–99.

Scheidegger, J.M., Bense, V.F. & Grasby, S.E., 2012. Transient nature of Arctic spring systems driven by subglacial meltwater. *Geophysical Research Letters*, 39, L12405, doi:10.1029/2012GL051445.

Schindler, C., Rothlisberger, H. & Gyger, M., 1978. Glaziale Stauungen in den Niederterrassen-Schottern des Aadorfer Feldes und ihre Deutung, vn C. Schindler, H. Röthlisberger und M. Gyger. *Ecologae Geologie Helvetiae*, 71, 159–174.

Scholz, H. & Baumann, M., 1997. An 'open system pingo' near Kangerlussuaq (Søndre Strømfjord), West Greenland. *Geology of Greenland Survey Bulletin*, 176, 104–108.

Schomacker, A., 2008. What controls dead-ice melting under different climate conditions? A discussion. *Earth-Science Reviews*, 90, 103–113.

Schomacker, A. & Kjær, K.H., 2007. Origin & de-icing of multiple generations of ice-cored moraines at Brúarjökull, Iceland. *Boreas*, 36, 411–425.

Schomacker, A. & Kjær, K.H., 2008. Quantification of dead-ice melting in ice-cored moraines at the high-arctic glacier Holmströmbreen, Svalbard. *Boreas*, 37, 211–225.

Schomacker, A., Krüger, J. & Kjær, K.H., 2006. Ice-cored drumlins at the surge-type glacier Brúarjökull, Iceland: a transitional-state landform. *Journal of Quaternary Science*, 21, 85–93.

Schumm, S.A., 1991. *To Interpret the Earth: Ten Ways to be Wrong*. Cambridge University Press.

Schytt, V., 1956. Lateral drainage channels along the northern side of the Moltke Glacier, Northwest Greenland. *Geografiska Annaler*, 38, 64–77.

Segal, R.A., Lantz, T.C. & Kokelj, S.V., 2016. Acceleration of thaw slump activity in glaciated landscapes of the Western Canadian Arctic. *Environmental Research Letters*, 11, 034025.

Shaw, J., 1977a. Tills deposited in arid polar environments. *Canadian Journal of Earth Sciences*, 14, 1239–1245.

Shaw, J., 1977b. Till body morphology and structure related to glacier flow. *Boreas*, 6, 189–201.

Shaw, J., 1982. Melt-out till in the Edmonton area, Alberta, Canada. *Canadian Journal of Earth Sciences*, 19, 1548–1569.

Slaymaker, O. & Embleton-Hamann, C., 2018. Advances in global mountain geomorphology. *Geomorphology*, 308, 230–264.

Sletten, K., Lyså, A. & Lønne, I., 2001. Formation and distintegration of a high arctic ice-cored moraine complex, Scott Turnerbreen. *Boreas*, 30, 272–284.

Sollid, J.L. & Sørbel, L., 1988. Influence of temperature conditions information of end moraines in Fennoscandia and Svalbard. *Boreas*, 17, 553–558.

Sollid, J.L. & Sørbel, L., 1994. Distribution of glacial landforms in southern Norway in relation to the thermal regime of the last continental ice sheet. *Geografiska Annaler: Series A, Physical Geography*, 76, 25–35.

St. Onge, D.A. & McMartin, I., 1999. La moraine du Lac Bluenose (Territories du Nord-Ouest), une moraine à noyau de glace de glacier. *Géographie Physique et Quaternaire*, 53, 287–295.

Stalker, A.M., 1976. Megablocks, or the enormous erratics of the Albertan Prairies. *Geological Survey of Canada Paper, 76-1C*, 185–188.

Sugden, D.E., 1968. The selectivity of glacial erosion in the Cairngorm Mountains, Scotland. *Transactions of the Institute of British Geographers*, 79–92.

Sugden, D.E. & John, B.S., 1976. *Glaciers and Landscape: A Geomorphological Approach*. Edward Arnold, London.

Sugden, D.E. & Watts, S.H., 1977. Tors, felsenmeer, and glaciation in northern Cumberland Peninsula, Baffin Island. *Canadian Journal of Earth Sciences*, 14, 2817–2823.

Sugden, D.E., Marchant, D.R., Potter, N. Jr., *et al.*, 1985. Preservation of Miocene glacier ice in East Antarctica. *Nature*, 376, 412–414.

Sugden, D.E., Knight, P.G., Livesey, N., *et al.*, 1987. Evidence for two zones of debris entrainment beneath the Greenland ice sheet. *Nature*, 328, 238–241.

Swinzow, G.K., 1962. Investigation of shear zones in the ice sheet margin, Thule area, Greenland. *Journal of Glaciology*, 4, 215–229.

Syverson, K.M. & Mickelson, D.M., 2009. Origin and significance of lateral meltwater channels formed along a temperate glacier margin, Glacier Bay, Alaska. *Boreas*, 38, 132–145.

Tonkin, T.N., Midgley, N.G., Cook, S.J. & Graham, D.J., 2016. Ice-cored moraine degradation mapped and quantified using an unmanned aerial vehicle: a case study from a polythermal glacier in Svalbard. *Geomorphology*, 258, 1–10.

Tsytovich, N.A., 1975. *The Mechanics of Frozen Ground*. McGraw-Hill, New York.

Vilímek, V., Klimeš, E.E. & Benešová, M., 2015. Geomorphic impacts of the glacial lake outburst flood from Lake No. 513 (Peru). *Environmental Earth Science*, 73, 5233–5244.

Viola, D., McEwen, A.S., Dundas, C.M. & Byrne, S., 2015. Expanded secondary craters in the Arcadia Planitia region, Mars: evidence for tens of Myr-old shallow subsurface ice. *Icarus*, 248, 190–204.

Vtyurin, B.I. & Glazovskiy, A.F., 1986. Composition and structure of the Ledyanaya Gora tabular ground ice body on the Yenisey. *Polar Geography*, 10, 273–285.

Wainstein, P., Moorman, B. & Whitehead, K., 2008. Importance of glacier–permafrost interactions in the preservation of a proglacial icing: Fountain Glacier, Bylot Island, Canada. *Ninth International Conference on Permafrost*, 29, 1881–1886.

Wainstein, P., Moorman, B. & Whitehead, K., 2014. Glacial conditions that contribute to the regeneration of Fountain Glacier proglacial icing, Bylot Island, Canada. *Hydrological Processes*, 28, 2749–2760.

Waller, R.I. & Tuckwell, G.W., 2005. Glacier-permafrost interactions and glaciotectonic landform generation at the margin of the Leverett Glacier, West Greenland. *Geological Society, London, Special Publications*, 242, 39–50.

Waller, R.I., Hart, J.K. & Knight, P.G., 2000. The influence of tectonic deformation on facies variability in stratified debris-rich basal ice. *Quaternary Science Reviews*, 19, 775–786.

Waller, R.I., Murton, J.B. & Knight, P.G., 2009a. Basal glacier ice and massive ground ice: different scientists, same science? In: Knight, J. & Harrison, S. (Eds.), *Periglacial and Paraglacial Processes and Environments*. Geological Society, London, Special Publication, 320, 57–69.

Waller, R.I., Murton, J.B. & Whiteman, C., 2009b. Geological evidence for subglacial deformation of Pleistocene permafrost. *Proceedings of the Geologists' Association*, 120, 155–162.

Waller, R.I., Phillips, E., Murton, J., Lee, J. & Whiteman, C., 2011. Sand intraclasts as evidence of subglacial deformation of Middle Pleistocene permafrost, North Norfolk, UK. *Quaternary Science Reviews*, 30, 3481–3500.

Weertman, J., 1961. Mechanism for the formation of inner moraines found near the edge of cold ice caps and ice sheets. *Journal of Glaciology*, 3, 965–978.

West, R.G., 1980. *The Pre-Glacial Pleistocene of the Norfolk and Suffolk Coasts*. Cambridge University Press, Cambridge.

Whiteman, C., 2002. Implications of a Middle Pleistocene ice-wedge cast at Trimingham, Norfolk, eastern England. *Permafrost and Periglacial Processes*, 13, 163–170.

Williams, P.J. & Smith, M.W., 1989. *The Frozen Earth: Fundamentals of Geocryology*. Cambridge University Press.

6

Future Research Directions

6.1 Basal Boundary Conditions and Processes Beneath Cold-Based Glaciers

The increasing recognition that cold-based glaciers can actively interact with subglacial permafrost to produce dynamically coupled systems is perhaps one of the most important advances in our understanding of the nature and significance of glacier–permafrost interactions. Whilst basal boundary conditions associated with 'frozen beds' were previously assumed to result in passive substrates and a cessation of subglacial processes (Section 3.4), field observations within modern and ancient glacial environments have revealed the extensive, distinctive and highly complex interactions that can take place (Section 3.5). These have the potential to significantly influence a range of glaciological attributes including their dynamic behaviour (e.g. Astakhov *et al.*, 1996; Echelmeyer & Zhongxiang, 1987), their ability to erode, entrain and transport sediment (e.g. Etzelmüller & Hagen, 2005) and their ability to generate distinctive geomorphic end products (e.g. Atkins, 2013; Dyke, 1993).

In challenging existing orthodoxies concerning the inaction of subglacial processes beneath non-temperate glaciers, the research findings resulting from a relatively small number of existing studies have provided early insights into the complexity of cryotic subglacial environments and seemingly contradictory results that serve to illustrate existing limitations in our current state of understanding. As noted by Fitzsimons *et al.* (2008), the resolution of conflicting views regarding the efficacy of glacial erosion beneath cold-based ice for example may require a clearer recognition and renewed research focus on the varied nature and influence of the glacier substrate as an independent control. Whilst glaciers resting on substrates comprising crystalline bedrock or coarse-grained sediments (hard-frozen permafrost; Tsytovich, 1975) may display limited rates of erosion, those resting on fine-grained substrates with more significant quantities of liquid water may be far more susceptible to deformation and entrainment.

As a case in point, the testing of this hypothesis would require the systematic observation of the nature and behaviour of the substrates beneath cold-based glaciers. Detailed field investigations on modern-day glaciers in Antarctica (e.g. Fitzsimons *et al.*, 1999), China (Echelmeyer & Zhongxiang *et al.*, 1987) and Svalbard have demonstrated the persistence of

Glacier–Permafrost Interactions, First Edition. Richard I. Waller.
© 2024 John Wiley & Sons Ltd. Published 2024 by John Wiley & Sons Ltd.

key basal processes under cryotic conditions and provided initial insights into their influence on ice-flow velocities (Section 6.5) and sediment fluxes (Section 6.4). However, when compared to the literature on warm-based glaciers, the paucity of targeted studies that have specifically examined the basal processes associated with glacier–permafrost interactions inevitably means that our understanding of the spatial and temporal variations in basal boundary conditions and processes beneath cold-based glaciers is limited.

There is a simple need therefore first and foremost to expand on the number of studies undertaken on these types of glaciers, particularly in relation to cold-based glaciers with soft and deformable beds and the greatest potential for glacier–permafrost interactions (Waller, 2001). In view of the rapid ongoing rates of cryospheric change, there is a need in particular to improve our understanding of the ways in which basal boundary conditions and subglacial processes are likely to change in response to glacier fluctuations (Section 6.6). One possibility here could be to use field evidence to test the hypothesis that glacier–permafrost interactions are particularly extensive during ice advances, with the evidence currently being limited to modelling studies (Cutler *et al.*, 2000; Marshall & Clark, 2002). The predominance of glacier recession however means that any such work is likely to be limited to short lived (e.g. surge-related) advances. With respect to the impacts of glacier recession, there is the opportunity to critically test a number of potentially significant recent research findings, the bases for which are currently limited to individual sites. One example includes the hypothesised connections between cold-based ice, soft-bed regelation and the accretion of basal ice and frozen subglacial sediments (e.g. Christoffersen *et al.*, 2006; Lovell *et al.*, 2015). If these processes and changing basal boundary conditions have a significant impact on dynamic behaviour and glacigenic sediment fluxes, then they could indicate the changes in glacier behaviour likely to occur in response to ongoing climate change and glacier recession (Section 6.5). The recent technological development of a range of remote sensing techniques has enabled the location of warm and cold ice to be mapped over wide areas and to infer the nature of basal boundary conditions and the possible extent of subglacial permafrost (e.g. Ruskeeniemi *et al.*, 2018). The wider application of these techniques therefore has the potential to elucidate the changing nature of basal boundary conditions across a range of glaciological environments and scenarios.

Any revised understanding of the nature of subglacial processes beneath cold-based glaciers may potentially have wide-ranging consequences. As an illustration, if strong ice-bed coupling combined with high subglacial porewater pressures beneath cold-based ice can facilitate the erosion, entrainment and subsequent comminution of large bedrock fragments (Section 3.4.2), then the associated increase in the available surface area could increase weathering rates and the release of trace gases from subglacial environments (MacDonald *et al.*, 2018). These may in turn provide an energy source for subglacial microbial populations whilst also suggesting that subglacial environments could constitute an active component of the global carbon cycle through methanogenesis and the release of methane during deglaciation (Wadham *et al.*, 2008).

An ongoing and broader challenge relates to the significant research costs entailed in undertaking field research in what are commonly remote locations. Alongside technological advances, this has resulted in a rapid increase in the number of studies relying solely on the application of remote sensing, modelling or a combination of the two. Whilst these have proven pivotal to a number of major research advances (e.g. Zwally *et al.*, 2002), generating more detailed understandings of the changing nature of basal boundary conditions and the role of subglacial processes is still likely to require detailed field investigations. What is perhaps most needed here and in other areas of glaciology are sustained investigations spanning multiple years that provide the opportunity to examine change over time and to explore the nature of interannual variability. The associated costs however mean that studies of the like undertaken on the John Evans Glacier in the Canadian Arctic (Section 4.3.2) remain extremely rare. However, the location of a number of relatively easily accessible cold and polythermal glaciers in places like Svalbard suggests that these localities will continue to provide an ideal locus for future sustained research on glacier–permafrost interactions in environments experiencing rapid climatic and landscape change (Section 6.6).

6.2 The Rheology of Subglacial Permafrost

Following the recognition that the basal processes are not restricted to warm-based glaciers and that subglacial sediments can continue to deform at temperatures well below the bulk freezing point (Section 3.5), it is clear that glaciologists, glacial geomorphologists and glacial geologists need to recognise the complexity of permafrost rheology and the implications for the basal processes operating beneath cold-based glaciers with soft beds. Permafrost researchers have made considerable advances in understanding the rheological properties of ice–sediment–water mixtures in this regard. Whilst early experimental studies suggested that ice–sediment mixtures display higher shear strengths than either pure ice or unfrozen sediments on account primarily of the increased cohesion, more recent studies (e.g. Arenson & Springman, 2005a; Yang *et al.*, 2010b) have highlighted the varied controls on shear strength, an appreciation of which is central to understanding their complex rheological behaviours.

The most influential control on permafrost rheology and therefore its likely response to the application of glacitectonic stresses is the abundance and distribution of unfrozen water. In his review of the deformation of ice–sediment mixtures, Moore (2014) identifies unfrozen water content as a key control that can influence the rheology of permafrost at any sediment concentration. At temperatures below the bulk freezing point of water, this is dependent upon the formation of interfacial films that can persist to temperatures below $-10\,°C$ and whose thickness varies in response to temperature, particle size distribution and solute content (Section 2.2.2). Their rheological significance was identified by Tsytovich (1975) who defined threshold temperatures above which the permafrost was 'plastic frozen' and deformable. Moreover, the association of interfacial films with frozen fringes and the migration of porewater has revealed potential connections between basal thermal regimes, basal ice formation, the porewater pressures in subglacial sediments and the dynamic behaviour of the overlying ice mass (e.g. Christoffersen & Tulaczyk, 2003a,b,c).

The measurement of the pre-melted water contents of glacigenic sediments with respect to temperature, particle size and solute content therefore provides the opportunity to generate systematic datasets on an attribute central to permafrost rheology. The wider application of time domain reflectometry that can be used to estimate the volumetric water content in unconsolidated sediments (e.g. Boike & Roth, 1997) in this specific context therefore provides a potential research opportunity. The ability of these techniques to monitor changes in liquid water concentration over seasonal freeze-thaw cycles and to demonstrate the persistence of at liquid water at temperatures below −15 °C illustrates their potential application in real world, field-based settings (Roth & Boike, 2001).

Unfortunately, there currently appears to be a limited awareness of these research developments within the glaciological literature, which still has a tendency to assume that the rheology of subglacial sediments, and the operation of basal processes more generally, can be conceptualised as a binary system (i.e. unfrozen, weak and permeable versus frozen, rigid and impermeable) separated by the bulk freezing point. In reality, the small number of investigations that have been undertaken in subglacial environments beneath cold-based glaciers have confirmed the influence of interfacial water in both reducing the viscosity of ice-rich sediments (Echelmeyer & Zhongxiang, 1987) and facilitating the continued operation of basal processes (Cuffey *et al.*, 2000). Therefore, with the bulk freezing point commonly marking the commencement of a freezing process spanning a considerable cryotic temperature range, it is more likely that the rheological properties of fine-grained subglacial sediments will change progressively as temperatures fall, the quantity of residual liquid water diminishes and the bond strength between ice and sediment increases (Williams & Smith, 1989).

The development of a rheological model applicable to subglacial permafrost is likely to constitute a significant research challenge for a number of reasons. First, there is still an active and as yet unresolved broader debate regarding the type of flow law that provides the best fit for the rheological behaviour for subglacial till (e.g. Kavanaugh & Clarke, 2006). The inclusion of temperature and state as additional variables, the associated variations in ice and liquid water contents and the potential migration of fluids through frozen fringes is likely to make this even more challenging. This is particularly the case when one considers the potential interrelationships and feedback loops, with the wide range of particle size distributions displayed by glacial tills providing a key determinant of the unfrozen water content and their rheological behaviour, which in turn influences their deformation, comminution and grain-size evolution (e.g. Hooke & Iverson, 1995). Second, there is a tendency for the existing empirical relationships derived for permafrost materials to break down at relatively 'warm' temperatures just below the pressure melting point (Williams & Smith, 1989). This means that the thermal regimes most likely to be encountered in subglacial environments that are of most potential relevance to glacier dynamics could well prove to be the most intractable.

An additional control of relevance to permafrost rheology relates to the role and dynamic nature of the effective bed - the interface below which glacier-induced motion ceases (Section 3.5.5). In situations where glaciers overlie unconsolidated beds, the advent of soft-bed glaciology has resulted in a shift in thinking from clear-cut 'ice–rock interfaces' to more dynamic 'ice-bed interfaces' in which the ice can couple with subglacial material. Mechanical coupling between cold-based ice and subglacial permafrost in turn promotes the transmission of basal shear stresses into the subglacial materials with ice-rich or clay-rich horizons providing potential lines of weakness and décollement surfaces. Further research is however required to better understand the associations between the characteristics of the glacier (e.g. basal shear stress and overburden pressure), the characteristics of the subglacial permafrost (e.g. temperature, grain size and water content) and their implications for the likely thickness of the resultant deforming layer and the style of deformation. Developing such an understanding is essential to understanding the dynamics of glaciers resting on permafrost (Section 6.5) and their geomorphological and geological signatures (Sections 6.3 and 6.4, respectively).

6.3 Using Geomorphological Evidence to Reconstruct Basal Thermal Regime

A recurrent theme in this book relates to the current lack of knowledge of non-temperate glaciers in comparison to their much better studied temperate counterparts. This stems from their inaccessibility, the cost of field-based research and these glacier types having previously been considered of limited intrinsic research interest. In reviewing existing studies in order to devise a landsystems model for polar continental ice margins, Fitzsimons (2003) pointed to 'a striking gap in our knowledge of polar landform and sediment assemblages' (p110). Whilst there has been some positive progress in the form of a series of studies that have demonstrated their distinctive geomorphic signatures (e.g. Hambrey & Fitzsimons, 2010), the number of studies and therefore our knowledge and understanding of both the nature and diversity of the related landsystems and the key controls remains rudimentary.

The development of accurate and realistic connections linking environment, process and form are fundamental to palaeoglaciological reconstructions in general and to the inverse-modelling approaches based around the interpretation of the landform record in particular. As discussed within Section 5.2, the subtle landscape impacts of cold-based glaciers and the absence of classic 'textbook' landforms indicative of glaciation has proven particularly challenging in differentiating sites glaciated by cold-based ice from those that were never glaciated, with significant implications for the reconstruction of former ice sheets in high-latitude regions (e.g. Dyke, 1999). In the same way that conceptual models describing the operation of basal processes and the rheological behaviour of subglacial sediments discussed in the previous sections have tended to be overly simplistic in their consideration of basal thermal regime, there is a similar need for inverse models to recognise and integrate the complex behaviours emerging from the investigation of glacier–permafrost interactions in ancient and modern environments. With process-based studies revealing an increasingly complex spectrum of processes and behaviours operating across a broad range of cryotic temperatures (Section 3.5), it seems reasonable to assume that the associated geomorphic end products could be equally diverse.

The formation of striations provides an interesting case in point and an illustration of the potential research opportunities that could help to elucidate the connections between basal thermal regime, basal processes and the resultant forms. Striations constitute very well known, small-scale erosional features that are routinely used to infer the former presence of warm-based ice, particularly when developed on exposed bedrock surfaces (e.g. Gellatly *et al.*, 1988). Recent research however has demonstrated the potential for basal sliding considered responsible for their formation to persist beneath cold-based glaciers at temperatures as low as −17 °C (Cuffey *et al.*, 2000). In addition, geomorphological investigations of cold glaciers have identified 'scrapes', 'striae' and 'grooves' as part of a suite of features considered indicative of clast–clast and clast–bedrock interactions within these extreme environments

(Atkins, 2013; Atkins *et al.,* 2002; Lloyd Davies *et al.*, 2009). In actively forming beneath some of the coldest glaciers on Earth, this indicates that these types of features cannot be used to infer the presence of warm or wet-based ice. Looking to the future research opportunities, Atkins *et al.* (2002) have raised the possibility that it is the orientation of these features rather than their presence or absence that could help to distinguish between warm and cold-based ice. Having examined the erosional features beneath the Manhaul Bay Glacier in Antarctica, they concluded that, 'the [abrasion] marks are variable in shape, size, and grouping, and are unlike the more consistent, uniform sets of parallel striae and grooves found commonly on bedrock abraded by warm-based sliding ice' (Atkins *et al.*, 2002, p662). The systematic examination of the variability in striation orientations at glaciers with contrasting thermal regimes provides an opportunity to test this intriguing hypothesis (Figure 6.1).

Geomorphological research undertaken to date suggests that it is highly unlikely that the recognition of any individual landform will provide a reliable basis for inferring the former presence of cold-based ice. Whilst some studies have suggested that lateral meltwater channels may provide the most convincing evidence, their observation along the margins of degrading, stagnant warm-based glaciers indicates that their presence is not necessarily diagnostic of cold-based ice (Syverson & Mickelson, 2009). This provides a reminder of the dangers inherent in basing reconstructions on individual lines of geomorphic evidence and the need instead to systematically map and examine entire glacial landsystems. Landsystem investigations undertaken to date have provided valuable insights into the distinctive characteristics of the landform-sediment assemblages associated with cold-based glaciers and with glacier–permafrost interactions. In addition to the small-scale abrasional features mentioned above, geomorphological observations of the forefields of retreating cold-based

Figure 6.1   Boulder featuring an abraded surface with striations with a range of different orientations located in front of a receding polythermal glacier on Bylot Island, Eastern Canadian Arctic. Pen for scale (Image: Richard Waller).

glaciers in Antarctica for example have highlighted the development of distinctive landform suites comprising erosional lineations (Atkins *et al.*, 2002), brecciated till deposits (Lloyd-Davies *et al.*, 2009), thrust-block moraines (Fitzsimons, 1996), lateral meltwater channels (Atkins & Dickinson, 2007) and areas of compacted ground (Atkins 2013). Meanwhile, an expanding range of studies in formerly glaciated permafrost regions have highlighted other important landform elements including broad belts of ice-cored moraine (e.g. Dyke & Savelle, 2000; Dyke & Evans, 2003). The number of published studies however remains limited and therefore there is again a simple need to expand this research in terms of the number of studies, their geographical coverage and the glaciological settings in which they have been undertaken. It is worth noting here that there are a number of regions in which glacier–permafrost interactions are likely to be common but which have yet to be explored in detail. This includes Novaya Zemlya in the Russian Arctic for example that contains 685 predominantly cold-based glaciers, some of which are thought to display surge-type behaviours (e.g. Grant *et al.*, 2009). The financial and logistical challenges associated with the acquisition of these observations however provide a reminder of why they remain so limited.

With much of the work on glacier–permafrost interactions highlighting the continued operation of basal processes and geomorphic activity at cryotic temperatures, it is the ability of cold-based glaciers overlying permafrost to preserve pre-existing features and surfaces and to leave little if any trace of their presence that is one of the most extraordinary elements of their behaviour that remains difficult to explain. It also provides a reminder that one of the most fundamental questions that emerged with the development of glacial theory – namely whether glaciers on balance promote the erosion or protection of landscapes – is still the subject of active and ongoing debate within geomorphology (e.g. Slaymaker & Embleton-Hamann, 2018, p237). The association of 'dry-based' glaciers with a cessation of basal processes and a hiatus in landform development is a key component of geomorphological inverse models (section 5.3.3, Kleman & Borgström, 1996) with the application of cosmogenic isotope dating techniques confirming that ancient surfaces can be preserved under these conditions (e.g. Fabel *et al.*, 2002). There is however a need to better understand the processes, mechanisms and basal boundary conditions that can allow some glaciers to advance and cover land surfaces whilst preserving delicate and fragile features, including vegetation that can remain biologically viable and able to re-grow once the ice has receded (Section 5.2; Figures 5.1 and 5.2). It is the ability of the same glaciers to simultaneously modify and erode some areas whilst leaving adjacent areas intact that is particularly difficult to explain. Further work on the nature, timing and mechanisms of ice advance may help to reveal the specific circumstances most likely to lead to this remarkable preservation, which may for example include advances over snow-covered surfaces or pre-existing features such as an ice-marginal aprons or aufeis.

6.4 Sedimentological and Structural Criteria for Identifying Glacier–Permafrost Interactions

In addition to the geomorphological evidence considered in the preceding section, there is a need to develop sedimentological and structural criteria that can be used to provide independent and robust evidence of the former occurrence of glacier–permafrost interactions.

Ideally, these should be combined with the geomorphological evidence discussed within the previous section as part of a landsystems approach. However, with the geomorphological record displaying an inherent bias towards wet-based conditions (section 2.4.2, Kleman & Glasser, 2007) and the landform record being subject to removal, modification or the preservation of pre-existing features rather than the generation of distinctive landforms, sedimentological and structural criteria are likely to be particularly important.

Focusing initially on sedimentological criteria, Hambrey & Glasser's (2012) systematic description of the contrasting sedimentary facies associated with glaciers from around the world created with the explicit aim of facilitating the reconstruction of thermal regimes and dynamic behaviours from the sedimentary record provides an excellent foundation for further work. The authors identify two key challenges associated with the development of discriminatory criteria. First, with the differing sedimentary processes identified occurring within all glacier regimes, it is their relative importance rather than their simple presence or absence that is required to infer the thermal regime (Figure 6.2). Second, any distinctive characteristics evident within the glacial sediments will be subject to post-depositional

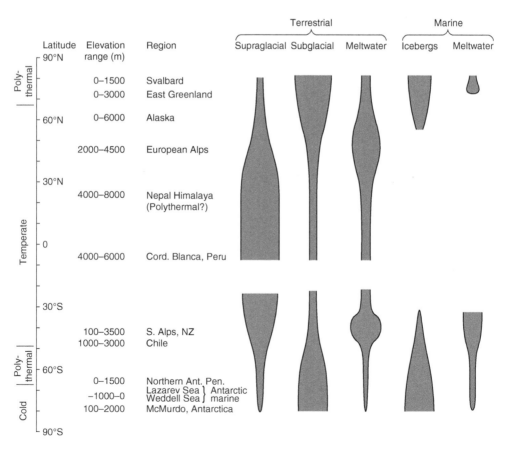

Figure 6.2 Conceptual model illustrating the varied climatic settings under which the principal glacial sedimentary processes take place in relation to place, region and latitude. The width of the shaded zones indicates the relative importance of each process.

reworking, with mass movement, fluvial and lacustrine processes associated with degrading ice-cored terrain being likely to be particularly problematic within warming permafrost environments (Section 5.3.2).

One avenue worthy of additional study relates to the clast roundness characteristics of subglacial diamictons and more specifically, the potential inclusion of greater quantities of angular material within subglacial diamictons developed beneath non-temperate glaciers. Early theoretical models of glacial erosion argued that the generation of strong ice–bed coupling can significantly enhance the process of plucking and the entrainment of large bedrock rafts (Boulton, 1972, 1974) with Clayton & Moran (1974) suggesting that the majority of glacial erosion occurs within these 'quarrying' zones (see Section 3.4.2). In addition, with the formation of subglacial sediments commonly involving significant extraglacial inputs and with the frost weathering of porous lithologies in permafrost environments resulting in segregated ice lens formation, fracture development and the production of angular debris (Matsuoka & Murton, 2008; Figure 4.19), it seems reasonable to assume that this will promote the incorporation of substantial quantities of material that is initially angular in nature (Lukas *et al.*, 2013).

Preliminary research has suggested that the glacigenic sediment generated beneath non-temperate glaciers can indeed display a surprising degree of angularity, related in large part to the predominance of bedrock shattering in these environments (Waller *et al.*, 2016). With studies on clast shape and roundness having examined the influence of lithology in detail (e.g. Bennett *et al.*, 1997; Lukas *et al.,* 2013), there is a similar need to systematically investigate the influence of glacier thermal regime as a potential additional control. One of the potential problems here is the potential mis-identification of glacial diamictons with high levels of clast angularity (and RA indices) as the products of supraglacial transport and deposition, which is currently the default interpretation (Benn & Ballantyne, 1994). In examining samples from small niche glaciers on James Ross Island in Antarctica, Hambrey & Glasser (2012) note that the basal debris displayed little evidence of modification from the source materials (rockfalls of volcanic breccia) with both featuring low C_{40} and high RA indices. One possibility here could relate to the identification of striae on the constituent clasts as a way of distinguishing sediments subject to basal traction and an element of clast–clast or clast–bed interactions, with early studies reporting the presence of striated clasts around cold-based glaciers in Antarctica (e.g. Mercer, 1971; Holdsworth, 1974). In contrast, Hambrey & Glasser (2012) state that none of the clasts they observed around glaciers in the Dry Valleys or the northern Antarctic Peninsula featured striations and therefore their presence and ability to act as a diagnostic tool clearly requires further investigation.

Looking at the textural characteristics of glacigenic sediments more broadly, there is a need to more open minded about their characteristics within cold climatic regimes. As with the landform record, glacigenic sediments associated with glacier–permafrost interactions commonly lack the classic 'textbook' features with Hambrey & Glasser (2012) commenting that 'recognising the deposits of cold glaciers in the geological record is challenging because of the near absence of clear indications of glacial transport' (p13). Within the extreme environments of the Dry Valleys, sediment entrained and transported by glaciers therefore simply reflects the reworking of pre-existing materials with little if any modification during transport, leading to fine sediment fractions being rare (Hambrey & Fitzsimons, 2010).

As noted by the same authors, this has resulted in uncertainties that have significant implications for palaeoglaciological reconstructions but that remain as yet unresolved,

> *From a geological perspective, the largest uncertainties that concern cold-based glaciers are the internal structure of ice-contact landforms and the attributes of the lithofacies that make up the landforms associated with cold-based glaciers. Specifically, little is known concerning the texture and clast characteristics of these sediments. Yet without such data, recognition of the former presence of cold-based ice in the geological record will remain speculative* (Hambrey & Fitzsimons, 2010, p858).

Sedimentological and structural investigations of pre-existing permafrost deformed by Pleistocene Ice Sheets within western Siberia and the Canadian Arctic have revealed a suite of distinctive features, many of which are composed of ice, thereby providing clear evidence of deformation under cryotic conditions. This has in turn helped to elucidate the maximum depth of glacitectonic deformation, the associated mechanisms and the potential thermal evolution of the deforming permafrost during phases of advance, retreat and stagnation. In view of the ongoing challenges and uncertainties associated with the development of sedimentological criteria, it may be that structural criteria could prove more promising in providing evidence of glacier–permafrost interactions that can be identified within the geological record. The formation and preservation of coarse-grained intraclasts with intact sedimentary structures contained within subglacial sediments displaying clear evidence of glacitectonic deformation provides one particular feature that could prove diagnostic (Waller *et al.*, 2011, Figure 5.26). However, whilst their formation is entirely consistent with expected variations in interfacial water content and sediment rheology, there is a need to critically test this hypothesis and to identify whether these intraclasts can in fact survive subglacial deformation under unfrozen conditions as an observation that would falsify their association with glacier–permafrost conditions.

6.5   Determining the Influence of Permafrost on Glacier Dynamics

In considering the potential influence of permafrost on the dynamic behaviour of glaciers and ice sheets, it is worth noting a broader shift in the focus of research examining the principal controls on velocity in general and states of fast ice flow in particular (Section 4.3). Research in the late twentieth century tended to focus on the influence of basal thermal regimes, with the investigation of surge events for example exploring the potential correspondence of active surge fronts with thermal transitions from warm to cold-based ice (e.g. Clarke *et al.*, 1984). This reflected a widespread recognition of the influence of thermal controls on the mechanisms of ice flow and the related assumption that basal processes, key to states of fast ice flow, are inoperative beneath cold-based ice (Section 6.1). More recently, however, research has increasingly focused on the influence of glacier hydrology as the primary control on glacier dynamics, with the influential paper of Zwally *et al.* (2002) and subsequent studies highlighting the critically important roles played by supraglacial meltwater production, its transfer to the glacier bed, its subsequent discharge (or storage) and ultimately its influence on basal water pressures. Whilst this undoubtedly represents a significant development in our understanding of glacier dynamics, it is unfortunate that this seems to have diminished the research attention afforded to the potential influence of basal thermal regimes (Section 4.3.3).

The few field studies that have examined the dynamic behaviour of relatively small-scale, non-temperate glaciers in detail have clearly demonstrated important process connections between a glacier's thermal regime, its hydrological behaviour and its velocity. The multi-year research studies undertaken on the John Evans Glacier on Ellesmere Island in the 1990s and 2000s in particular (Copland & Sharp, 2001; Bingham *et al.*, 2003, 2006; Copland *et al.*, 2003), provided unique insights into the mechanisms associated with dramatic seasonal variations in velocity. The ability of a thermal dam of cold, marginal ice to influence the capacity and the evolution of the subglacial drainage system, which in turn controls the surface velocities, provides an indication of the potential role played by glacier–permafrost interactions in regulating meltwater outputs and subglacial storage (Section 4.3.2). As discussed in Section 6.1, there is a need therefore to replicate these studies and to determine the extent to which these mechanisms might translate to larger-scale ice sheets, particularly with recent geophysical studies having demonstrated the extension of subglacial permafrost beneath the margin of the Greenland Ice Sheet (Ruskeeniemi *et al.*, 2018). These studies could help to resolve outstanding research questions concerning the influence of basal thermal transitions on glacier flow. These have been advocated to explain the entrainment of debris through thrusting (Section 5.3.2) and the formation of ribbed moraine (Section 5.3.3), although targeted investigations have questioned their validity, suggesting that alternative controls including subglacial topography and hydrology may play a more influential role (Moore *et al.,* 2011).

In presenting enthalpy budgets and enthalpy gradients as a means of explaining glacier surges, Benn *et al.* (2019a,b) have provided a unifying model with the potential to reconcile these ongoing debates in its consideration of the combined influence of temperature and water content on glacier dynamics. The seasonal acceleration of non-temperate glaciers observed in numerous studies can in this respect be viewed as a dynamic response to an increase in enthalpy associated initially with the production, transfer and temporary storage of meltwater to the bed, followed by the instigation of basal processes that result in strain heating of the basal ice. Subsequently, it is the breaching of ice-marginal thermal dams and the re-establishment of a channelised drainage systems in the summer months that results in the release of stored water, a loss of enthalpy and a reduction in velocity. The formation of a barrier of cold ice within the lower tongue of Morsnevbreen during the lead up to a surge event provides a tantalising insight into the potential influence of glacier–permafrost interactions (Benn *et al.*, 2019b). Downwasting and a reduction in ice thickness in the lower part of the glacier resulted in a state of negative enthalpy, causing basal freezing and therefore the potential formation of subglacial permafrost. In steepening the enthalpy gradient, this influenced the evolution of the surge with the associated formation of a hydraulic dam and the storage of large volumes of pressurised meltwater up-glacier coinciding with a peak in the surge velocities. A wider application of this enthalpy model to other studies of non-temperate glaciers could therefore prompt further consideration of the complex connections between the thermal character of glaciers and permafrost, hydrological systems and glacier dynamics, the importance of which is likely to become more apparent with the glaciological impacts of climate change becoming increasingly evident (Section 6.6).

Recent technological advances have the potential to expand the limited empirical data available on the dynamic behaviour of non-temperate glaciers, many of which are located within remote and inaccessible locations. The development of SAR (synthetic aperture radar) interferometry in particular has substantially enhanced our knowledge of glacier dynamics. This research has enabled large scale, synoptic investigations of the ice dynamics of the Earth's remaining ice sheets (e.g. Fahnestock *et al.*, 1993; Mohr *et al.*, 1998) as well as more detailed investigation of changes in the terminal position and the velocities of outlet glaciers as they respond to polar climate change (e.g. Luckman *et al.*, 2006; Goliber *et al.*, 2022). Millan *et al.*'s (2017) investigation of the changing ice velocities of tidewater glaciers in the Canadian Arctic archipelago illustrates their potential application to polar environments characterised by permafrost. Such studies have however tended to focus on the fast-flowing glaciers contributing to sea-level rise rather than slow flowing terrestrial glaciers in which glacier–permafrost interactions are likely to play a more influential role. Whilst it would necessitate site visits and entail the associated set up costs, the application of newly developed ground-based (GB-SAR) interferometry could enable smaller-scale and more detailed investigations of individual sites (e.g. Luzi *et al.*, 2007). In providing real-time data over extended time periods on both the surface elevation (ice volume) and ice surface velocity, this could help to clarify the dynamic impacts of thermal transitions (cf. Moore *et al.*, 2011) and to monitor the impacts of glacier recession and associated changes in thermal regime on their dynamic behaviour (Section 6.6).

There is a need finally to combine these data on the dynamic behaviour of non-temperate glaciers with an improved understanding of basal processes beneath cold-based ice (Section 6.1) and the rheology of cryotic sediments (Section 6.2) in order to develop more accurate representations of these boundary conditions within numerical ice sheet models. With the incorporation of basal processes within such models and the development of basal sliding laws for example representing an ongoing area of research development (e.g. Gudmundsson, 2011), this is likely to prove challenging. What is increasingly clear however is that conceptualisations based around a binary system (frozen bed and zero basal velocity versus unfrozen bed with active basal processes) separated by a clearly defined threshold (the pressure melting point) are likely to be overly simplistic. Identification of the circumstances in which basal processes can remain active at temperatures below the bulk freezing point, combined with the collation of larger datasets demonstrating the relationship between basal thermal regimes and ice surface velocities in different settings, can be considered key starting points. These will hopefully facilitate progress on some of the most consequential research questions, including the degree to which active basal processes beneath cold-based glaciers significantly enhance their surface velocities and whether ice-marginal, glacier–permafrost interactions are more likely to limit flow velocities and promote ice-sheet stability or, by contrast, to promote states of fast ice flow (Clayton *et al.*, 2001).

6.6 Glacier–Permafrost Interactions and Future Climate Change

Previous and ongoing work has highlighted a range of important connections between glaciers, permafrost and climate. It is important to recognise at the outset that long-term climate histories and past climate changes as well as modern-day climate change can play an important role in determining a glacier's current behaviour. This time dependency is illustrated by the persisting influence of the Little Ice Age (LIA) on the thermal character of a number of Arctic glaciers, with the polythermal Laika Glacier in the Canadian Arctic retaining a zone of warm-based ice thought to represent a thermal legacy of its larger LIA extent (Blatter & Hutter, 1991; section 3.2). Similarly, the comparatively fast flow of Erikbreen, its active subglacial drainage and the abundant subglacial sediment store that enables high sediment fluxes are similarly considered legacies of its LIA form (Sollid *et al.*, 1994; section 4.3.2).

Etzelmüller & Hagen (2005) provide a valuable conceptual framework through which the impacts of climate change on glaciers, permafrost and glacier–permafrost interactions can be considered. With the response time of glaciers to climate change typically being much more rapid than for permafrost, equilibrium line altitudes are likely to change more rapidly than mean permafrost altitudes, potentially leading to a more a complex response of the coupled system. Glacier advances caused by climatic cooling for example tend to lead to the warming of the overridden permafrost as the glacier ice both insulates the ground from the atmospheric temperatures and results in enhanced advective heat transport. The considerable thermal inertia of permafrost can however allow it to persist for millennia even under temperate glaciers (e.g. Cutler *et al.*, 2000; section 2.4.3). A climatically induced glacier retreat can conversely lead to a renewed phase of permafrost aggradation within newly exposed glacier forelands.

One of the most significant and widely reported changes relating to the ongoing recession of glaciers in the Arctic in response to rapid climate warming is a transition from polythermal basal thermal regimes to entirely cold-based regimes. This glacier cooling occurring in response to climate warming is one of the most striking paradoxes caused primarily by a progressive thinning and deceleration of the receding glaciers (e.g. Björnsson *et al.*, 1996; Nowak & Hodson, 2014; Lovell *et al.*, 2015; section 2.3.2). Work by Reinhardy *et al.* (2019) suggests this thermal transition can also occur in lower latitude areas with temperate glaciers becoming polythermal in nature, which in turn suggests that glacier–permafrost interactions could become more extensive, at least temporarily, as deglaciation progresses. In much the same way that the treatment of basal processes has been oversimplified (Section 6.1), Irvine-Fynn *et al.* (2011) caution that the use of the umbrella term 'polythermal' hides a complex diversity of glacier thermal regimes. They argue that there is a consequent need for more detailed research on these ongoing thermal transitions to better understand the rates and patterns of change in the distribution of temperate and cold ice as a pre-condition to understanding their hydrological and dynamic implications.

Nonetheless, ongoing research has provided insights into the dynamic and hydrological impacts one might expect to result from such thermal changes. Numerous studies have considered the hydrological implications, with glacier thinning and deceleration potentially leading to a reduction in subglacial meltwater production, the refreezing of taliks, a reduction in groundwater recharge and the shutdown of proglacial spring systems (e.g. Haldorsen & Heim, 1999; Etzelmüller & Hagen, 2005). With the hydrological systems of glaciers and their dynamic behaviour being intrinsically linked, improving our understanding of the cascade of potential changes is clearly a research priority. Through their multi-year investigation of the hydrological changes relating to the recession of Austre Brøggerbreen and its transition from a polythermal to an entirely cold-based thermal regime, Nowak & Hodson (2013, 2014) provide an example of the long-term studies required. In showing that 'the simple concept wherein warming means more glacial melt and therefore more glacial discharge does not apply' (Nowak & Hodson, 2014) and revealing complexities that have yet to be incorporated into models, this provides a reminder of the importance of these studies that remain all too rare in the literature. An illustration of this complexity is the ability of increased precipitation to compensate for a reduction in subglacial meltwater fluxes, with Bogen & Bønsnes (2003) demonstrating the influence of summer storm events on suspended sediment fluxes. An increase in the retention of meltwater relating to the constriction of drainage across the cold-temperate transition zone could in theory increase melt rates and trigger a dynamic response (Irvine-Fynn *et al.*, 2011; section 4.2.3) although Benn (2019a) has suggested that a reduction in accumulation rates and an associated loss of enthalpy could conversely result in an end to surging on Svalbard.

The geomorphological implications of a thermal evolution to more extensive cold-based conditions are potentially broad ranging, thereby providing additional opportunities for further research. Initial work has suggested that such a transition is likely to result in reduced rates of subglacial erosion whilst facilitating the basal freeze-on of formerly unfrozen sediment to generate thick sequences of debris-rich basal ice (e.g. Etzelmüller & Hagen, 2005; Lovell *et al.*, 2015). A related question is whether such debris entrainment is more appropriately conceptualised as a mechanism of basal ice formation or as the aggradation of permafrost beneath the thinning glaciers. The theoretical models on the influence of subglacial temperatures on fluid flow and debris entrainment produced by Menzies (1981) provide a good starting point for any such studies, with the author highlighting the lack of data on the temperatures of subglacial sediments as a major research gap. Whilst there has been some progress here (e.g. Alexander *et al.*, 2020), these more recent publications have confirmed that our knowledge of the thermal character of subglacial permafrost remains poor due to a lack of direct observation and measurement.

A range of studies has demonstrated how the impact of climate change on ice-cored terrains that can in turn be considered the product of ice-marginal glacier–permafrost interactions (Dyke & Evans, 2003; Evans, 2009) is resulting in their widespread destabilisation and a concomitant increase in sediment fluxes into the wider catchments in which they occur (e.g Etzelmüller & Hagen, 2005; Schomacker, 2008; Midgley *et al.*, 2018). Within modern-day environments, recent research has indicated that it is an acceleration of the processes responsible for the removal of the protective debris cover, including precipitation-induced mass movement and thermal erosion by channels (Etzelmüller, 2000) that are the key drivers. In line with a greater recognition of the role of groundwater within glacial environments, the accumulation of meltwater on dead-ice surfaces and within ice-walled lakes may also accelerate the rates of ice loss (Schomacker & Kjær, 2008; section 5.6.3).

The landscape impacts of these processes are most dramatically illustrated by the exponential increase in the areas of ice-cored terrains affected by retrogressive thaw slump activity in areas such as the western Canadian Arctic; a rapid change that represents a re-initiation of the process of deglaciation that remains as yet incomplete (Kokelj *et al.*, 2015, 2017; Lewkowicz & Way, 2019). In combination with the increased magnitude and frequency of rock slope failures being observed in mountainous areas with glaciers and alpine permafrost, this demonstrates the influence of permafrost on the timescales of paraglacial landscape adjustment, the resultant environmental impacts and the escalating risks posed by natural hazards (e.g. Haeberli *et al.*, 2017). More detailed studies of both the extent and landsystems context of the relict glacier ice vulnerable to future degradation, the resulting process cascades and the risks posed to the environment and resident populations all represent important research priorities.

Finally, with glaciers and ice sheets themselves potentially playing a more active role within the carbon cycle than was previously thought, this represents an additional area likely to generate increased research interest. Zeng (2003) has proposed a 'glacial burial hypothesis' in which ice sheets bury vegetation and soil carbon during their advance phases to create a hypothetical carbon store of c. 500 Gt, with subsequent deglaciation resulting in its release back into the atmosphere. Zeng (2007) hypothesised that these feedbacks might have contributed to the development of 100 ka glacial–interglacial cycles with the release of carbon upon deglaciation driving some terminations. Of particular relevance to glacier–permafrost interactions is the suggestion that the advance of a cold-based ice sheet over permafrozen terrain is more likely to lead to the burial of carbon and the generation of the subglacial store central to the mechanism. Recent work illustrating the remarkable preservation of vegetation beneath cold-based glaciers on Ellesmere Island has certainly illustrated the potential for carbon storage, although in these modern-day situations the vegetation has subsequently provided the locus for re-growth following deglaciation and exposure to sunlight (e.g. La Farge *et al.*, 2013; section 5.2).

6.7 Towards an Integrated Analysis of Glaciers and Permafrost

In conclusion, it is worth returning to the point made at the start of this book (Section 1.1), namely that glaciologists and permafrost scientists have tended to work in isolation from each other as largely distinct research communities, in spite of sharing the study of ice as a shared common ground. A greater recognition of the nature and importance of the wide-ranging interactions between glaciers and permafrost that has formed the focus of this contribution inevitably provides an invitation and an opportunity for these two research communities to work in closer collaboration. With the separation of the research activities having arguably been to the detriment of both disciplines, their closer integration has the potential to deliver mutual benefits.

First, this will help to promote and embed analyses and conceptual frameworks that treat glaciers and permafrost as integrated components of the cryosphere, the importance of which has been demonstrated from a range of perspectives. From a thermal perspective, thin and slow moving glaciers can be viewed as a constituent part of the permafrost horizon with cold-based conditions being likely if the local permafrost thickness exceeds the glacier thickness (Dyke, 1993), with Etzelmüller & Hagen (2005) further emphasising the thermal interconnections through their integration of the concepts of equilibrium line altitude and mountain permafrost altitude (Section 3.2; Figure 3.2). The wider significance of these fundamental thermal connections is illustrated by the increasing risks posed by large rock slope failures in high mountain environments, triggered by a combination of glacier recession and increasing rock wall temperatures and permafrost degradation (Section 5.6.2). It is encouraging to note here that the urgent need to assess the risks posed by ice-rock avalanches in deglaciating terrains has resulted in increased communication and collaborative work between glaciologists and permafrost researchers (Haeberli *et al.*, 2017). Hydrological investigations focusing on both glaciers and permafrost have similarly highlighted their interconnected nature. The pioneering work of Olav Liestøl demonstrated the influential role played by glaciers as the principal source of groundwater recharge in catchments characterised by continuous permafrost (Figure 4.11; Liestøl, 1977), with the resultant groundwater fluxes facilitating active spring discharge (Section 4.2.5) and the associated formation of open-system pingos (Section 5.3.5). In regulating the capacity of subglacial drainage systems and the storage of water in glaciers, glacier–permafrost interactions and the formation of thermal dams can also influence their dynamic behaviour, with the new enthalpy model of glacier surging (Benn *et al.*, 2019a,b) illustrating the additional connections between temperature and water content (Sections 4.3.2 and 6.5). Meanwhile, the strong ice-bed coupling combined with the enhanced cohesion of ice-rich permafrost demonstrates the mechanical interconnections facilitated by glacier–permafrost interactions, enabling the propagation of the stress field into proglacial areas and the formation of large push moraine complexes (Section 5.3.2).

Second, the closer integration of the glaciological and permafrost research communities will facilitate progress on a range of specific research questions the resolution of which is dependent upon active collaboration. These include the accurate characterisation of the rheology of ice-rich subglacial diamictons and their influence on glacier flow (Section 6.2), the interpretation of the origin and palaeoglaciological significance of massive ice (Section 5.3.4) and the potential application of frost heave theory to the formation of basal ice, subglacial pore pressures and glacier dynamics (Section 4.3.3).

In recognising both the diversity and the significance of the connections between glaciers and permafrost, it is hoped that this book will act as a stimulus for further dialogue and collaborative research between the associated research communities and the opportunity to open up the 'new scientific land' envisioned by Haeberli in 2005.

References

Alexander, A., Obu, J., Schuler, T.V., Kääb, A. & Christiansen, H.H., 2020. Subglacial permafrost dynamics and erosion inside subglacial channels driven by surface events in Svalbard. *The Cryosphere*, 14, 4217–4231.

Arenson, L.U. & Springman, S.M., 2005a. Mathematical descriptions for the behaviour of ice-rich frozen soils at temperatures close to 0°C. *Canadian Geotechnical Journal*, 42, 431–442.

Astakhov, V.I., Kaplyanskaya, F.A. & Tarnogradsky, V.D., 1996. Pleistocene permafrost of West Siberia as a deformable glacier bed. *Permafrost and Periglacial Processes*, 7, 165–191.

Atkins, C.B., 2013. Geomorphological evidence of cold-based glacier activity in South Victoria Land, Antarctica. *Geological Society, London, Special Publications*, 381, 299–318.

Atkins, C.B. & Dickinson, W.W., 2007. Landscape modification by meltwater channels at margins of cold-based glaciers, Dry Valleys, Antarctica. *Boreas*, 36, 47–55.

Atkins, C.B., Barrett, P.J. & Hicock, S.R., 2002. Cold glaciers erode and deposit: evidence from Allan Hills, Antarctica. *Geology*, 30, 659–662.

Benn, D.I. & Ballantyne, C.K., 1994. Reconstructing the transport history of glacigenic sediments: a new approach based on the co-variance of clast form indices. *Sedimentary Geology*, 91, 215–227.

Benn, D.I., Fowler, A.C., Hewitt, I. & Sevestre, H., 2019a. A general theory of glacier surges. *Journal of Glaciology*, 65, 701–716.

Benn, D.I., Jones, R.L., Luckman, A., *et al.*, 2019b. Mass and enthalpy budget evolution during the surge of a polythermal glacier: a test of theory. *Journal of Glaciology*, 65, 717–731.

Bennett, M.R., Hambrey, M.J. & Huddart, D., 1997. Modification of clast shape in high-Arctic glacial environments. *Journal of Sedimentary Research*, 67, 550–559.

Bingham, R.G., Nienow, P.W. & Sharp, M.J., 2003. Intra-annual and intra-seasonal flow dynamics of a High Arctic polythermal valley glacier. *Annals of Glaciology*, 37, 181–188.

Bingham, R.G., Nienow, P.W., Sharp, M.J. & Copland, L., 2006. Hydrology and dynamics of a polythermal (mostly cold) High Arctic glacier. *Earth Surface Processes and Landforms*, 31, 1463–1479.

Björnsson, H., Gjessing, Y., Hamran, S.E., *et al.*, 1996. The thermal regime of sub-polar glaciers mapped by multi-frequency radio-echo sounding. *Journal of Glaciology*, 42, 23–32.

Blatter, H. & Hutter, K., 1991. Polythermal conditions in Arctic glaciers. *Journal of Glaciology*, 37, 261–269.

Bogen, J. & Bønsnes, T.E., 2003. Erosion and sediment transport in High Arctic rivers, Svalbard. *Polar Research*, 22, 175–189.

Boike, J. & Roth, K., 1998. Time domain reflectometry as a field method for measuring water content and soil water electrical conductivity at a continuous permafrost site. *Permafrost & Periglacial Processes*, 8, 359–370.

Boulton, G.S., 1972. The role of thermal regime in glacial sedimentation. In: Price, R.J. & Sugden, D.E. (Eds.), *Polar Geomorphology*. Institute of British Geographers Special Publication No. 4, 1–19.

Boulton, G.S., 1974. Processes and patterns of glacial erosion. In: Coates, D.R. (Ed.), *Glacial Geomorphology*. A proceedings volume of the Fifth Annual Geomorphology Symposia series. Springer, New York, 41–87.

Christoffersen, P. & Tulaczyk, S., 2003a. Thermodynamics of basal freeze-on: predicting basal and subglacial signatures of stopped ice streams and interstream ridges. *Annals of Glaciology*, 36, 233–243.

Christoffersen, P. & Tulaczyk, S., 2003b. Response of subglacial sediments to basal freeze-on 1. Theory and comparison to observations from beneath the West Antarctic Ice Sheet. *Journal of Geophysical Research: Solid Earth*, 108(B4).

Christoffersen, P. & Tulaczyk, S., 2003c. Signature of palaeo-ice-stream stagnation: till consolidation induced by basal freeze-on. *Boreas*, 32, 114–129.

Christoffersen, P., Tulaczyk, S., Carsey, F.D. & Behar, A.E., 2006. A quantitative framework for interpretation of basal ice facies formed by ice accretion over subglacial sediment. *Journal of Geophysical Research: Earth Surface*, 111(F1), doi:10.1029/2005JF000363.

Clarke, G.K., Collins, S.G. & Thompson, D.E., 1984. Flow, thermal structure, and subglacial conditions of a surge-type glacier. *Canadian Journal of Earth Sciences*, 21, 232–240.

Clayton, L. & Moran, S.R., 1974. A glacial process-form model. In: Coates, D.R. (Ed.), *Glacial Geomorphology*. A proceedings volume of the Fifth Annual Geomorphology Symposia series. Springer, Dordrecht, Netherlands, 89–119.

Clayton, L., Attig, J.W. & Mickelson, D.M., 2001. Effects of late Pleistocene permafrost on the landscape of Wisconsin, USA. *Boreas*, 30, 173–188.

Copland, L. & Sharp, M., 2001. Mapping thermal and hydrological conditions beneath a polythermal glacier with radio-echo sounding. *Journal of Glaciology*, 47, 232–242.

Copland, L., Sharp, M.J. & Nienow, P.W., 2003. Links between short-term velocity variations and the subglacial hydrology of a predominantly cold polythermal glacier. *Journal of Glaciology*, 49, 337–348.

Cuffey, K.M., Conway, H., Gades, A.M., *et al.*, 2000. Entrainment at cold glacier beds. *Geology*, 28, 351–354.

Cutler, P.M., MacAyeal, D.R., Mickelson, D.M., Parizek, B.R. & Colgan, P.M., 2000. A numerical investigation of ice-lobe–permafrost interaction around the southern Laurentide ice sheet. *Journal of Glaciology*, 46, 311–325.

Davies, M.L., Atkins, C.B., van der Meer, J.J., Barrett, P.J. & Hicock, S.R., 2009. Evidence for cold-based glacial activity in the Allan Hills, Antarctica. *Quaternary Science Reviews*, 28, 3124–3137.

Dyke, A.S., 1993. Landscapes of cold-centred Late Wisconsinan ice caps, Arctic Canada. *Progress in Physical Geography*, 17, 223–247.

Dyke, A.S., 1999. Last glacial maximum and deglaciation of Devon Island, Arctic Canada: support for an Innuitian Ice Sheet. *Quaternary Science Reviews*, 18, 393–420.

Dyke, A.S. & Evans, D.J.A., 2003. Ice-marginal terrestrial landsystems: northern Laurentide and Innuitian Ice Sheet margins. In: Evans, D.J.A. (Ed.), *Glacial Landsystems*. Arnold, London, 143–165.

Dyke, A.S. & Savelle, J.M., 2000. Major end moraines of Younger Dryas age on Wollaston Peninsula, Victoria Island, Canadian Arctic: implications for paleoclimate and for formation of hummocky moraine. Canadian Journal of Earth Sciences, 37, 601–619.

Echelmeyer, K. & Zhongxiang, W., 1987. Direct observation of basal sliding and deformation of basal drift at sub-freezing temperatures. *Journal of Glaciology*, 33, 83–98.

Etzelmüller, B., 2000. Quantification of thermo-erosion in pro-glacial areas - examples from Svalbard. *Zeitschrift für Geomorphologie*, 44(3), 343–361.

Etzelmüller, B. & Hagen, J.O., 2005. Glacier-permafrost interaction in Arctic and alpine mountain environments with examples from southern Norway and Svalbard. In: Harris, C. & Murton, J.B. (Eds.), *Cryospheric Systems: Glaciers & Permafrost*. Geological Society, London, Special Publications, 242, 11–27.

Evans, D.J.A., 2009. Controlled moraines: origins, characteristics and palaeoglaciological implications. *Quaternary Science Reviews*, 28, 183–208.

Fabel, D., Stroeven, A.P., Harbor, J., *et al.*, 2002. Landscape preservation under Fennoscandian ice sheets determined from in situ produced 10Be and 26Al. *Earth and Planetary Science Letters*, 201, 397–406.

Fahnestock, M., Bindschadler, R., Kwok, R. & Jezek, K., 1993. Greenland ice sheet surface properties and ice dynamics from ERS-1 SAR imagery. *Science*, 262, 1530–1534.

Fitzsimons, S.J., 1996. Formation of thrust-block moraines at the margins of dry-based glaciers, south Victoria Land, Antarctica. *Annals of Glaciology*, 22, 68–74.

Fitzsimons, S.J., 2003. Ice-marginal landsystems: polar continental margins. In: Evans, D.J.A. (Ed.), *Glacial Landsystems*. Arnold, London, 89–110.

Fitzsimons, S.J., McManus, K.J. & Lorrain, R.D., 1999. Structure and strength of basal ice and substrate of a dry-based glacier: evidence for substrate deformation at sub-freezing temperatures. *Annals of Glaciology*, 28, 236–240.

Fitzsimons, S., Webb, N., Mager, S., *et al.*, 2008. Mechanisms of basal ice formation in polar glaciers: an evaluation of the apron entrainment model. *Journal of Geophysical Research: Earth Surface*, 113(F2), doi:10.1029/2006JF000698.

Gellatly, A.F., Gordon, J.E., Whalley, W.B. & Hansom, J.D., 1988. Thermal regime and geomorphology of plateau ice caps in northern Norway: observations and implications. *Geology*, 16(11), 983–986.

Goliber, S., Black, T., Catania, G., *et al.*, 2022. TermPicks: a century of Greenland glacier terminus data for use in scientific and machine learning applications. *The Cryosphere*, 16, 3215–3233.

Grant, K.L., Stokes, C.R. & Evans, I.S., 2009. Identification and characteristics of surge-type glaciers on Novaya Zemlya, Russian Arctic. *Journal of Glaciology*, 55, 960–972.

Gudmundsson, G.H., 2011. Ice-stream response to ocean tides and the form of the basal sliding law. *The Cryosphere*, 5, 259–270.

Haeberli, W., Schaub, Y. & Huggel, C., 2017. Increasing risks related to landslides from degrading permafrost into new lakes in de-glaciating mountain ranges. *Geomorphology*, 293, 405–417.

Haldorsen, S. & Heim, M., 1999. An Arctic groundwater system and its dependence upon climatic change: an example from Svalbard. *Permafrost and Periglacial Processes*, 10, 137–149.

Hambrey, M.J. & Fitzsimons, S.J., 2010. Development of sediment–landform associations at cold glacier margins, Dry Valleys, Antarctica. *Sedimentology*, 57, 857–882.

Hambrey, M.J. & Glasser, N.F., 2012. Discriminating glacier thermal and dynamic regimes in the sedimentary record. *Sedimentary Geology*, 251, 1–33.

Holdsworth, G., 1974. *Meserve Glacier Wright Valley, Antarctica: Part I. Basal Processes*. Research Foundation and the Institute of Polar Studies, Report No. 37, The Ohio State University Research Foundation.

Hooke, R.L. & Iverson, N.R., 1995. Grain-size distribution in deforming subglacial tills: role of grain fracture. *Geology*, 23, 57–60.

Irvine-Fynn, T.D., Hodson, A.J., Moorman, B.J., Vatne, G. & Hubbard, A.L., 2011. Polythermal glacier hydrology: a review. *Reviews of Geophysics*, 49, RG4002.

Kavanaugh, J.L. & Clarke, G.K., 2006. Discrimination of the flow law for subglacial sediment using in situ measurements and an interpretation model. *Journal of Geophysical Research: Earth Surface*, 111(F1), doi:10.1029/2005JF000346.

Kleman, J. & Borgström, I., 1996. Reconstruction of palaeo-ice sheets: the use of geomorphological data. *Earth Surface Processes and Landforms*, 21, 893–909.

Kleman, J. & Glasser, N.F., 2007. The subglacial thermal organisation (STO) of ice sheets. *Quaternary Science Reviews*, 26, 585–597.

Kokelj, S.V., Tunnicliffe, J., Lacelle, D., *et al.*, 2015. Increased precipitation drives mega slump development and destabilization of ice-rich permafrost terrain, northwestern Canada. *Global and Planetary Change*, 129, 56–68.

Kokelj, S.V., Lantz, T.C., Tunnicliffe, T., Segal, R. & Lacelle, D., 2017. Climate-driven thaw of permafrost preserved glacial landscapes, northwestern Canada. *Geology*, 45(4), 371–374.

La Farge, C., Williams, K.H. & England, J.H., 2013. Regeneration of Little Ice Age bryophytes emerging from a polar glacier with implications of totipotency in extreme environments. *Proceedings of the National Academy of Sciences*, 110, 9839–9844.

Lewkowicz, A.G. & Way, R.G., 2019. Extremes of summer climate trigger thousands of thermokarst landslides in a high arctic environment. *Nature Communications*, 10, 1–11.

Liestøl, O., 1977. Pingos, springs and permafrost in Spitsbergen. *Norsk Polarinstitutt Årbok*, 1975, 7–29.

Lovell, H., Fleming, E.J., Benn, D.I., *et al.*, 2015. Former dynamic behaviour of a cold-based valley glacier on Svalbard revealed by basal ice and structural glaciology investigations. *Journal of Glaciology*, 61, 309–328.

Luckman, A., Murray, T., De Lange, R. & Hanna, E., 2006. Rapid and synchronous ice-dynamic changes in East Greenland. *Geophysical Research Letters*, 33, L03503.

Lukas, S., Benn, D.I., Boston, C.M., *et al.*, 2013. Clast shape analysis and clast transport paths in glacial environments: A critical review of methods and the role of lithology. *Earth-Science Reviews*, 121, 96–116.

Luzi, G., Pieraccini, M., Mecatti, D., *et al.*, 2007. Monitoring of an alpine glacier by means of ground-based SAR interferometry. *IEEE Geoscience and Remote Sensing Letters*, 4, 495–499.

Macdonald, M.L., Wadham, J.L., Telling, J. & Skidmore, M.L., 2018. Glacial erosion liberates lithologic energy sources for microbes and acidity for chemical weathering beneath glaciers and ice sheets. *Frontiers in Earth Science*, 6, 212.

Marshall, S.J. & Clark, P.U., 2002. Basal temperature evolution of North American ice sheets and implications for the 100-kyr cycle. *Geophysical Research Letters*, 29, 67–61, doi:10.1029/2002GL015192.

Matsuoka, N. & Murton, J., 2008. Frost weathering: recent advances and future directions. *Permafrost and Periglacial Processes*, 19, 195–210.

Menzies, J., 1981. Freezing fronts and their possible influence upon processes of subglacial erosion and deposition. *Annals of Glaciology*, 2, 52–56.

Mercer, J.H., 1971. Correspondence - Cold glaciers in the central Transantarctic Mountains, Antarctica: Dry ablation areas and subglacial erosion. *Journal of Glaciology*, 10, 319–321.

Midgley, N.G., Tonkin, T.N., Graham, D.J. & Cook, S.J., 2018. Evolution of high-Arctic glacial landforms during deglaciation. *Geomorphology*, 311, 63–75.

Millan, R., Mouginot, J. & Rignot, E., 2017. Mass budget of the glaciers and ice caps of the Queen Elizabeth Islands, Canada, from 1991 to 2015. *Environmental Research Letters*, 12, 024016.

Mohr, J.J., Reeh, N. & Madsen, S.N., 1998. Three-dimensional glacial flow and surface elevation measured with radar interferometry. *Nature*, 391, 273–276.

Moore, P.L., 2014. Deformation of debris-ice mixtures. *Reviews of Geophysics*, 52, 435–467.

Moore, P.L., Iverson, N.R., Brugger, K.A., *et al.*, 2011. Effect of a cold margin on ice flow at the terminus of Storglaciären, Sweden: implications for sediment transport. *Journal of Glaciology*, 57, 77–87.

Nowak, A. & Hodson, A., 2013. Hydrological response of a High-Arctic catchment to changing climate over the past 35 years: a case study of Bayelva watershed, Svalbard. *Polar Research*, 32, 19691.

Nowak, A. & Hodson, A., 2014. Changes in meltwater chemistry over a 20-year period following a thermal regime switch from polythermal to cold-based glaciation at Austre Brøggerbreen, Svalbard. *Polar Research*, 33, 22779.

Reinhardy, B.T.I., Booth, A.D., Hughes, A.L.C., *et al.*, 2019. Pervasive cold ice within a temperate glacier - implications for glacier thermal regimes, sediment transport and foreland geomorphology. *The Cryosphere*, 13, 827–843.

Roth, K. & Boike, J., 2001. Quantifying the thermal dynamics of a permafrost site near Ny-Ålesund, Svalbard. *Water Resources Research*, 37, 2901–2914.

Ruskeeniemi, T., Engström, J., Lehtimäki, J., *et al.*, 2018. Subglacial permafrost evidencing re-advance of the Greenland Ice Sheet over frozen ground. *Quaternary Science Reviews*, 199, 174–187.

Schomacker, A., 2008. What controls dead-ice melting under different climate conditions? A discussion. *Earth-Science Reviews*, 90, 103–113.

Schomacker, A. & Kjær, K.H., 2008. Quantification of dead-ice melting in ice-cored moraines at the high-arctic glacier Holmströmbreen, Svalbard. *Boreas*, 37, 211–225.

Slaymaker, O. & Embleton-Hamann, C., 2018. Advances in global mountain geomorphology. *Geomorphology*, 308, 230–264.

Sollid, J.L., Etzelmüller, B., Vatne, G. & Ødegard, R.S., 1994. Glacial dynamics, material transfer and sedimentation of Erikbreen and Hannabreen, Liefdefjorden, northern Spitsbergen. *Zeitschrift für Geomorphologie, Supplementband*, 97, 123–144.

Syverson, K.M. & Mickelson, D.M., 2009. Origin and significance of lateral meltwater channels formed along a temperate glacier margin, Glacier Bay, Alaska. *Boreas*, 38, 132–145.

Tsytovich, N.A., 1975. *The Mechanics of Frozen Ground*. McGraw-Hill, New York.

Wadham, J.L., Tranter, M., Tulaczyk, S. & Sharp, M., 2008. Subglacial methanogenesis: a potential climatic amplifier? *Global Biogeochemical Cycles*, 22, 2008, doi:10.1029/2007GB002951.

Waller, R.I., 2001. The influence of basal processes on the dynamic behaviour of cold-based glaciers. *Quaternary International*, 86, 117–128.

Waller, R.I., Phillips, E., Murton, J., Lee, J. & Whiteman, C., 2011. Sand intraclasts as evidence of subglacial deformation of Middle Pleistocene permafrost, North Norfolk, UK. *Quaternary Science Reviews*, 30, 3481–3500.

Waller, R.I., Hambrey, M.J. & Moorman, B., 2016. Glacier-Permafrost interactions, debris transfer mechanisms and the development of distinctive sediment-landform associations in a High Arctic glacial environment: the case of Fountain Glacier, Bylot Island, Nunavut, Canada. In: *XI. International Conference On Permafrost–Book of Abstracts, 20–24 June 2016, Potsdam, Germany*. International Permafrost Association, pp. 106–107.

Williams, P.J. & Smith, M.W., 1989. *The Frozen Earth: Fundamentals of Geocryology*. Cambridge University Press.

Yang, Y., Lai, Y. & Chang, X., 2010b. Laboratory and theoretical investigations on the deformation and strength behaviours of artificial frozen soil. *Cold Regions Science and Technology*, 64, 39–45.

Zeng, N., 2003. Glacial-interglacial atmospheric CO_2 change—The glacial burial hypothesis. *Advances in Atmospheric Sciences*, 20, 677–693.

Zeng, N., 2007. Quasi-100 ky glacial-interglacial cycles triggered by subglacial burial carbon release. *Climate of the Past*, 3, 135–153.

Zwally, H.J., Abdalati, W., Herring, T., *et al.*, 2002. Surface melt-induced acceleration of Greenland ice-sheet flow. *Science*, 297, 218–222.

Index

Glacier–Permafrost Interactions, First Edition. Richard I. Waller.
© 2024 John Wiley & Sons Ltd. Published 2024 by John Wiley & Sons Ltd.